Music Production

We're all able to record music; a smartphone will get you quick results. But for a good sound, a lot more is involved. Acoustics, microphone placement, and effects have a huge influence on the resulting sound.

Music Production: Learn How to Record, Mix, and Master Music will teach you how to record, mix, and master music. With accessible language for both beginner and advanced readers, the book contains countless illustrations, includes tips and tricks for popular digital audio workstations, and provides coverage of common plugins and processors. Also included is a section dedicated to mastering in a home studio. With hundreds of tips and techniques for both the starting and advanced music producer, this is your must-have guide.

Hans Weekhout is one of the best-known names in the Dutch studio scene. Besides collaborating with countless Dutch artists, he has worked with international artists like Prince, Falco, and Girls Aloud. Under the moniker of Capricorn, he scored a worldwide hit with the dance track "20 Hz." As a lecturer in the Pop department of the Conservatory of Amsterdam, he teaches his students the intricacies of music production.

Music Production

Learn How to Record, Mix, and Master Music

Third Edition

Hans Weekhout

Routledge
Taylor & Francis Group
NEW YORK AND LONDON

First edition published in Dutch by Django Music & Publishing 2016

Second edition published in Dutch by Django Music & Publishing 2018

Third edition published in Dutch by Django Music & Publishing 2019

Third edition published 2019
by Routledge
605 Third Avenue, New York, NY 10017

and by Routledge
2 Park Square, Milton Park, Abingdon, Oxon, OX14 4RN

Routledge is an imprint of the Taylor & Francis Group, an informa business

Library of Congress Cataloging-in-Publication Data
Names: Weekhout, Hans, 1965– author.
Title: Music production : learn how to record, mix, and master music / Hans Weekhout.
Description: Third edition. | New York, NY : Routledge, 2019. | Includes index.
Identifiers: LCCN 2019004452 (print) | LCCN 2019004757 (ebook) | ISBN 9780429860911 (pdf) | ISBN 9780429860898 (mobi) | ISBN 9780429860904 (epub) | ISBN 9781138626096 (hardback : alk. paper) | ISBN 9781138626102 (pbk. : alk. paper) | ISBN 9780429459504 (ebook)
Subjects: LCSH: Sound recordings—Production and direction. | Popular music—Production and direction.
Classification: LCC ML3790 (ebook) | LCC ML3790 .W386 2019 (print) | DDC 781.49—dc23
LC record available at https://lccn.loc.gov/2019004452

ISBN: 978-1-138-62609-6 (hbk)
ISBN: 978-1-138-62610-2 (pbk)
ISBN: 978-0-429-45950-4 (ebk)

Typeset in Giovanni
by Apex CoVantage, LLC

Visit the companion website: www.musicproduction-pro.com

DOI: 10.4324/9780429459504

Contents

Preface

We've been able to record music for almost a century now. But it's only since the 1950s that technology has added a new dimension to music, namely, *sound*. Sound is never the same, and that's exactly what makes recording music a fascinating process—and unpredictable at the same time. I've been working as an engineer, producer and artist for more than 25 years, but my fascination for sound is still growing.

Recording is the most critical aspect in a musician's career. A hit reaches far and can reach a wide audience. Every time it's played, the work of musician, producer and engineer is under the microscope. So it better be good! But what is a good record? Of course, it starts with a good song and a good voice, but in pop music, **production value** is often just as important. If you listen to either The Beatles, Prince or Beck, the emotional impact of the song is interwoven with its sound; they are inseparable. This book is about the process of sound-making: it shows you how to apply both analog and digital techniques in order to get a good sound. It will show you how to use software for making professional productions in your living room.

Music Production was not written for aspiring engineers/producers only. It's just as valuable for musicians. Despite the fact that longtime, successful artists like Pharrell, Björk, Radiohead and Lady Gaga generally don't flick the switches themselves, they know exactly how to instruct the people surrounding them. So anyone aspiring a long-term career in the music business should learn about the technical aspects of music production. This book can be the guide!

Thanks to Jan van der Plas.
Hans Weekhout, Amsterdam, May 2019

Introduction

At the end of the book, you'll find Appendices about the characteristics of sound and our hearing. I encourage you to read these first, as it will give you a better understanding of the techniques described earlier in the book.

TECHNIQUES SHOWN IN PRO TOOLS AND LOGIC PRO

I've chosen to demonstrate techniques in two of the most popular programs: **Pro Tools** (Avid) and **Logic Pro** (Apple). In case you work with other software, that shouldn't be too much of a problem. Most programs offer similar functions, although they may be located in a different menu, under a similar name.

FAMOUS HARDWARE . . . IN SOFTWARE

With the classic pop recordings, analog gear has been largely responsible for the sound. Some of the vintage devices have gained legendary status and are modeled into so-called **plugins**. Plugins are like little apps inside your project. They behave like the real hardware device, adapting both look and character. That character can then be imparted on a specific instrument. In the book, I will show you plugins with good reputations and demonstrate which settings could work as a starting point.

Note for Logic Users

Before applying bespoke techniques in **Logic**, don't forget to switch on "Advanced Tools" in Logic Pro X/Preferences/Advanced Tools (see Figure 0.1). Otherwise, many functions are unavailable.

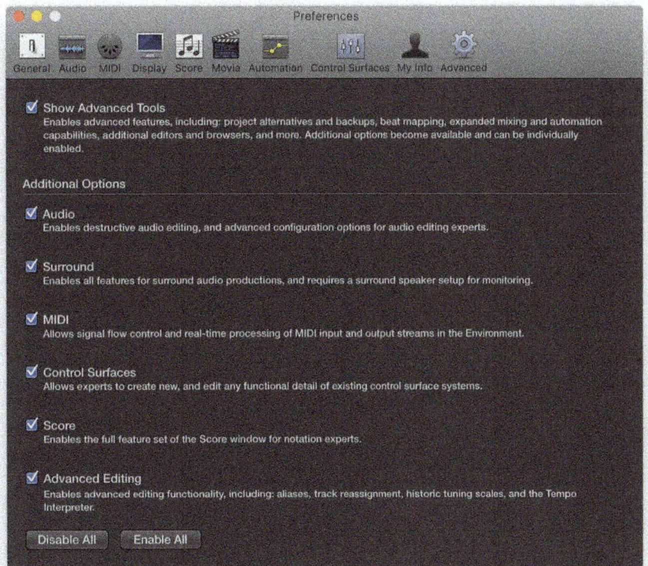

FIGURE 0.1
Logic Pro: "Advanced Tools."

Check www.musicproduction-pro.com for extra audio and video.

MUSICPRODUCTION-PRO

Note for Pro Tools Users

On the Mac, secondary functions are commonly activated by holding the "Control" button, followed by a left click. In **Pro Tools**, however, some functions require a "real" right click. On a laptop, a right click can be executed by double tapping the trackpad with two fingers. On a desktop Mac, right-clicking must be activated in System Preferences/Mouse.

PART I
Recording

CHAPTER 1

Production

The Studio as an Instrument

The records of The Beach Boys, The Beatles, Michael Jackson, Trevor Horn, Nirvana and Max Martin not only contained seminal songs but sounded fantastic too. For us, listeners, the sound of these classics is inextricably linked to the music forever. Organizing sound waves can cause the experience of pop music to be total. The questions with music production are, How can a song be translated into a record? and How can the listener's experience be enhanced with the help of technology? Producer Mark Ronson found the answer by adding a kick drum sample to Amy Winehouse's organic-sounding "Rehab." Nirvana-producer Butch Vig doubled guitars and vocals on "Smells like Teen Spirit" without Kurt Cobain knowing. Producer Shawn Everett made Alabama Shake's Britney Howard sing through obscure mics from Craigslist—with a cloth in her mouth, in order to force an extreme performance. These examples effectively demonstrate that there are innumerable ways for production to get to a sound that's specific.

▶ PRODUCTION VALUE

To gain insight into the production value of a given song, it can be useful to assign it one of four categories: documentation, coloration, organic/electronic or fully electronic (see Figure 1.1).

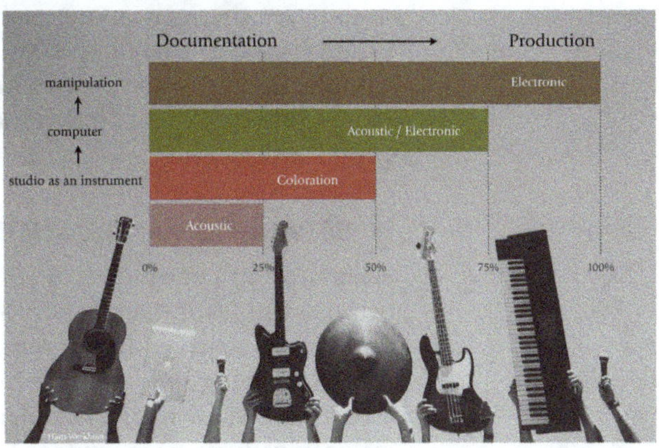

FIGURE 1.1
From documentation to production: farther to the right, means more sound manipulation to take place.

First Category: Documentation

Production always starts with the recording of instruments. If the producer wants an instrument to sound as natural as possible, he'll choose the best-quality microphones and record to the best medium available.

When mixing, the goal is to preserve a natural sound, as if the listener is part of the soundstage. Recordings in this category are classical, jazz, live recordings and many singer-songwriters' recordings. From a pop perspective, no production is involved.

Second Category: The Studio as an Instrument

Instead of the purist's approach, a producer might decide to record a beautiful acoustic guitar in the bathroom, by using a pawnshop mic. Doing so, the resulting coloration becomes an integral part of the music and evokes an emotion with the listener. The studio is not used for documentation only but, rather, as an instrument. Coloration can be achieved by using atypical microphones or acoustics, or with tape machines, equalizers, compressors, distortion units, guitar amplifiers and other sound manipulation tools.

Third Category: Computer

Organizing sound evokes **emotion** with the listener.

Third category productions typically mix acoustic and electronic ingredients. With the use of a computer, it is now possible to intervene **musically**. For example, cutting up musical phrases and reversing them, changing timing (such as *quantizing* drums) or correcting the pitch of the vocal for example. Actually, a fair share of modern pop music is in the third category, although it may **appear** as if it is from the second (or even first) category.

Fourth Category: All Electronic

Fourth category productions might not use any acoustic element at all. In case it does get used, the producer is free to manipulate it so much that little or nothing of the instrument can be recognized in the end product. The composition and its ingredients are initiated (or at least inspired) by technology. Conventional song structures may be abandoned. When assessing a song, it's impossible to rely on "old values," such as the quality of the composition or good musicianship.

▶ SEPARATION AND CONTRAST

Separation of instruments is usually an important goal for production. That's because you want the listener to connect with every single instrument. Separation is best achieved in the recording stage, by choosing the right instrument, the best playing style and the musical notes that best support the song. These aspects are decisive for the general direction of the record. Only after all musical options have been explored, the studio and its technology come into play to build a good sound and to further improve separation and contrast. This involves finding the best mic and its position, using the best acoustics and, last but not least, making the best possible mix.

▶ DOUBLE-TRACKING

Double-tracking is an often-used studio technique in order to achieve a thicker and wider sound. Doubling happens when you record the same part on a new track, possibly with another instrument. It is commonly used on backing vocals, lead vocals and guitars. Phil Spector used to double-track every single instrument, including drums and bass. As this technique causes individual characteristics and differences to flatten, the resulting sound is somewhat "depersonalized" and less defined. That can be positive. Doubling reinforces a performance, makes it more abstract and increases the stereo width of a mix. These are important qualities in production. Doubling doesn't add any new notes to the song; that's why this technique perfectly fits the maxim of "*less is more.*" Soundwise, however, doubling means "more is more."

With **double-tracking**, the musical content of the song remains the same.

Examples of common doubles:

- A fuzz guitar on the left is doubled with a second fuzz guitar on the right
- An acoustic guitar on the left is doubled with a second acoustic guitar on the right

FIGURE 1.2
Crescente Studio, Tokyo, Japan, with a (rare) 72-channel Focusrite mixer.

- A string pad (chords) from one synthesizer is doubled with a sound from another synthesizer
- The lead (and/or) backing vocals are recorded multiple times on individual tracks
- An acoustic piano and an electric piano play the same part

All in all, production is about reinforcing the emotion that's contained within the song. When the sound is "right," the listener can be affected emotionally. This book focuses on production in the second and third category, with the occasional trip to the left or the right.

CHAPTER 2
In the Studio

Most of the great pop classics wouldn't have been possible without the experience and talent of many skilled people working together in a large studio. What exactly are the roles of those people, and how did they get there? How have the big temples of sound been responsible for the sound of pop music? And what does modern-day recording look like?

The evolution of studios and technology has been of great influence on the sound of pop music. Since the 1950s, sound has changed roughly per decade.

BEFORE THE 1960S

Orchestral recordings were made in large rooms with one or more microphones. The natural reverb of the acoustic space was an important ingredient of the sound. After recording, the balance between the instruments could not be altered anymore.

THE 1960S: MULTITRACKING

By using a 4- to 8-track **multitrack recorder**, it became possible to record multiple microphones separately. This allowed balancing signals after recording, rerecording of individual instruments and changing the sound of an individual instrument. People discovered that manipulating instruments was more effective by separating them acoustically. To minimize **crosstalk**, acoustic screens (*gobos*) were positioned between instruments. Multitracking became the standard.

THE 1970S: ISOLATION

Technology advanced quickly; multitrack recorders with 24 tracks became the norm. This allowed a separate recording of just about every single instrument. To minimize leakage between microphones, studios built iso-rooms (or booths; see Figure 2.2). In these acoustically damped spaces, musicians were physically separated, hearing the music through headphones, and maintained visual contact through windows. As the booths were acoustically almost dead, the instruments

FIGURE 2.1
Control room A at Sound Emporium Studios, Nashville, USA. The wooden frame on the back wall is a "QRD diffuser," which is used to scatter sound waves. Behind the walls are "bass traps": spaces that are partly filled with sound-absorbing materials such as rockwool or glass wool. Note the slanted walls.

Source: Photo courtesy Jeff Carpenter/Readylight Media & Sound Emporium Studios.

FIGURE 2.2
Recording room at Sound Emporium Studios. Both asymmetrical surfaces and different floor and wall materials help sound to become diffuse. Acoustic screens are used to acoustically isolate musicians.

Source: Photo courtesy Jeff Carpenter/Readylight Media & Sound Emporium Studios.

sounded dry. In case reverb was needed, plate reverbs or echo chambers were used (see Chapter 14, "Effects | Reverb"). If you listen to typical 1970s' bands like Thin Lizzy, Steely Dan, Pink Floyd, The Eagles and Chic, the close and dry sound clearly illustrates the big influence of the studio and its technology on pop music.

THE 1980S: MORE, MORE, MORE

It was discovered that with more reverb, productions could be made to sound bigger. The invention of **digital reverb** made reverb so accessible and easy to use that it was used in great amounts. For more separation on the drums, **noise gates** (see Chapter 27, "Advanced Mixing Techniques: Drums") were used to separate the individual microphone signals of a drum kit. In the mix, these artificially separated signals had to be forged into a whole again. This caused the sound to sometimes turn out unnatural and artificial. By synchronizing two or three 24-track machines, track counts increased to 48 or 72.

Some well-known producers of the 20th century include Les Paul, Sam Philips, Phil Spector, Joe Meek, George Martin, Brian Wilson, Glyn Johns, Gamble & Huff, Chinn & Chapman, Phil Ramone, Roy Thomas Baker, Todd Rundgren, Lee "Scratch" Perry, Stock, Aitken & Waterman, Eddie Kramer, Jeff Lynne, Arif Mardin, John "Mutt" Lange, Chris Thomas, Quincy Jones, Toni Visconti, Giorgio Moroder, Steve Lillywhite, John Leckie, Andy Wallace, Stephen Street and Trevor Horn.

THE 1990S: DIGITAL TECHNOLOGY

Digital technology entered the studio. It became possible to record on a 24-, 32- or 48-track multitrack machines digitally. Drum sounds were enhanced or even replaced with samples, drummers were replaced by drum loops. Gradually, more and more production tasks could be executed with a computer.

2000S: IN THE BOX

Why hire a big studio if you can achieve the same result conveniently, at home? Expensive hardware from the traditional studio is made redundant by computers. Multitracking, mixing and applying effects can be done efficiently "in the box" with an infinite number of tracks. Session recall is only a mouse click away, and projects can be exchanged quickly thru the internet.

▶ The Producer

Traditionally, the job of a music producer can be compared to that of a film director. A producer is in charge of the recording and brings in ideas for the music. He develops a vision about how the final product should sound. How can the artist be guided to express himself in the best creative way? The producer decides on the studio and the engineer. A good producer creates a pleasant atmosphere and inspires all people in the room to bring out their best. Experience, psychological qualities and a certain authority are needed to guide the artist through the often-complicated process of record making. That's why it is impossible to become a full-fledged producer through study or courses.

A producer is also the connecting factor between the artist and the record company. The record company will want a vote in the songs to be recorded and the sound for the final product. They allocate a budget for renting the studio, the engineer and the musicians. It's the producer's job to manage that budget.

Most producers are accomplished engineers themselves. Yet there are also producers who may never touch a knob, such as Rick Rubin, Trevor Horn, Quincy Jones, George Martin and Phil Spector. Although they might know the ins and outs of the equipment, they leave the technical tasks for the engineer. The producer needs full attention for guiding the artist, as getting a good performance is the highest goal.

LARGE STUDIOS

From the 1960s to the 1990s, the record industry was peaking. Large, beautiful studios were built in which the great classic pop albums were recorded. Although most of these temples of sound have ceased operation, some of the legendary places are still in operation, like The Power Station, Abbey Road Studios, Sunset Sound, Electric Lady Studios, Air Studios, Conway Recording, Eastwest Studios, Larrabee Studios, Ocean Way, Record Plant and Sound City Studios. Large-scale studios offer important advantages over the typical project studio:

- Sound isolation. Built as a box-in-a-box, no sound can come in or go out of the studio.
- Professionally designed control room acoustics and good-quality speakers result in better sounding mixes.
- The natural reverb of a beautiful room is superior to artificial reverb.
- Analog equipment, such as the mixer, effects, tape and vintage microphones, is considered superior in sound.
- Professional infrastructure and staff can support the artist.

Apart from quality and convenience, recording in the same room that spawned great pop classics is undeniably exciting!

▶ The Engineer

A good engineer knows how to translate the producer's vision into sound. He chooses the microphones and their positions and manages the equipment in the control room. He knows how a certain device will affect the sound. The relationship between engineer and producer is crucial; the goal is a superb recording, both artistically and technically.

Studio days can take long, and the life of an engineer is not always easy. It can be challenging to perform all technical operations without making mistakes and in the meantime push sonic barriers. It helps if the engineer is organized, patient and stays calm in stressful situations. Last, the will to experiment and being open to new ideas are definitely bonuses, as well as a good sense of humor.

THE TEA BOY

Similar to runners in the movie industry, tea boys were responsible for getting food, drinks and groceries. Sooner or later during the session, the engineer might call off, allowing the tea boy to quickly take his place. Historically, many tea boys have made the transition to engineer and/or producer that way: Steve Lillywhite (U2, The Killers), Eddie Kramer (Jimi Hendrix), Chris Thomas (Beatles, Pink Floyd), Jim Abbiss (Arctic Monkeys, Placebo), Michael Brauer (Coldplay, Jeff Buckley), Nigel Godrich (Radiohead, Beck) and Alan Moulder (Smashing Pumpkins, Nine Inch Nails).

Modern Recording

THE 2000S

After 2000, revenues in the record industry started to decline. This caused recording budgets to drop too. At the same time, technology became accessible to everyone, making large studios redundant. Large studios are no longer required, as pop music has become more and more electronic. These factors have caused recording and mixing to move to home studios and project studios.

GIRL POWER

Although the studio may seem like a men-only affair, there are definitely women working too: Lenise Bent (Blondie, Steely Dan and Supertramp), Susan Rogers (Prince), Marcella Araica (Britney Spears, Timbaland, Mariah Carey), Ann Mincieli (Alicia Keys), Leslie Ann Jones (Alice in Chains, Marcus Miller) and Sylvia Massy (Tool, Prince, Johnny Cash), to name a few.

Well-known female mastering engineers are Darcy Proper, Mandy Parnell and Emily Lazar.

▶ The Project Studio

Project studios are typically privately owned, smaller spaces. Although the minimum outfit of such a studio consists of a computer, microphone and a pair of speakers, many project studios have added analog equipment and a (small) mixer. It is perfectly possible to work in a project studio the traditional way, with the roles of engineer and producer separated between two persons. But often the budget will not allow for this. Therefore, it has become common for the producer to work alone and handle both tasks of engineering and producing. This is just as efficient as it is demanding. All in all, record making has changed drastically since the 2000s.

Well-known 21st-century producers are Steve Albini, Mark Ronson, Timbaland, Dr. Dre, Guy Sigsworth, Dangermouse, Paul Epworth, Bloodshy & Avant, Dave Fridmann, Brendan O'Brian, Flood, Alan Moulder, David Bendeth, John Congleton, Pharrell Williams, Dr. Luke, Eric Valentine, Nigel Godrich and Butch Vig.

Brian Eno, Nile Rodgers, Rick Rubin and Max Martin have achieved relevance for more than 25 years!

▶ The Mixer

Especially in the US, engineers have specialized in recording; others, in mixing. They'll get sent the files from the recording studio and perform the mix, often without the producer or artist being present. Once the mix is finished, it is sent for review. After the producer or artist has made his comments, the mixer makes a new revision of the mix. Sometimes, a streaming connection allows the artist and producer to become part of the mixing process. Well-known mixers are Manny Marroquin (John Mayer, Alicia Keys), Michael Brauer (Coldplay, Athlete), Phil Tan (Rihanna), Serban Ghenea (The Weeknd, Ariana Grande), Tchad Blake (Arctic Monkeys, The Black Keys), Toni Maserati (Beyoncé), Chris Lord Alge (Biffy Clyro, Snow Patrol) and Mark "Spike" Stent (5 Seconds of Summer, Muse, Björk).

Many successful **producers** started out as a **tea boy**.

THE MIX ASSISTANT

As modern projects may contain 100 tracks or more, just organizing the project can take a long time. That's why many mixers work with a mix assistant. A mix assistant will sort tracks, group them, name and color them and set up basic effects. When those time-consuming jobs are done, the mix engineer can fully concentrate on the creative part of the mix.

CHAPTER 3

Analog Recording

Traditionally, recording was done on tape. Nowadays, we record on the computer, at home and in project studios. But the technology and terminology of today are based on the old methods and equipment. So before we start recording, it is essential to learn about the old equipment and working methods. How did the Beatles manage to record so many instruments on just four tracks? What exactly is the sound of tape? Can the sound of tape be faithfully reproduced in software? And is recording on tape worth the effort?

BING CROSBY

Bing Crosby was the man who gave analog tape recording an important push. In the late 1940s he hosted his own show on American radio. But he actually hated doing this live. Once he heard about the Ampex company manufacturing one of the first good-quality tape machines, he immediately purchased a unit and became a devoted user. After producing the best-selling single ever, "White Christmas" (more than 50 million copies), he bought a large portion of shares in the Ampex company. That turned out to be visionary, as Ampex was to become one of the largest tape and tape machine manufacturers in the world. To our standards, music recording in this era was laborious. The mix of a few microphones was recorded on a mono tape recorder. In case one of the musicians made a mistake, or if the mix was lacking, the musicians had to perform the whole piece again.

LES PAUL

Les Paul is the man who gave the Gibson guitar its iconic name. Together with his wife Mary Ford, he recorded "How High the Moon" in 1951 on an Ampex 3-track multitrack recorder. By using *sound on sound* (see Figure 3.1), he managed to record no fewer than 24 guitar and vocal parts. Even by today's standards, the result is a great sounding song. Les Paul has also been crucial for the development of both tape delay (Chapter 13) and the legendary Fairchild compressor (Chapter 15).

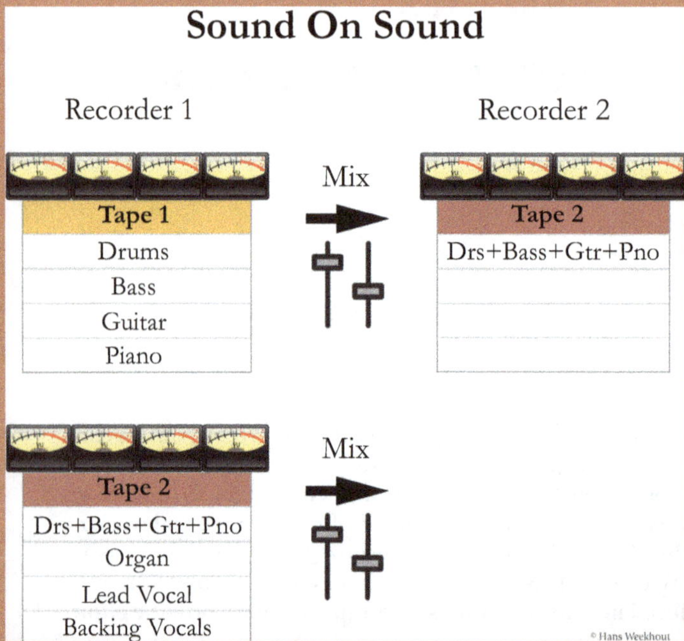

FIGURE 3.1
In order to record many different instruments on a limited number of tracks, Les Paul invented the process that has become known as "sound on sound." For sound on sound, two multitrack recorders are needed. In the case of 4-track recorders, the first four instruments can be recorded on machine 1. Then, a proper mix of these instruments is recorded on the first track of the other machine. On this recorder, three new instruments can be recorded on the remaining tracks. After loading machine number 1 with a new tape, a good mix of machine 2 can then be recorded on the first track of the first machine. This can be repeated endlessly. A great advantage of sound on sound is the almost infinite amount of instruments that can be recorded. However, there are also disadvantages. Revisiting an earlier mix is impossible, as the previous four instruments are burned into a single track. So in case one instrument appears too soft in the mix, this would not only require remixing the corresponding tape but also rerecording all the following instruments. Another disadvantage is the degradation of audio quality: noise, distortion and high-frequency loss are doubled at each step!

THE BEATLES AND THE BEACH BOYS

In the 1960s, studio engineers and producers became masters in sound on sound. If you listen to The Beatles and The Beach Boys, it's evident that their records sound rich and well balanced, despite the luscious arrangements and multitude of instruments. Between 1965 and 1969, when The Beatles worked with producer George Martin, a lot of energy went into pushing the limits of the studio. Production techniques that are common nowadays, such as feedback, reverse playback, flanging and the doubling of guitar and vocals were introduced to a large audience. On the other side of the ocean, The Beach Boys were breaking new ground as well. While the rest of the band was on tour, composer and mastermind Brian Wilson recorded complex productions in the studio with session musicians. Back and forth, The Beatles and The Beach Boys inspired each other to great achievements, requiring the utmost of their engineers.

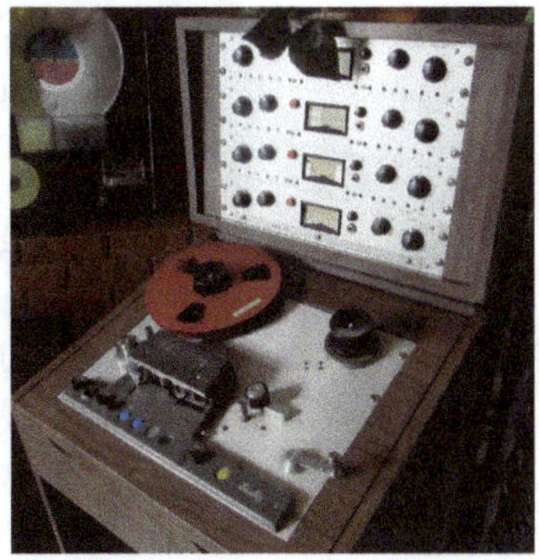

FIGURE 3.2
Scully 280 4-track recorder. Brian Wilson recorded *Pet Sounds* on the almost identical Scully 288.

PHIL SPECTOR

From the early 1960s, producer Phil Spector scored big hits with artists such as The Ronettes, The Raveonettes, Ike & Tina Turner and The Beatles. By using 3-track machines from Ampex, he recorded richly instrumented songs, containing atypical pop instruments such as castanets and the glockenspiel. Although doubling of vocals and guitars has always been common in pop music, Spector doubled drums, bass, piano and orchestral instruments too. This resulted in the famous "Wall of Sound." In Spector's vision, doubling also served another goal, namely, elimination of the musician's personality. Remarkably, the lead vocal was usually excepted from doubling.

FIGURE 3.3
Studer A800: probably the most popular 24-track recorder ever.

FROM 4 TO 24

Although The Beatles recorded on four tracks at the beginning of their career, track counts quickly went up with the advance of technology. Ampex started manufacturing the first 16-track machine in 1966. Only a few years later, MCI came up with the first 24-track recorder. When artists like Pink Floyd, Todd Rundgren and Queen started recording their albums on 24 tracks, the 2-inch, 24-track format was established. It was destined to become the industry standard for the next 30 years.

The Production Process

With multitrack recording, the production process has four distinct stages: **recording**, **overdubbing**, **mixing** and **mastering**.

RECORDING

When multitracking, individual microphones are recorded on separate tracks (see Figure 3.4). Only when the signals are completely separated will changes on one instrument not affect others. It is important to realize that a multitrack recording doesn't contain a **mix** of the music.

OVERDUBBING

For musicians, it's no longer necessary to perform at the same time. While hearing previously recorded instruments on headphones, they can add their performance to spare tracks. In case of a mistake, the tape is rewound so that the same part can be rerecorded. This can be repeated endlessly. Even single notes can be punched in and punched out, although it requires full concentration of the engineer. By erroneously pressing the record button, any existing performance would be gone forever. Fortunately, we now have *undo* buttons.

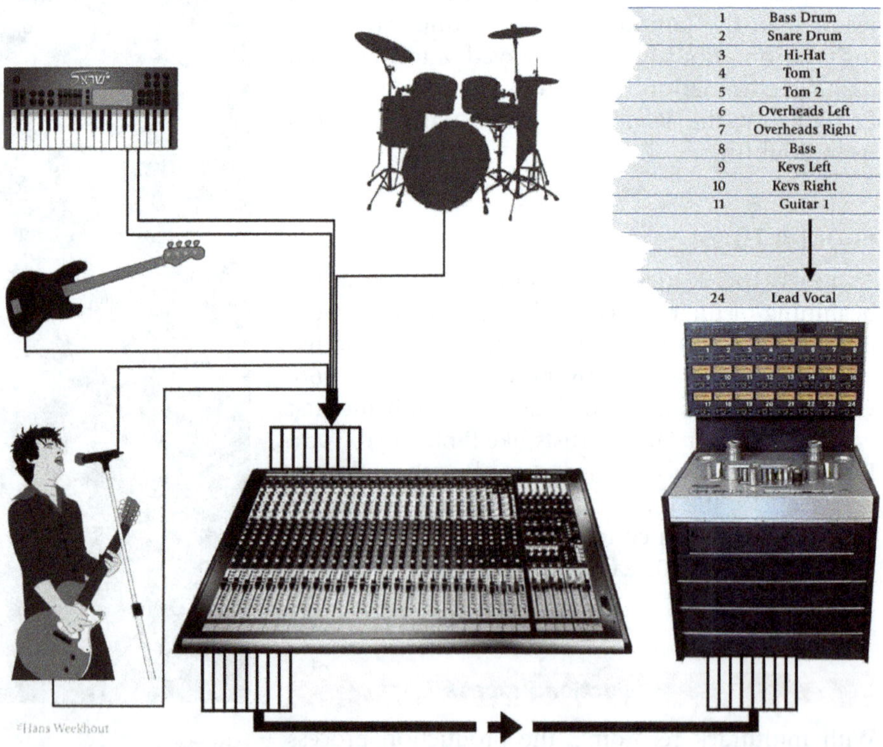

FIGURE 3.4
Recording on a 24-track tape recorder. Note that a multitrack recording doesn't contain a mix of the music but rather the unprocessed microphone signals.

MIXING

Once recording is done, the individual tracks of the multitrack feed the mixer (see Figure 3.5) and can then be treated with equalizing (EQ), reverb, or other effects. By use of the *pan-pot* (panorama knob), an instrument can be positioned in the stereo image. The resulting stereo mix can then be recorded on a 2-track analog recorder or on the computer. It's only now that the music has taken on the left–right stereo format that is suited for distribution to the consumer.

MASTERING

During mastering (see Chapter 31, "Mastering"), the finished mix of a song is treated with EQ, compression and other effects to make it sound as good as possible. The mastering studio produces different formats for distribution on CD, vinyl, internet streaming, movie or game audio.

Tape in Practice

While digital records a signal "as is," tape adds a certain **sound**. Due to the loss of high frequencies, tape is perceived as "fat" and "warm." It has increased clarity and density due to third harmonic distortion. Tape *saturates* the signal and chops off the peaks. The effect is similar to compression, which causes the signal to be denser and more compact. As long as you stick to normal recording levels, these by-effects are subtle. By pushing the levels, saturation and nonlinear artifacts increase proportionally.

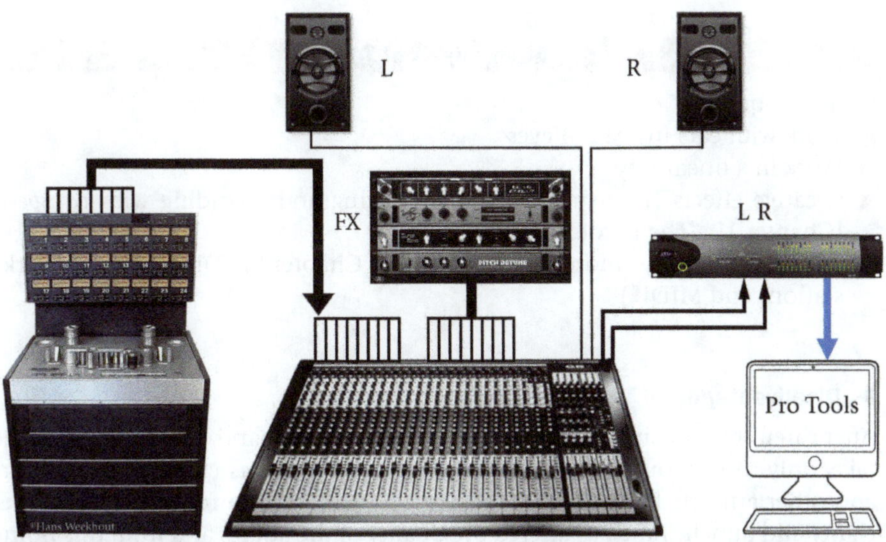

FIGURE 3.5
Analog mix down: while the multitrack machine plays, a 2-track tape machine (or computer) records the stereo mix from the console.

Workflow

Although the information on a computer screen is very informative, it could distract and even lead to the wrong decisions being taken, for example, by erasing noise on a guitar track because "it cleans up the project nicely" or when you think that blocks that are shown on the left should be copied to the right or when you spot timing errors and then visually line up waveforms. These are all actions that may seem logical to our symmetrical-working brain but do not necessarily lead to **better music**. Working with tape forces you to work with your **ears** instead of your eyes.

Attitude

In the computer, literally anything can be changed, even in the last stages of a mix. Want to recall the version of half a year ago? It's only a mouse click away. Although this is, of course, very powerful and flexible, it allows for a certain comfort that could lead to decisions being postponed. Typical symptoms include too many tracks and a lack of direction. Tape, however, urges you to commit and is therefore "dangerous." It has limited tracks, and moving or copying parts takes time or causes generation loss. Redoing a take may mean erasing the previous take. The linear workflow of tape requires decisions to be taken, which results in focus and direction for the production. As such, choosing for tape is not only a matter of sound but also of **attitude**.

Advantages of Tape

- Sound quality
- Work with ears instead of eyes
- Work in a linear way
- Creative effects (reversing tape, tape-flanging and recording with *varispeed* [Chapter 11, "The Recording Session"])
- Absence of *latency* (more about latency in Chapter 9, "Digital Audio Workstations and MIDI")

▶ Disadvantages of Tape

After carefully choosing microphone positions for the band, you record the first take, only to find out that after recording, the sound has changed. Distortion and other artifacts have affected the signal, resulting in a recording that lacks clarity and punch. Tape has altered the quality of the audio, and from this point of view, the acclaimed **sound** of tape is counterproductive.

Tape Wear

As a result of physical contact with heads and roller mechanism, tape wears out. Magnetic particles that are glued to the plastic coating, gradually loosen. This causes sound to slowly deteriorate and a brown sticky sludge to build up on both the heads and tape deck. Fortunately, tape will store your precious sonics for a long time, even with day-in, day-out usage. But there have been infamous sessions where the tape was literally falling apart. In those cases, quickly copying the tracks to a second multitrack machine saved the song, be it at the cost of generation loss.

Workflow

Working with tape needs patience from all people involved. Rewinding a song may easily take 20 sec. Although it is possible to make edits by use of a razor blade and sticky tape, this is time-consuming. Depending on tape speed, either three or six songs can be recorded on one reel. This means that reels need to be changed regularly during a session, which takes time.

Disadvantages of Tape

- Sound quality
- Noise, *wow and flutter* (fluctuations in playback speed)
- Editing is time-consuming
- Changing reels and rewinding take time
- Expensive
- Limited number of tracks
- Tape wears out, and sound quality slowly deteriorates
- Generation loss when copying: side effects double
- Tape machines need regular cleaning and maintenance

As you can see, tape has strong arguments on both sides of the coin. A decision for the recording medium probably comes down to sacrificing flexibility on the computer in return for better sound and an organic workflow.

▶ Tape in the Computer

Computer technology has progressed so far that tape artifacts can be mimicked in software. Waves (Kramer Tape [see Figure 3.6] and J37), Universal Audio (Studer A800, Ampex ATR-102), Slate Digital (Virtual Tape Machines), U-He (Satin), Fabfilter (Saturn), Cranesong (Phoenix) and Empirical Labs (Fatso) are virtual tape machines that can be used to warm up either single instruments or the total mix. Other than real tape, it is now possible to apply the tape effect on selected instruments only.

How good is software? Well, purists could argue that software doesn't sound the same, but it is getting better and better. Besides, two analog tape machines that

FIGURE 3.6
Waves "Kramer Tape." Based on the 1958-released Ampex 351 recorder, this virtual tape machine was developed in cooperation with Eddie Kramer (Jimi Hendrix). Increasing the "flux level" simulates a stronger magnetic field in the recording head, which increases tape saturation. Turning up the "Record level" instead will cause the electronic circuits to distort as well.

sound exactly the "same" are yet to be found. Variation is caused by the age of the machine, the way it is calibrated, the tape brand and recording levels. All in all, plugins may help achieve a certain sound, without the need for a real tape machine. This can at least be called attractive.

CHAPTER 4
Speakers and Acoustics

In the studio, speakers are the extension of your ears. But every speaker will, more or less, color the sound, and this can lead to the wrong decisions being taken. With better speakers, a recording or mix will come out better. Coloration is caused not only by the speakers but also by less-than-ideal acoustics. In an untreated room, even expensive speakers can't get you a proper sound. How can you improve your own room? What do you listen for when buying speakers? And is it possible to make a good mix on headphones?

Before reading this chapter, it is advised to first learn about the characteristics of sound in Appendix 1.

Acoustics

Acoustics play a big role in every single stage of the production, be it recording, mixing or mastering. Outside, in the free field, an instrument or speaker will sound completely natural, as sound waves can travel to the ear directly. With a sound source **in a space**, reflections occur. Reflections can be seen as **bad copies** of the original wave, as they lose energy after bouncing off a surface. Due to the longer distance traveled, reflections arrive **later** than direct sound. This will **color** sound when recording. When mixing, the control room will color the sound too. Our ear picks up both direct sound from the speakers and reflections from the room. This blurs the image and could lead to the wrong decisions being taken.

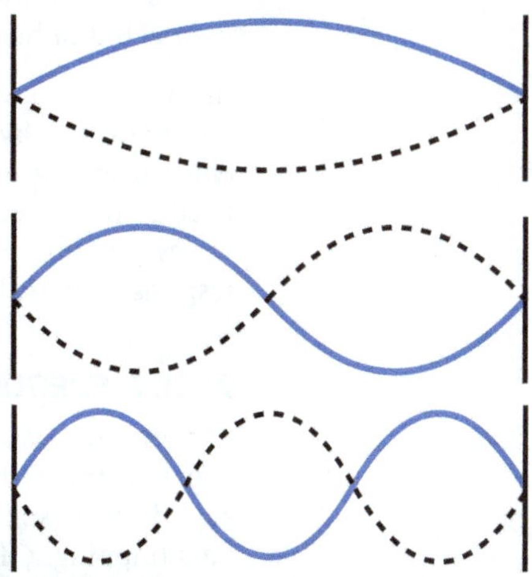

FIGURE 4.1
Standing waves.

Although "good" reflections may add a nice character to a room or recording, "bad" reflections lead to bad sounding recordings and mixes. So the questions are, When are reflections bad? And how can they be tamed? To answer this we'll separate the spectrum into low, mid and high frequencies.

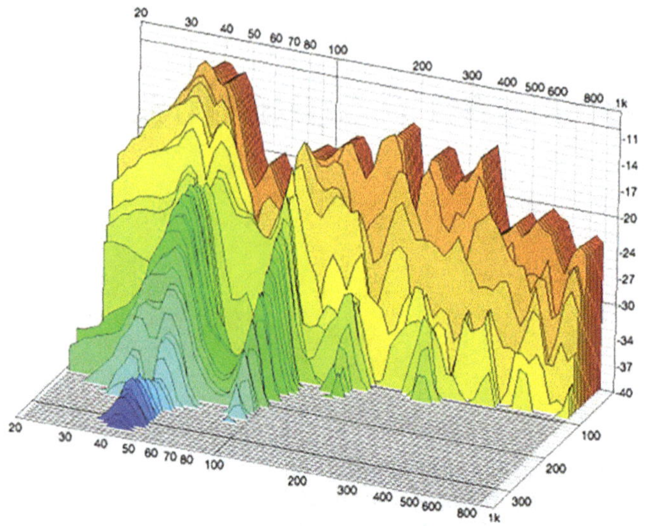

FIGURE 4.2
In the *waterfall plot* of a room, it's easy to see how standing waves lead to resonances. In this room, it appears that 45 Hz (and the octaves at 90 Hz and 180 Hz) are resonating frequencies. The increased decay time blurs the sound.

▶ LOW FREQUENCIES: STANDING WAVES

Low-frequency problems of a room can be predicted by dividing length, width or height by the wavelength. For example, the wave of a 50-Hz tone fits a room with a length of 22 feet. As the wave bounces off the wall symmetrically, only little energy is lost, causing it to bounce back and forth in the room. This will not only make it sound **louder** than other waves, but it will also take **longer** to decay (see Figure 4.2). So, standing waves (see Figure 4.1) are actually resonances of a room. By the way, frequencies of 100 Hz and 150 Hz will prove problematic as well, as they fit exactly two or more times.

To make matters worse, bass response varies with position and frequency. At one position, a node meets another node, resulting in loud bass. At another position, a node meets an antinode, causing that frequency to (almost) cancel. Different frequencies either build up or cancel out, depending on the position in the room. When mixing, this poses us with a big problem. When the bass plays an E, the bass might sound earth-shaking, while an A could sound thin. In case you grab an EQ in order to boost the A's corresponding frequency, you'll mess up the mix. It's not the bass being wrong, but rather the acoustics!

In an untreated room, it's not uncommon for the frequency response in the bass area to deviate by a staggering 20 to 30 dB. Of course, such a room cannot present you with a reliable picture of the frequency spectrum. So, standing waves need to be tamed by treating the acoustics. As low frequencies consist of large waves containing a lot of energy, this is not easy. So in practice, getting the bass response right is the hardest part of getting good room acoustics.

▶ LOW-FREQUENCY ABSORPTION

Superfluous bass and low-mid energy can be **absorbed** by **bass traps**. After this energy is trapped, it will no longer interfere with direct sound. Then, the pure sound of the speakers will prevail. For the low frequencies, it's hard to overdo bass trapping. Of course, speakers aren't linear in the bass area either, but the deviations are way smaller than the acoustics ruining the response.

In professional studios, bass traps appear as cavities behind walls or ceiling. They are (partly) filled with damping materials, such as mineral wool. For existing rooms, bass traps are often barrel-like objects or panels that can be attached to the walls. As bass waves are large, bass traps are also large. Professional studios may sacrifice 25% to 50% of their volume for bass traps.

Mid-High Frequencies: Absorption and Diffusion

Absorption

Imagine being in an empty, concrete cubic space. By clapping your hands, specific **mid-** and **high** frequencies will start resonating. These are also standing waves, and they sound like an echo with a long decay, called ***flutter echo***. So, a single material room with parallel walls will present you with the worst possible acoustic conditions. Fortunately, your own studio room might already be decorated with curtains, carpets and other objects. Those materials will help absorb the smaller sound waves of the mids and highs. Any excessive energy can be absorbed with thin panels, similar to bass traps. Note that mids and highs should not be completely absorbed. Although a completely dead room would allow us to listen to the pure (and rather linear) response of the speakers, this wouldn't represent a realistic listening situation. A control room must therefore retain a certain amount of reflections and liveliness.

Diffusion

To prevent the remaining mid-high frequencies from being too singular and focused, they should be dispersed as randomly as possible. The aim is an even distribution, especially at the listening position. This can be achieved with **diffusers**. Diffusers are panels that come in all forms and shapes, from irregularly shaped wooden constructions to polystyrene skylines (see Figure 4.3).

FIGURE 4.3
Diffusers: sound waves of the mids and highs reflect randomly, causing an even response throughout the room.
Source: Photo courtesy vicoustic.com.

▶ IMPROVING THE ACOUSTICS OF YOUR OWN ROOM

In case the room is still to be built, the best results can be achieved by avoiding parallel walls. With (slightly) slanted walls, sound waves reflect diagonally, reducing the risk for standing waves.

In case of an existing room, acoustic panels can be attached to the walls and ceiling (see Figure 4.4). Many companies offer acoustic sets nowadays (see Figure 4.5), at various price points. Brands with a good reputation are Primacoustic, Vicoustic, EQAcoustics, Gikacoustics and Hofa en Auralex. A set may contain bass traps, mid-high frequency panels and diffusers.

Note:

- Controlling bass will lay the biggest claim on both space and budget.
- It's almost impossible to overdo bass trapping. The more bass absorbed, the more you get to listen to the pure and direct sound of the speakers.

FIGURE 4.4
EQAcoustics mid-hi absorber and bass trap.
Source: Photo courtesy of eqacoustics.com.

- By leaving a space between wall and panel, the effective frequency range of the panel can be extended downward. Let's say, the panel is 2 in. thick and absorbs frequencies down to 400 Hz. By leaving a space of 2 in., the effective range can be extended to 200 Hz.

It's easy to check the bass response of your own room with **sine waves**. Every DAW (*Digital Audio Workstation*) has a tone generator plugin built in and there are also tone-generator apps for smartphones. At the listening position, all frequencies should sound equally loud. By walking through the room, positions can be found where bass response is louder than others. As bass frequencies usually pile up in the corners or along the edges of the ceiling, that's where you'll want to attach the bass traps. Mid-high panels can be fixed anywhere along walls or ceiling but especially at the listening position.

How Many Panels Are Needed?

It all depends on budget, more is better. Starter sets contain four bass traps, four mid-high panels and two to four diffusers; a medium set may double this. An advanced set might contain 12 to 16 bass traps, 10 to 14 mid-high panels and 8 diffusers.

FIGURE 4.5
Acoustically treated home studio. Bass traps are piled up in every corner. Every wall that causes early reflections at the listening position is treated with mid-high absorbers, including the ceiling.
Source: Photo courtesy of hofa-akustik.de.

Do It Yourself

Large acoustic sets may set you back considerably. Building your own panels instead saves money, and is not too difficult. On the internet, many do-it-yourself projects can be found. Panels often consist of a wooden frame covered with acoustically transparent (!) cloth and filled with mineral wool. The thicker the panel, the lower the effective frequency range. Mid and high panels are often 1 to 2 in. thick, while bass panels might be anything from 5 to 10 in. For diffusion, the cheapest solution is a big classic bookcase with books spaced as randomly as possible. www.realtraps.com and the sites of previously mentioned manufacturers contain additional information on how to improve acoustics.

Speakers and Headphones

▶ BIG MONITORS

Traditional studios generally offer big speakers built into the wall (see Figure 4.6). Why should they be built in? Well, in case the speakers were positioned in front of the wall, sound waves reflecting from the back wall would arrive later at the listening position than the direct sound of the speakers. This would cause coloration. With built-in speakers, back-wall reflections arrive in time with direct sound.

Big monitors offer a good bass performance, not only because of their size but also because the construction of the building prevents resonances to become part of the sound field. In a good studio, bass response could be linear up to 20 Hz, which is the lowest frequency we can hear. This allows for fine-tuning bass frequencies in the mix with great precision. Big monitors are also **analytic** in the mids and the highs, revealing the smallest details. Last, but not least, they allow for getting a kick in the studio by playing loud every now and then—essential when working on a creative product!

If you want to know how the mix sounds on your aunt's stereo, big speakers may not be of great reference. First of all, consumer speakers are not capable of reproducing the lowest and highest frequencies. But there's also another aspect. Imagine standing close to a big painting. Although it's perfectly possible to zoom in on the smallest details, you won't get a proper overview.

FIGURE 4.6
Big monitors in the wall at Onkio Haus Tokyo, Japan.

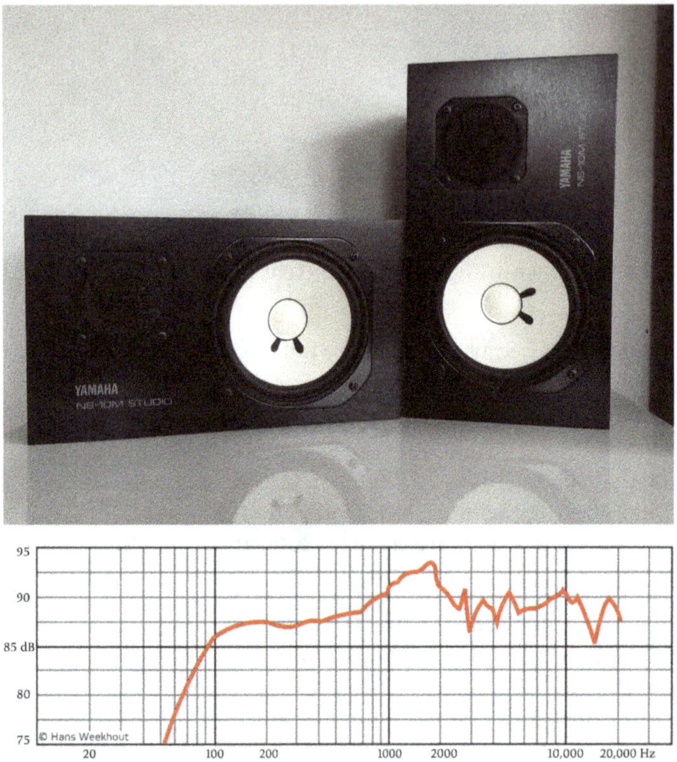

FIGURE 4.7
Yamaha NS10s have a frequency response that's not even close to the ideal curve: the lowest octaves (20–80 Hz) are hardly there. The peak around 1.8 kHz is causing the NS10s to sound slightly aggressive in the mids.

Standing close to a small painting, however, let's you oversee the global picture easily. Well, the same goes for speakers, and that's exactly the reason why studios also offer *nearfield monitors*.

▶ NEARFIELD MONITORS

Nearfield monitors allow for a good impression of music in the average living room. Positioned close to the listener, direct sound is louder (and earlier) than sound reflected from the walls and ceiling. Therefore, acoustics are less critical.

Yamaha NS10

By looking at classic studio pictures, chances are that you'll spot those small black boxes with white cones on top of the console. These are the legendary Yamaha NS10s (see Figure 4.7). Although the mid emphasis of these speakers has caused some people to not be fans of its sound, it has been *the* nearfield monitor of choice since the beginning of the 1980s. Once the mix sounds good on the Yamaha's, you can rest assured that it will be fine on consumer speakers too. Unfortunately, Yamaha has ceased producing this classic.

ACTIVE OR PASSIVE?

Speakers can be active or passive. Passive speakers are driven by a separate power amp. The (long) cable between amp and speaker has a disadvantage: losses in the cable reduce fidelity. Active speakers have a power amp built in, so the speaker cable can be (very) short. Besides, active speakers have their amp designed to match the speaker as closely as possible. What about (long) cables that connect a mixer or interface with active speakers then? Well, cable quality is always important, but the signal feeding an amplifier is more robust than the signal feeding a speaker. Therefore, cable quality is less critical when using active speakers.

Since the NS10, the sound of nearfields has improved. With the aid of computers and modern materials, recent nearfields offer a frequency response that's flatter than yesteryear's speakers. Plus, reproduction is often more **analytic**.

▶ SUBWOOFERS

A subwoofer is an enclosure containing a big speaker and a built-in amplifier. It can be used to extend the bass performance of nearfield speakers. Subwoofers have controls for both volume and cutoff frequency. Setting this control to 100 Hz for instance, will prevent higher frequencies to be reproduced, while any connected nearfields are fed with a signal that contains frequencies **higher** than 100 Hz. As it is hard for our ear to get directional information from bass frequencies, only one sub is needed. For some people, a subwoofer–nearfield setup fails to integrate the playback of low and high frequencies. Also, it can be hard to find the right position for the subwoofer in the room; this can only be done by trial and error.

▶ HOW SHOULD SPEAKERS BE POSITIONED?

The way speakers address a room is not only dependent on the shape of the room and the materials used, but also on their position. Herewith you'll find some general guidelines—experimentation will result in better sonics!

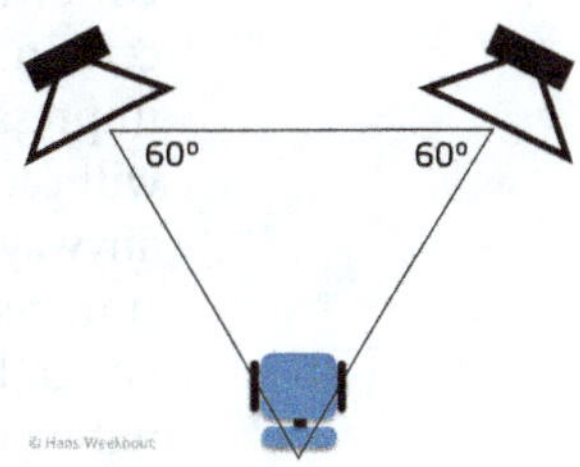

FIGURE 4.8
Speakers should form an equilateral triangle with the listener.

1. The speakers should form an equilateral triangle with the listener (see Figure 4.8): 5 ft per side is a good starting point.
2. Positioning the triangle symmetrically in the room will cause left–right reflections to bounce off the walls similarly: this will improve quality of the stereo image.
3. Avoid setting up in the exact middle of the room or close to the front or back wall: one-third in front of and two-third behind the listener offers best results generally.
4. Move the complete setup in order to find the best position in the room.
5. Low frequencies spread in all directions, while high frequencies are directional (see Appendix 1: "Characteristics of Sound"). That's why the tweeters (high-frequency units) should be pointed directly at your ears. Tilting the speakers may be necessary.
6. Keep speakers away from flat surfaces like walls, windows or the mixing console. This could cause reflections to interfere with direct sound from the speakers, resulting in coloration.
7. By positioning speakers on spikes, pads, heavy stands or boxes filled with sand, reduces resonances and vibration of surfaces.
8. Avoid positioning speakers on the same surface. When vibrations of one speaker influence the other, imaging will suffer.

PORTED OR CLOSED DESIGN?

Speaker manufacturers either use ported or closed designs. With ported designs, aka **bass reflex**, the interior of the speaker is connected to the outside world by means of a port. The port is tuned, in order to create a resonance frequency, similar to the effect of blowing on a bottle. This allows the bass performance of the speaker to be extended artificially. Below the resonance frequency, bass response decreases rapidly. Although the wider frequency range is an advantage, bass performance can be slightly unstable with this design. That's because the woofer (low-frequency unit) can move freely in the open enclosure. With a **closed design**, woofer movement is confined by air pressure in the box. This results in a tighter and more controlled bass response, be it over a limited frequency range.

▶ ARE CONSUMER SPEAKERS SUITED FOR USE IN THE STUDIO?

Unfortunately, no. Consumer speakers are generally designed to produce a certain **sound**. With emphasized bass for instance, small speakers may sound impressive at first sight. Using these impressive sounding speakers for mixing will cause you to be reluctant with kick and bass as the low end will sound big anyway. But after playing the mix on a better system, it will sound thin. Consumer speakers, especially the cheaper ones, are not designed to sound **analytic** either. But in the studio, details are of great importance, for instance when evaluating a mic setup or when establishing the exact right volume for a subtle effect in the mix. On consumer speakers, details can go by unnoticed. Last, consumer speakers have limited **power**. They're designed to playback the limited dynamics of a final product. Checking the bass drum on a high volume will cause consumer speakers to distort.

▶ CHOOSING SPEAKERS

Although product tests might help you decide on a certain speaker, the ultimate choice is a personal one. Visiting a store and listening to music for a longer time is highly recommended. What should you listen for?

1. Listen to **details**. How much detail of a recording is revealed by the speaker? How analytic is it?
2. Listen to the **sound** of a speaker. Although the effect is probably subtler than with consumer speakers, studio speakers may sound slightly impressive too. At first sight, it will make the speaker appealing, but a "dull," and at the same time, natural-sounding monitor might cause you to work harder on sound. Once that sound is captured, the mix will sound good on other speakers too. As we are used to the sound of human voices in everyday

life, our ear is very sensitive for its characteristic. That's why coloration is revealed best on vocals that are meant to sound natural.

3. The **price**. In case your budget is limited (and whose budget isn't?), decisions can be made by setting priorities. With studio gear, some items are more important to spend money on than others. A good speaker is certainly one of them, as it is the extension of your ears: better speakers will lead to better sounding productions. Apart from this, speakers have a longer life than computers and software, for instance, so they are a better investment. Professional brands with a solid reputation are Focal, Adam, Dynaudio, Neumann, PMC, Quested, ATC, PSI, HEDD and Genelec.
4. Reserve budget for **improving room acoustics**. Only then can good speakers reach their full potential.

Good speaker reviews can be found at www.soundonsound.com.

▶ ARE HEADPHONES SUITED FOR USE IN THE STUDIO?

Yes! Headphones lock out **room acoustics**. Plus, any sirens, traffic, fridges and neighbors. At the same time, they prevent the neighbors from hearing **your** music. Good-quality headphones can reveal **details** that speakers don't, like little crackles, noises or other imperfections. Even in professional studios, the final mix gets checked on headphones. Last but not least, studio headphones have a (very) **flat frequency response**, also in the bottom end. As long as you can't afford proper speakers, it is therefore a good idea to work on headphones instead.

Do Headphones Have Disadvantages?

Yes. But as long as you are aware of them, they can be overcome. Herewith are the most important:

- Headphones reveal so many details, that you tend to lower them in the mix. Like delay and reverb. After playing the mix on speakers, it might turn out too dry;
- Lead vocals or instrumental leads might seem loud in the mix, which will cause you to turn them down;
- Exaggerated stereo-image. This may cause you to pan instruments closer to the middle, resulting in a mix that's not wide enough;
- Wearing headphones can be uncomfortable, especially for a longer time.

Similar to speakers, headphones are either for consumer- or studio use. Professional brands like Beyerdynamic, AKG, Sennheiser (see Figure 4.9), Audio Technica and Sony produce good-quality headphones for studio use.

FIGURE 4.9
Sennheiser HD650 open headset.
Source: Photo courtesy of sennheiser.com.

CHAPTER 5

Microphones

The sound quality of a recording is determined by the acoustics, the microphone and its position. Now that you know of acoustics, it's time to talk about microphones. Which mics did they use on the classic pop recordings? Are modern mics just as good? Which mic produces the most natural sound? This chapter provides you with an answer, and we'll also find out how the artifacts of certain mics can work in your favor!

▶ TRANSDUCER

A device that converts a specific type of energy into another, is called a **transducer**. A microphone for example, converts the pressure of sound waves into an electrical current (see Figure 5.1). Other transducers in music production are speakers, tape recorders, audio interfaces, turntables and guitar pickups. An amplifier or a mixer is not a transducer, as the form of the energy stays the same. Losses caused by amplifiers are therefore lower than those of transducers.

Due to the laws of physics, energy is lost during the transducer's conversion process. It will therefore never be possible to record according to the linear goal, that is, a straight line from 20 to 20,000 Hz. Apart from frequency loss, byproducts such as distortion and noise are added to the signal. The specifications of a transducer indicate how much it deviates from the linear goal. For example,

FIGURE 5.1
A microphone transduces soundwaves into an electrical current that can be recorded.

Frequency response: 20–20,000 Hz ± 1 dB (deviations in the frequency response are no more than 1 dB)
Distortion: 0.01% THD (Total Harmonic Distortion)
Noise: −98d B (if 0 dB is the loudest signal that can be recorded, then a signal that's softer by 98 dB will drown in noise)

In music production, at least three transducers are needed before we can listen to the recording of an instrument. Let's look at the process in detail.

The sound waves of an instrument set the membrane of a microphone in motion. This movement generates a current that is sent to a tape machine. Before this electrical signal can be recorded, it needs to be converted into a magnetic field, which, in turn, magnetizes the iron particles of tape. On playback, the magnetic charge is converted to an electrical current in the replay head. In the speaker, this current is converted to a magnetic force that moves the cone. Now, the original sound waves of the instrument are reconstructed, and the pressure of sound waves sets our eardrum into motion.

As you can see, the original sound waves have been converted at least **six times** before arriving at our ear. It's nothing less than a miracle that a recorded instrument can sound so realistic!

In pop music, we often look for the most natural and pure sound. But just as often it's the shortcomings and artifacts of a recording that make the sound interesting. The trick is using these "shortcomings" to your advantage. **Knowing** the artifacts of your gear prevents making mistakes and will help you achieve a certain **sound**. So let's start taking in knowledge!

Microphones

The way a microphone picks up sound is reflected in its **polar pattern. Omni** mics pick up sound from all directions, **cardioid** mics pick up sound from their front side and "**figure-8**" (or **bidirectional)** microphones pick up sound from the front and the back (see Figure 5.2). This *directionality* should not be

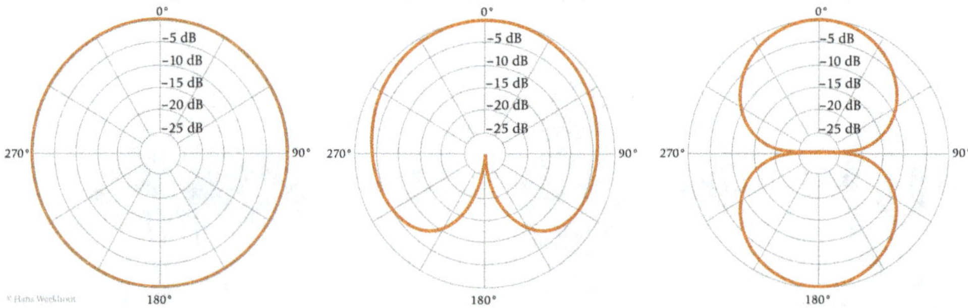

FIGURE 5.2
Polar patterns: omni, cardioid, 8.

taken too literally; every mic picks up a certain amount of sound from the sides or back, although high frequencies may be lacking. This means that directional mics capture bleed of other instruments in a colored fashion.

Different microphones use different methods for converting sound pressure into an electrical current. Most popular types in the studio are **dynamic**, **condenser** and **ribbon**. The construction of these types is reflected in their sound character!

Dynamic Microphones

Dynamic microphones have an electric coil, that moves in the magnetic field of a magnet. Once the sound waves of an instrument set the membrane into motion, the coil starts to move and generates a current. Energy is lost in this process, due to the rigid membrane and the relatively heavy coil. That's why dynamic microphones are less sensitive than condensers or ribbons and produce a sound that's less natural. At the same time, they're robust and capable of capturing sound at a high volume.

SHURE SM57

The Shure SM57 is probably the most popular mic ever, and it's also one of the cheapest. It's the preferred choice for recording snare and guitar amps. However, it can also be useful when looking for an alternative sounding vocal or acoustic guitar. There are even producers who record nine out of ten instruments with it, like Tony Hoffer (Beck, M83).

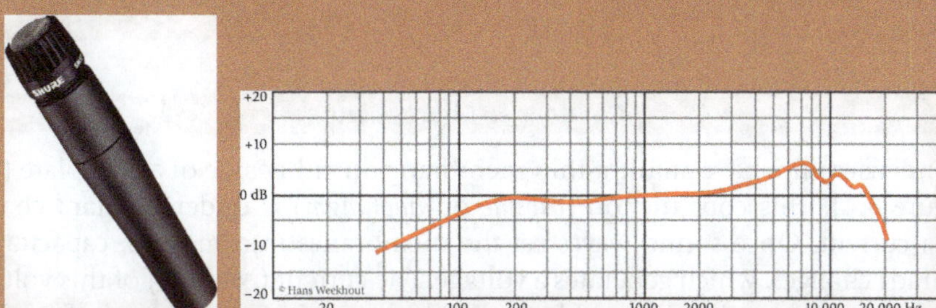

FIGURE 5.3
If linear were the ideal, then this mic should sound horrible! Notice the reduced response in the lows and top highs. The peak around 6 kHz is responsible for the bright sound of the SM57. The SM57 is not an exemplary mic because of its natural response, but rather a good example of how an uneven response can lead to an attractive and effective sound!

Source: Photo courtesy of shure.com.

DYNAMIC MICROPHONES:

- Have a cardioid response
- Cannot handle sub-low or top-high frequencies properly
- Show an uneven frequency response
- Can handle high SPLs without distorting
- Work without phantom power (48 V)
- Are usually cheaper
- Are robust
- Have the proximity effect
- Generate a low electrical voltage
- Record crosstalk with coloration

Some well-known classic dynamic mics are:

Shure: SM7, SM57 (see Figure 5.3), SM58. Apart from the detachable, ball-shaped pop shield, the SM57 and SM58 are identical.
Sennheiser: MD421 (the "razor"), MD409, E604, E609, E90
AKG D12/112 (the famous bass drum microphone)
Electrovoice RE20, PL20

WHAT'S THE PROXIMITY EFFECT?

With **cardioid** and **figure 8** mics, recording closer to a source causes a low-frequency boost. This effect can turn out either positive or negative, depending on the sound needed. For instance, by recording a vocalist closer to the mic during a softer verse, the added low frequencies may get you a warm impression of the vocal. However, when close miking an acoustic guitar, the low frequencies of the hollow body could occupy space in the mix that's needed for other instruments. Omnis don't exhibit the proximity effect.

Condenser Microphones

Condenser mics have an ultra-thin membrane on either side of a backplate (see Figure 5.4). This construction (*capsule* or *diaphragm*) is under constant charge of a current. Once sound waves set the membranes into motion, capacitance (Farad) changes, which generates a voltage. The current (Ampère) of this voltage is so low that the signal would have died before arriving at the end of a cable. That's why an impedance converter is built in; it is responsible for the typical high electrical output of condensers. Actually, when recording loud sources, such as a kick drum, snare or a loud vocal, the level may turn out so high that it causes distortion. Therefore, many condenser mics feature a –10- or – 20-dB *pad* switch.

FIGURE 5.4
Capsule of a condenser mic.

Source: Photo courtesy of Neumann Berlin.

To power both the membranes and the converter, **phantom power** (48 V) is used. Phantom power was invented in 1966 by Georg Neumann; most mixers and audio interfaces feature a 48-V switch. The voltage is not dangerous, as its power figure (Ampère) is relatively low. Although dynamic mics don't need phantom power, no harm is done by accidentally activating.

▶ SIZE MATTERS

Condenser mics come with either large or small diaphragms. Although both have a very linear response, the physical dimensions and weight of a large diaphragm prevent it to move as freely as a small diaphragm. That's why the **transient response** (the short impulse of a sound) of small diaphragm mics (**pencil mics**; Figure 5.5), is slightly better. Pencil condensers provide the most uncolored, natural sound, with a frequency response that extends beyond human hearing. Large diaphragm mics sound a little smoother and richer.

FIGURE 5.5
AKG C451 pencil mic (with CK1 cardioid capsule). Some pencil mics allow changing capsules, which is cheaper than buying two separate microphones.

Source: Photo courtesy of akg.com.

FIGURE 5.6
Neumann U87 switch.

Source: Photo courtesy of Neumann Berlin.

CONDENSER MICROPHONES:

- Often feature a polar pattern switch (see Figure 5.6)
- Have a very linear frequency response
- Subtly emphasize the highs
- Distort on (very) high sound pressure levels
- Require phantom power (48 V)
- Offer a high electrical output level
- Must be handled with care
- Are sensitive to humidity
- Pick up crosstalk without coloration (omni mode only)

Some well-known classic condenser mics are:

Neumann large diaphragm: U87 (see Figure 5.8), U67 (see Figure 5.12), U47 (see Figure 5.7), M149, TLM series
Neumann small diaphragm: KM 84/184 (cardioid), KM 83/183 (omni)
AKG large diaphragm: C414, C12
AKG small diaphragm: C451, C1000

FIGURE 5.7
Neumann U47-FET. This model has the original tube replaced by FET electronics. This allows louder sources to be recorded without distortion. It is a favorite for recording bass drum.

Source: Photo courtesy of Neumann Berlin.

FIGURE 5.8
Neumann U87, with frequency response (cardioid, low cut filter on/off). Note the 2-dB emphasis around 10 kHz.

Source: Photo courtesy of Neumann Berlin.

Sennheiser: E406, E906
Telefunken U47, CU12, ELA250/251
DPA: 4006 (omni), 4011 (cardioid), Schoeps MK2 (omni), MK4 (cardioid)

Newer and cheaper alternatives:

Røde: NT1 (cardioid), NT2 (switchable), NTK (with tube), NT5 (small diaphragm cardioid)
sE Electronics 2200A
Audio Technica AT2020
Warm Audio WA14, WA47, WA87

USB MICROPHONES

USB microphones are typically condenser mics fed by the USB socket of a computer. A great advantage of this mic is that no losses can occur in the cable, as the signal is converted to digital directly. The maximum length of a USB cable (about 10 ft) prevents it from use in a professional studio. USB mics with good reputations are Apogee Mic 96K (24 bit), Blue Snowball (16 bit) (see Figure 5.9), AT 2020 USB+ (16 bit), Sennheiser MK4 (24 bit) and Samson G-Track Pro (16 or 24 bit).

FIGURE 5.9
Blue Snowball.
Source: Photo courtesy of bluedesigns.com.

▶ CONDENSER MICS IN PRACTICE

As a result of the fragility of the membranes, condenser mics are very sensitive and capable of capturing the smallest nuances in sound. They pick up sub-low and top-high frequencies very well. At the same time, by accidentally dropping a condenser, it could die. It may also distort on sources with a high volume. Although the frequency response of condensers is very flat, they emphasize frequencies around 10 kHz. For certain applications this can cause the sound to turn out slightly brittle. But it can be an advantage too, for instance, when a vocal has to cut through the mix.

In general, condensers are the first choice when looking for a natural sound. For most engineers, they are a go-to choice for vocals, piano, drum overhead, percussion, acoustic guitar and orchestral instruments. Keep in mind that better acoustics of the recording room are required for working in omni-mode. With close miking, bass response of omnis is inferior to cardioids and 8s, as they lack the proximity effect. Further away, omnis offer a better bass response.

Ribbon Mics

In 1931, the first ribbon mics became available. Although sensitive, heavy and expensive traditionally, modern ribbons are cheaper, lighter and more robust. Ribbons are typically **bidirectional**, but there are exceptions, such as the cardioid Beyerdynamic M160.

Ribbon mics have a membrane that moves in the magnetic field of a permanent magnet. This ultra-thin aluminum folded **ribbon** is on both sides connected to a step-up transformer, which amplifies the low output voltage. Despite the transformer, the output level of older ribbon mics is so low, that a powerful and noise-free mic pre-amp is needed to record low-level sources, like finger-picked acoustic guitar or a soft vocal. As it is nowadays possible to manufacture more powerful magnets, modern ribbons offer a higher output level, that is often amplified by a built-in amplifier. Such an **active** ribbon allows recording even the quietest instruments. The high output voltage is less sensitive to cable loss and preserves signal quality, even at the end of a long cable. Although this type of ribbon requires phantom power, old **passive** ribbons might get fried by it. So be careful when changing cables or when engaging the 48-V button!

Although their frequency response is slightly uneven, ribbons are known for their natural sound, especially in the high frequencies. In case the 10 kHz emphasis of a condenser sounds too brittle, a ribbon could serve as a good alternative.

FIGURE 5.10
Fethead mic pre-amp.

Source: Photo courtesy of tritonaudio.com.

TRITON AUDIO FETHEAD

In order to boost the low electrical output of passive ribbon mics, phantom-powered amplifiers housed in XLR plugs have arrived on the market (see Figure 5.10). Such an amp not only circumvents the low output problem, but it will also prevent blowing up your precious old ribbon in case you accidentally engage phantom power.

RIBBON MICROPHONES:

- Feature a figure 8 characteristic
- Offer a (rather) flat frequency response
- Can deal with very high volumes
- Can be fried by phantom power (old ribbons)
- Are sensitive to wind (i.e., bass drum or vocals)
- Have a low electrical output voltage (passive ribbons)
- Are fragile and sensitive to handling
- Feature the proximity effect (even more so than cardioids)
- Record crosstalk with less coloration

Some well-known classic ribbon mics are:

Coles 4038 (see Figure 5.11)
Royer Labs R121/R122
Beyerdynamic M160
AEA R84/R44

Newer and cheaper alternatives:

Royer Labs R110
Golden Age Project R1

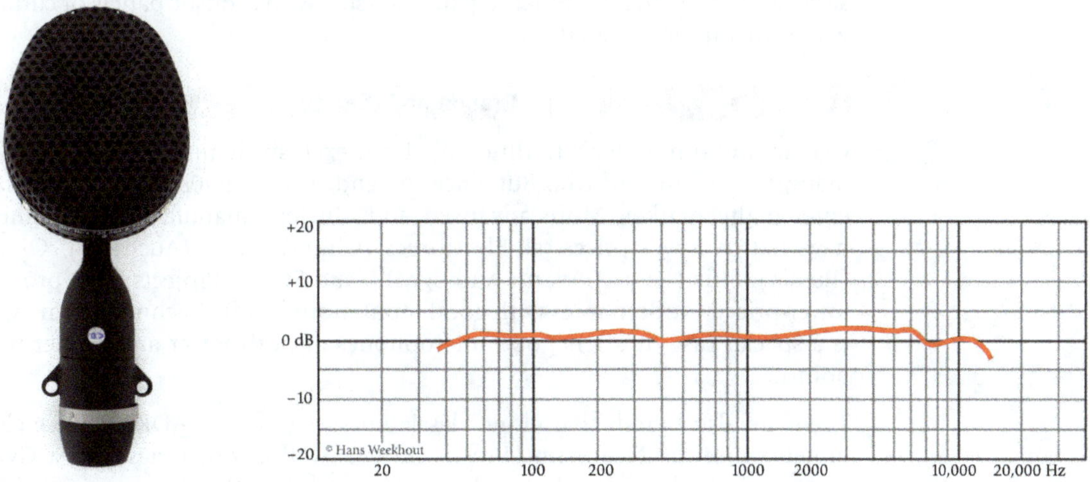

FIGURE 5.11
Coles 4038 ribbon, once BBC broadcast mic. Specifications generally reveal a lot . . . but not always. By looking at the frequency response of the average ribbon mic, you'd get the impression that it would sound unnatural. Actually, the opposite is true: ribbon mics are known for their natural and musical character.

Source: Photo courtesy of coleselectroacoustics.com.

▶ HOW COME RIBBON MICS ARE PERCEIVED AS MORE NATURAL SOUNDING THAN CONDENSERS?

To answer that question, we must look at three aspects:

Transient response: condenser mics tend to "overshoot." That happens when the diaphragm is overreacting to impulses. The transient response of ribbons is truer to the natural attack of an instrument.

Resonance (ringing): the diaphragm of a condenser has its own resonance frequency, which we have come to know as the typical emphasis around 10 kHz. This resonance is always present and colors the sound. Although ribbons have a resonating frequency too, it is beyond human hearing.

Directionality: although directional mics pick up fewer high frequencies off-axis, ribbons perform better. In other words, ribbons pick up crosstalk with less coloration than dynamic mics and condensers.

▶ RIBBON MICS IN PRACTICE

If a natural character and big low end are what you're looking for, then a ribbon can be the best choice. Its figure 8 characteristic is useful in case you want to capture beautiful room acoustics or when recording two opposite sound sources. That being said, ribbons are used less often than dynamic mics and condensers. That's because of their "polite" tone and their backside's sensitivity. In case you choose a ribbon for color but don't want the mic to capture sound at its backside, you can angle it away from other instruments or reflective surfaces (such as stone walls or windows). Shielding the backside with acoustic panels or curtains will further improve isolation.

Brands and Quality

Classic studio mics have traditionally been exclusively manufactured in Germany, the US and Austria. But since the end of the 1990s, new brands have entered the market. More advanced and efficient manufacturing methods have led to lower prices. sE-Electronics (China), Røde (Australia), Oktava (Russia), Blue (Latvia), Warm Audio, MXL and Studio Projects (US) produce lower-priced modern mics with good reputations. Audio Technica from Japan is a special case; it offers good microphones at both lower and higher price points.

There are also brands that **clone** classic mics. Peluso and ADK produce close imitations of the Neumann U-series mics as well as other classics. At Gyraf, you can find schemes and assembly instructions for DIY vintage mic projects. A completely new approach is taken by the "Virtual Microphone System" (VMS) by Slate Digital. It consists of a microphone/pre-amp combo with accompanying software. By means of the software, the VMS-mic can closely mimic the character of a selection of legendary mics. Not only can you change the virtual model during recording, but also in the mix! Similar systems are offered by Townsend Labs and Antelope Audio.

Refined production processes have caused great sounding microphones to come to the market. Although the classic mikes have the reputation and confidence, it is perfectly possible to make fantastic recordings with new mics.

FIGURE 5.12
The legendary Neumann U67 was produced from 1960 till 1971, while a reissue is produced since 2018.
Source: Photo courtesy of Neumann Berlin.

▶ HOW TO DECIDE ON A MIC

Apart from some exceptions (such as the SM57), quality comes at a price in music production. This means that you can't expect the best quality from cheap mics. The law of diminishing returns is also valid here: once you upgrade your $80 condenser mic to a model that costs $250, you'll add substantial quality. But for the next step, you'd suddenly need to invest significantly more money.

Even if the rest of your equipment is not up to par yet, it makes sense to invest in a decent mic. A microphone is the most important element in the recording chain, and good sound starts at the "ears" of the recording. If later on you decide on upgrading the rest of your studio, the uncolored and even sound of a good mic can flourish even better.

There are shops nowadays that allow buying an item with the possibility of returning it after some time. The best thing to do is to **try out** different microphones. The character of a microphone can only be understood after putting it to test for a longer time, in your own environment.

The Microphone Pre-amp

Before the microphone signal can be recorded, it must be amplified by a *microphone pre-amplifier*. Most audio interfaces (see Chapter 10, "Recording on the Computer") and mixers contain mic *pre-amps*. Those devices combine pre-amps with other functions, such as A/D (analogue-to-digital) and D/A (digital-to-analogue) converters and a headphone amplifier. To keep prices down, a compromise has been made regarding **sound quality**. That's why professionals commonly use dedicated (but expensive) pre-amps from brands like Neve, API, Universal Audio, BAE, EMI/Chandler, Manley, Avalon and Millennia Media. Most famous pre-amp ever must be the Neve 1073 (see Figure 5.13) from 1970. Nowadays, affordable clones are manufactured by Golden Age Projects, Warm Audio and others.

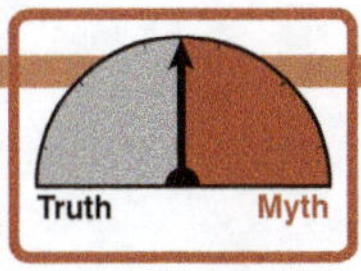

THE QUALITY OF A CABLE: TRUTH OR MYTH?

Music shops offer cables at different price points. As money can be spent only once, it is attractive to think that cable quality doesn't matter. But both microphone cables and guitar cables carry fragile currents. With better conductive properties, losses are reduced. There are OFC (Oxygen Free Copper) cables, silver cables and even gold cables. Better cables allow for a transparent sound, with more depth of field. Without a good microphone, it probably doesn't pay off to invest in expensive cables. But in case you do own a good mic, the fidelity of your recordings can be improved. Brands with a good reputation are Mogami, Monster Cable, Evidence Cable and Zaolla.

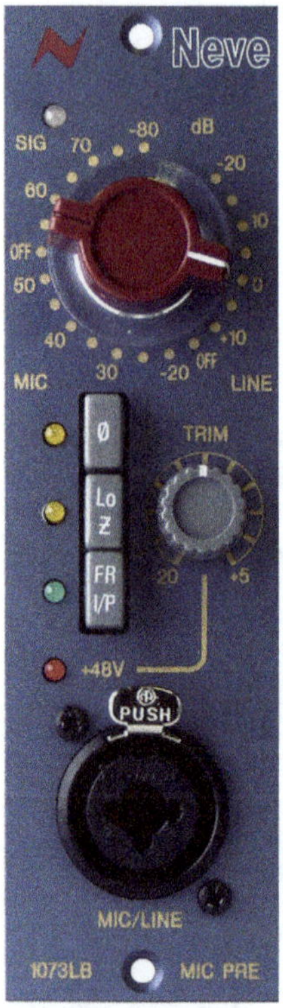

FIGURE 5.13
Legendary mic pre-amp: Neve 1073LB.

Source: Photo courtesy of ams-neve.com.

So how much sound quality can be gained when using a dedicated pre-amp instead of an audio interface? Well, probably not as much as changing microphones. That being said, the sound quality of a beautiful instrument and a proper mic can only flourish with a good-quality pre-amp. Similar to microphones, a good quality pre-amp buys you confidence. As an investment, a top-class model will still be top-class within 20 years. Something that can't be said of computer hardware, software or audio interfaces!

CHAPTER 6

Recording Techniques

Before starting to record specific instruments, we need to know more about positioning mics and recording in general. For example, what's the thought process for choosing mics and finding the best distance? What is phase, and when should you use two mics instead of one? And which stereo microphone techniques are useful in pop music?

With every recording, the signal passes several stages before sound reaches our ears:

musician -> instrument -> acoustics -> microphone -> recording medium -> mix

This means that the sound quality of a recording will never be better than the weakest link in the chain. When using an SM57 for the kick drum for example, the signal will lack sub lows, and there's no process that allows for re-creating missing information. In the recording stage, alterations to sound can be done way more effective than in the mix. Therefore, it's always better to first swap microphones or alter their positions.

Mic Techniques

CARDIOIDS

In pop music, **close miking with cardioids** is most popular. There are many good reasons for that:

- Cardioids feature the **proximity effect**. Only with sufficient bottom end can an instrument sound warm.
- Cardioids capture a **dry** sound, with less *ambience*. This is generally a better starting point for the mix: reverb can always be added, but it cannot be removed.
- Close miking captures more of the desired instrument and less of other instruments. **Less bleed** equals more control in the mix.

The sound quality of a recording will **never** be better than the weakest link in the chain.

- Close miking captures the most **punchy** tone, due to the sound waves being intact. Farther away, instruments sound duller due to high-frequency loss.

Remember, cardioid mics come as dynamic, condenser and ribbon!

DO GO ANY FARTHER

There are disadvantages to close miking too. The closer the mic's position, the more it will emphasize just one aspect of the instrument. As a rule of thumb, the full instrument can only be captured at a distance that's at least the size of the instrument's biggest diameter. An acoustic guitar mic would therefore need at least 2 ft distance in order to capture all tonal aspects. Generally, farther away yields more **tone** and a slightly longer **decay**. Choosing for either a close or distant mike position depends on the sound wanted. In case other instruments cause bleed, this limits your options to close miking.

FIGURE 8S

Although cardioids and omnis are most popular in the studio, figure 8 mics have attractive qualities too. In case you want to record the room's acoustics, 8s can effectively capture reflections from either the walls or the ceiling, for example, when recording a drum kit. Other times, you may not be looking specifically for the 8 characteristic but, rather, for the warm sound character of a ribbon. Then the trick is to minimize reflections by pointing the mic's backside in the least harmful direction. Which means, away from windows and other hard surfaces. Any remaining reflections can be damped with acoustic panels or curtains. Figure 8s are also used for M/S recording (see the later discussion in this chapter).

OMNIS

Recording with omnis has great advantages, too. They capture the most natural, uncolored version of an instrument. Instead of using cardioids for suppressing bleed, you could also decide to capture "beautiful" bleed by using omnis.

POINTING THE MIC'S BACKSIDE

A common mistake when positioning cardioids and 8s is considering their front side only. But finding the best direction for its *null* (insensitive side) is just as important. For example, aiming a cardioid mic at the center of the snare will cause hi-hat bleed to enter the mic sideways. Pointing the back of the cardioid at the hi-hat instead will minimize leakage.

> **Hi-Fi versus Lo-Fi**
>
> It's always easier to turn a hi-fi sound ("hi-fidelity") into lo-fi ("lo-fidelity") than vice versa. That means, as long as you're not sure about the instrument's function in the mix, it is better to capture the full spectrum of the instrument, for example, by choosing a better microphone.

PHASE

Although recording with one mic will get you great results for many applications, a second mic allows balancing two timbres separately in the mix. But recording with multiple mics has one big disadvantage: sound waves will arrive **later** in one microphone than in the other (see Figure 6.3). This results in a phase difference.

Phase compares **timing** between two sound waves and is measured in degrees, from 0° to 360°. When two identical sound waves start exactly at the same time, we call that **in-phase** (0°). Summing the two results in a louder signal. In case the phase difference is 180° ("out of phase"), the positive movement of one wave (node) is cancelled by the negative movement (antinode) of another. This will result in silence (see Figure 6.1).

In practice, we don't work with pure sine waves, but rather with **complex** sounds. Every sound that we know of consists of multiple sine waves, each with its own frequency, volume and phase. If one complex sound encounters another complex sound, certain frequencies become louder, while others are attenuated. This creates a spectrum with peaks and valleys, or a *comb filter* (see Figure 6.2).

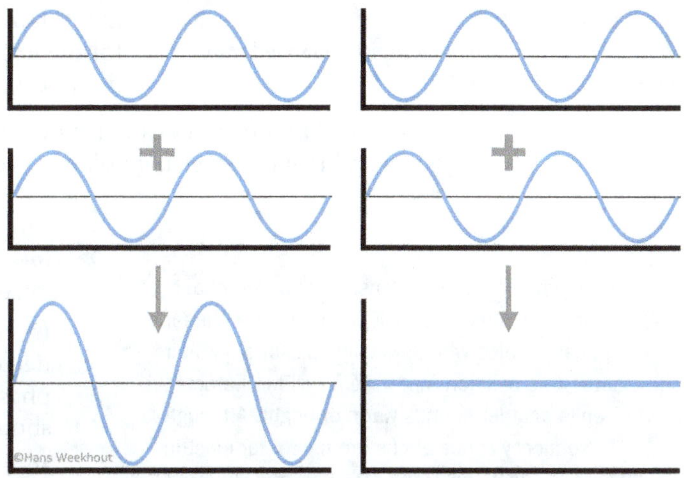

FIGURE 6.1
Summing in-phase signals results in higher volume; summing out-of-phase signals results in silence.

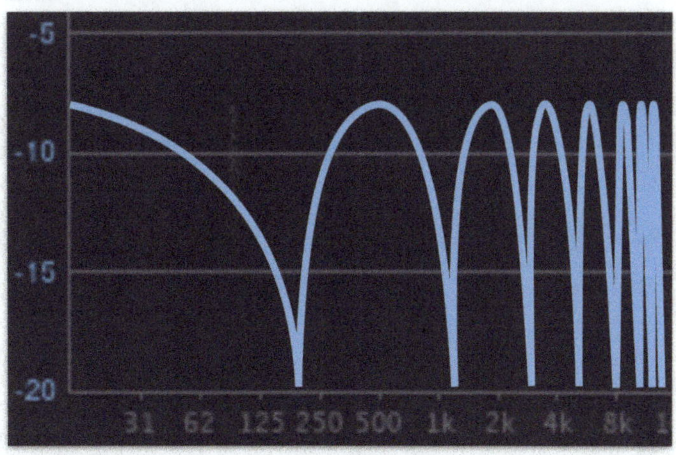

FIGURE 6.2
Comb filter: summing two (slightly) out-of-phase signals causes certain frequencies to (partly) cancel.

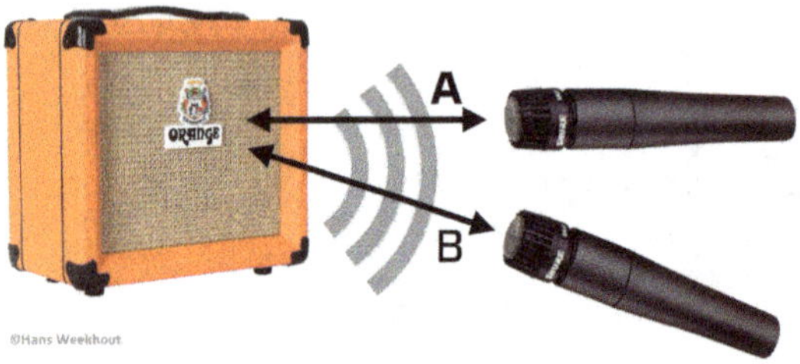

FIGURE 6.3
Phase difference: microphone A picks up sound waves earlier than microphone B.

THE SOUND OF PHASE DIFFERENCES

When miking a kick drum with both a mic inside and outside, you'll quickly hear how sound is affected by phase differences. After adding the second mic in the mix, certain aspects in the frequency spectrum become stronger while others disappear. Depending on the volume of the second mic, this may radically change the character of the instrument. How bad is that? Well, it's not uncommon for the low frequencies of a kick drum to (almost) disappear after adding the second mic!

By the way, phase issues can also occur when a cable is soldered incorrectly or when accidentally reversing polarity of one speaker.

> Guitarists are grateful consumers of the phase-canceling effect, not only when using a phaser pedal but also when switching pickups on their guitar. One position may have a hollow character, while another sounds warm or bright. Although frequency cancelations are **useful** for electric guitar, the effect can be **disastrous** when you want a piano or drum kit to sound natural.

▶ How Can Phase Differences Be Solved When Recording?

1. **Decrease microphone distance.** This decreases delay, which in turn, decreases phase differences. In the ideal world, distance should be zero, although physically this wouldn't be possible of course. But you can minimize the space between the membranes. As it is often hard to see the membrane with the eye, the "perfect" position can be found by reversing the phase of one of the mics. By playing music from a telephone and moving one mic slowly forward and backward, you will find the spot where the signal is (almost) cancelled. This position can then be fixed by tying the mics with tape.
2. **Increase microphone distance.** This will cause the signals to differentiate so much that the **3:1 rule** becomes applicable, for instance, when one microphone is used as a *close mic* (e.g., at 1 ft) and another as a *room mic* (e.g., at 3 ft or greater). At a greater distance, signals become less related, causing phase issues to diminish.

▶ How Can Phase Differences Be Solved When Mixing?

1. **Reverse phase:** this swaps polarity of the signal. All mixers, mic pre-amps and audio interfaces have phase-reverse buttons (see Figure 6.4). In **Logic,**

phase can be reversed with the "Gain"-plugin (in "Utilities"), **Pro Tools** has a phase reverse switch in "EQ1" or "EQ7." When the second signal is out of phase, flipping the phase might (partly) restore phase relationship.

2. The sweet art of deduction: find the **best sounding microphone** and mute others, for example, with multiple mics on a guitar amp.
3. Compensate for delay by **visually aligning** tracks on the computer, for example, when a bass guitar was recorded with both mic and Direct Injection box ("DI box," Chapter 6). Small movements (1 ms or less) may have a huge impact on sound.

FIGURE 6.4
Phase switch.

Stereo Microphone Techniques

In case you want to capture a soundstage, or you want the recording to reflect the size of a larger instrument, you'll need a stereo miking technique. Most stereo techniques work on the basis of two identical microphones, preferably a *matched pair*. With a matched pair, the manufacturer has selected two copies from the production line with minimized tolerances. Which stereo techniques are common in pop music?

An **X–Y** array (see Figures 6.5 and 6.6), consists of two **cardioids** positioned at an angle of 90°. A sound source positioned at the left will cause the **volume** in the right mic to be lower. An X–Y array produces a detailed representation of the stereo image. The close membranes allow for minimal phase differences; an X–Y has therefore very good *mono-compatibility*. The width of the stereo field can be widened by decreasing the angle (lower than 90°), and shrunk by increasing the angle. Due to the cardioids' proximity effect, an X–Y produces more bottom end when positioned close and less when positioned far.

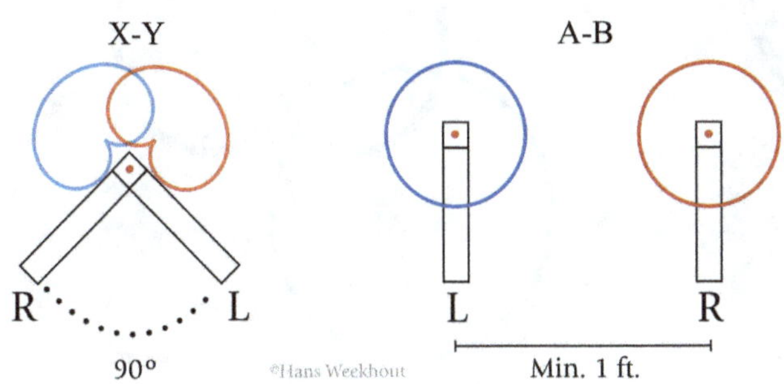

FIGURE 6.5
Left: two cardioids: X–Y. Right: two omnis: A–B.

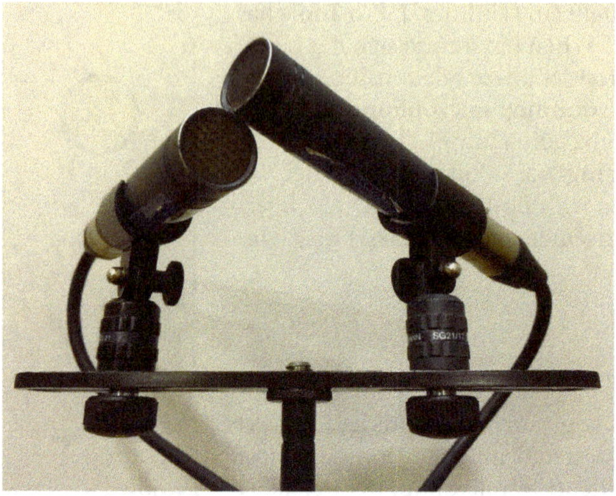

FIGURE 6.6
X–Y based on two Neumann KM184s.

An **A–B** array (or *spaced-pair*, see Figure 6.5) works with two **omnis** positioned in parallel. A–B's derive their stereo information from **timing differences**. A sound source positioned at the left will cause sound waves in the right mic to arrive later. The stereo image of an A–B is slightly blurrier than an X–Y and is less mono-compatible due to the distance between the membranes. There are no rules for the distance between the microphones, other than the minimum being 1 ft. A larger distance yields a wider stereo image but could result in a "hole in the middle." For an A–B, you may also use cardioids or 8s, although the proximity effect will cause the bottom end to suffer when positioned far. When sidewalls are close, or produce nasty reflections, cardioids or 8s can be a good choice.

An **M/S array** (*Mid-Side,* see Figure 6.7), consists of a forward-looking **cardioid**, positioned as close as possible to a sideways-looking figure 8 mic. The cardioid is connected to channel 1 of the mixer; this is the Mid-channel. The S-signal is sent to two identical channels of the mixer. These should be hard-panned left

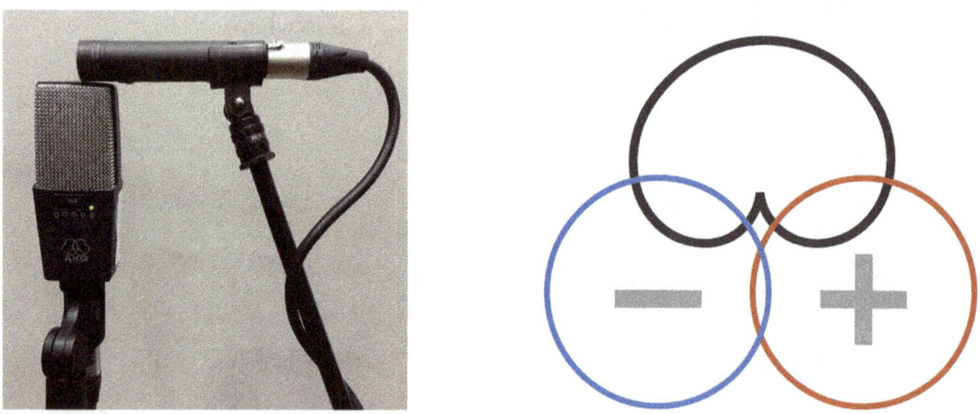

FIGURE 6.7
M/S array: "8" in combination with cardioid.

and right in the stereo image, while the phase of one of the channels should be reversed (see Figure 6.8). As the 8 signals don't meet electronically, they won't cancel.

How should the M/S signals be mixed? Start with the first channel, which represents the natural mono sound of the instrument. To widen the image, add channels 2 and 3 to taste. Playing an M/S recording in mono cancels the (opposite phase) 8 signals. As the uncolored sound of the cardioid will remain, M/S is an elegant way to make stereo recordings with perfect mono-compatibility. M/S can be used for a variety of instruments, like grand piano, drum kit, or acoustic guitar.

Note: With microphone setups, always trust what you hear, not what you see. A critical ear is perfectly capable to identify the weak spots of a specific setup.

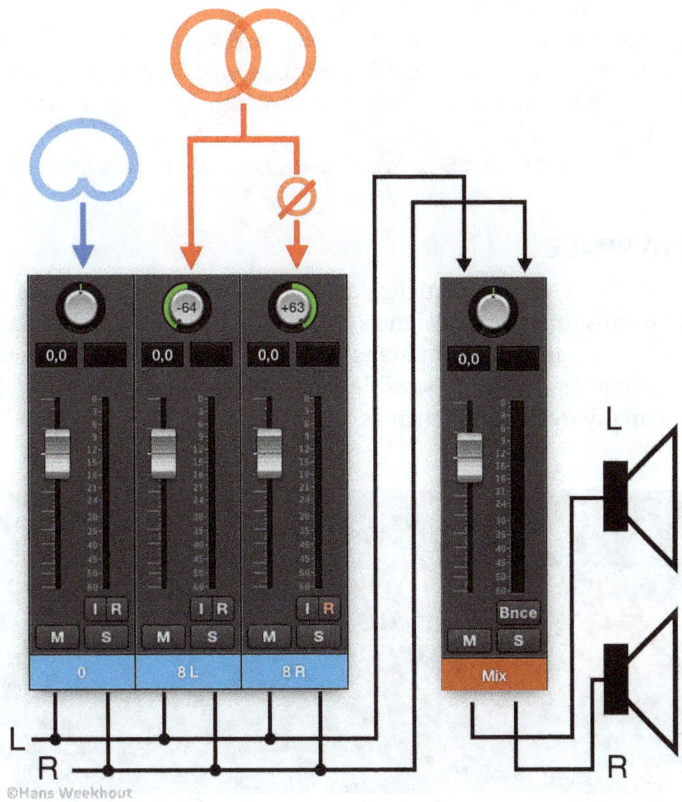

FIGURE 6.8
M/S in the mixer: the 8 signal is split; one of the channel's polarity is reversed.

LEARN HOW TO HEAR PHASE ISSUES

Step 1. In your DAW (Chapter 9, "Digital Audio Workstation and MIDI"), drag a mono audio file to a track and pan it hard left (a full mix usually works best). Copy the file to a second track that's panned hard right, and reverse the phase of (either) one of the tracks. Now, by listening with your head exactly in the middle of the speakers, it may seem like your head gets "separated." Almost freaky. Why "freaky"? Well, normally, a positive wave (node) will cause the woofer in both speakers to move forward. This **pushes** molecules toward our eardrum. When phase is flipped, one woofer moves backward instead of forward, causing one eardrum to **pull**. In nature, this wouldn't be possible, and therefore, the experiment feels a little weird.

When recording, out-of-phase effects are more subtle. Now that you know how this weird pressure feels, it is possible to develop a sensitivity for it. Once detected during a session, alarm bells should ring.

Step 2. Pan both signals in the middle. When two perfect out-of-phase signals are summed, silence will be the result! Why didn't the signal cancel when played over speakers? Well, as we have two ears, there is no single point for the sound waves to merge and cancel.

VIEWING PHASE

Most DAWs have *correlation* meters on board that display phase as a horizontal line: "1" means the left/right signals correlate 100%, (which is effectively mono) and "–1" means the left/right signals are antiphase, while "0" means that the signals deviate as much as possible (maximum stereo width; see Figure 6.9). There are many free meters, such as Flux "Stereo Tool" or Voxengo "Span Free."

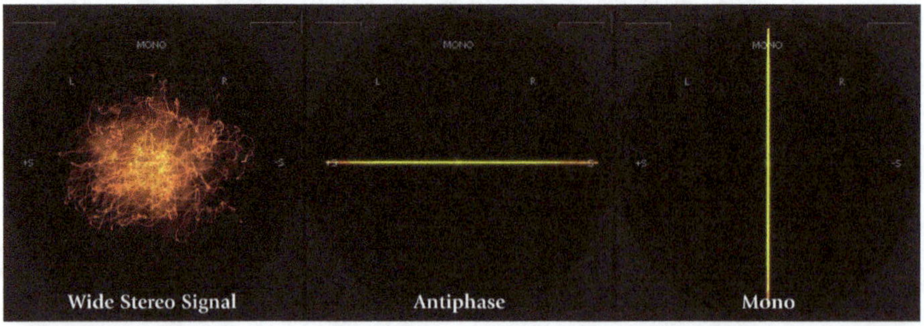

FIGURE 6.9
Goniometers' visualization of the stereo image.

▶ Mono

Although most playback systems are stereo nowadays, there are still many mono devices around. Like mono televisions, mono sound docks, bed radios, telephones or ceiling systems in shopping malls. Playing a mix with poor mono compatibility on such a system will cause changes to both volume and tone of the instrument with phase problems. Apart from this, a mix with too much out-of-phase information cannot be cut onto vinyl.

A mix with too many **out-of-phase** ingredients cannot be cut onto **vinyl**.

CHAPTER 7

Recording | Drums and Percussion

Of all instruments in pop music, drums are probably the most challenging to record. There are many variables, while changes to one mic will affect other parts of the kit. The good news: a good drum recording can be made with just three mics!

SOUND STARTS WITH THE INSTRUMENT AND THE ROOM

We may fiddle around with mics and their positions and process the mix endlessly, but really the musician, the instrument and the room make all the difference. Only a great sounding instrument can turn out great in the mix. So what can you do to improve sound at the source?

With drums, sound starts with proper **tuning** and **damping**. Although it may not be easy to hear a specific tone in a tom or kick drum, drums will sound better when they are in tune with the music. Often-used tunings for individual drums are the root, the fourth or the fifth. As far as **damping** is concerned, there are many options, from Gaffer's tape to tea towels and from O-rings to Moongel. Adding or removing the resonant head on kick or toms makes a big difference too. Damping is dependent on style: many alternative and vintage styles favor tone, while some pop-rock and metal styles prefer attack. Think Led Zeppelin/T-Bone Burnett versus Muse/Paramore.

Volume

Many drummers hit their drums too hard. When playing loud, the tonal balance of the instrument shifts toward the higher frequencies. This causes both the **tone** and the **bottom end** to suffer. It may sound counterintuitive, but playing softer often results in a bigger sound. For cymbals, this is even truer. Smacking cymbals causes extra bleed in all mics. In the mix, the volume of cymbals increases even further because of EQ and compression. Therefore, a drummer striking his cymbals softer is of great help for the production. In this respect, a higher position of the cymbals will also help, as distance (and therefore separation) with other drum mics increases. But in practice, both solutions require the drummer to adapt his style, which could restrain him too much.

Kit Position

The room and the instrument form a relationship. How the kit **addresses** the room is determined by its position. At specific positions, the kit will sound either better or worse. Although moving a complete kit may be a bit of a hassle, a tom, a snare or a kick drum can be used to find a specific position in the room that has a good tone. Clapping your hands at various positions will point your ear toward unpleasant flutter echoes. Knowing what exactly to listen for and recognizing the sweet spots can take a while, but it's time well spent.

Acoustics

With a low ceiling, reflections can cause the sound of the cymbals to suffer. It might be necessary to dampen the ceiling with acoustic foam, curtains or blankets. The same applies to other hard surfaces.

Assistance makes recording easier. With one person moving the mics and another listening in the control room, it will be easier to assess mic positions.

1, 2 or 3 Mic Recording

ONE MICROPHONE

Many classic drum recordings were made with just one mic. In case you're after a vintage sound or own just one mic, there is the great advantage that a one-mic recording always has perfect phase. A large diaphragm condenser or a ribbon mic is the preferred choice here because each has a warm sound. An omni will provide better bottom end, while a cardioid picks up less acoustics, which is handy in case the room's acoustics are anything but perfect.

As for positioning, a mic in front of the drum kit at a height of 3 to 6 ft or above the head of the drummer are frequently used *hot spots*. Best results for mic positioning can be had by testing positions by using your **ears** first. Keep in mind that the kick drum usually turns out soft and the cymbals loud. This could help you decide for a specific position. As one mic allows for the best possible natural sound, it is more important than ever for the drummer to play **balanced**.

TWO MICROPHONES

With a second mic, the kick drum can be *spot miked*. Not only will this improve definition, but it also allows for balancing *overheads* and kick afterward. With **two identical mics**, X–Y or A–B stereo recording becomes possible. Or, with a cardioid and an 8, an M/S array can be set up. Although the width of a stereo drumkit is a great improvement over a mono recording, a third mic might be needed for the kick drum to sound defined.

With **fewer** microphones, it is vital that the drummer plays **balanced**.

GLYN JOHNS TECHNIQUE

Early in the 1970s, producer Glyn Johns (The Who, The Eagles) developed a three-microphone technique to record the drums for Led Zeppelin's album *IV*. Despite the unconventional, asymmetric arrangement of the overhead mics, this technique yields a defined stereo image and has a natural sound. Plus, it is relatively easy to set up.

Point two cardioids at the snare, one about 3 ft directly above the instrument and the other 1 to 2 ft behind the floor tom (1–2 ft above the rim of the floor tom). Pan one mic left in the stereo image and the other right. The trick now is to position the mics at **equal distance** of (the center of) the snare by using a measuring tape (see Figure 7.1). This will cause the sound waves of the snare to arrive **in phase** on both mics. Other components of the kit will, of course, not be in phase, but given the importance of the snare in the average pop song, this is a great advantage. A third mic is used to spot mic the kick drum, we'll talk about that in a minute.

FIGURE 7.1
Glyn Johns technique: both overhead mics are at equal distance to the snare.

Johns himself was actually quite easygoing about the mics' distance to the snare. He used Neumann U67 cardioids, but you could use Beyerdynamic M160 ribbon mics (cardioid!) too. This mic's high-frequency roll-off will prevent cymbals from sounding brittle. Figure 8 ribbons such as the Coles 4038 are also popular for this application, although the room's acoustics will be more critical in that case.

Adding a fourth microphone for the snare allows for balancing this instrument afterwards. Such a mic can also be used to add reverb to the snare. If you were to send the

By playing an instrument **softer**, its sound may turn out bigger, with more bottom end, sustain and **tone**.

signals of Johns's main mics to a reverb, this could easily lead to a messy sound. You could even mute the mic's direct signal and use the reverb only.

Recording with Multiple Mics

Since the 1970s, the **multi-mic** technique (see Figure 7.2) has become the most popular technique for recording drums. When multi-miking, every mic is recorded on a separate track. The individual tracks contain a **deconstruction** of the drumkit, which must be recombined into a coherent sounding instrument in the mix. Although the drummer may have played in perfect **balance** acoustically, this balance must be re-created **artificially** when mixing.

OVERHEADS

For **overheads**, an X–Y or A–B array is most popular. Although an X–Y provides good phase behavior and stereo imaging, the asymmetrical setup of a drum kit limits options here. Because what's the exact center of a drum kit? The snare? Or the kick? Positioning an X–Y above the snare causes the kick to appear off-center and vice versa. Prioritizing the snare seems obvious, since you're not likely to capture a useful kick signal with an overhead array. Increasing the height of the array will cause kick and snare to appear more centered, however, at the expense of a less detailed stereo-image, less bottom end and more of the acoustics.

By **selecting** and **positioning** the microphones, **production** has started.

With an **A–B**, it might be easier to find mic positions that allow both kick and snare to appear in the middle. Plus, the use of omni mics allows for better bottom end and more room reflections to be captured.

Of all stereo arrays, **M/S** is best mono-compatible. Be aware that the 8 will look sideways only, so the acoustic properties of the sidewalls are more important than those of ceiling and floor.

Spot miking cymbals is the most controlled option, although it will result in a sound that's less organic. The closer the mic, the less bleed from the rest of the kit. Pointing the mic toward the middle of the cymbal captures hi frequencies and attack, while the edge produces a warmer tone with more bottom end. Too close to the edge and the sound will change during the up and down movement of the cymbal.

As for positioning, **higher** positions not only capture more room reflections but will help the kit to sound natural, with a slightly longer decay of the cymbals. **Lowering** the mics creates separation between the cymbals and the rest of the kit.

Pencil mics capture a very natural sound as they feature a fast transient response and an extended frequency spectrum. Large diaphragm condenser mics and ribbons sound warmer and prevent cymbals from sounding brittle.

> **TOP 3**
>
> **OVERHEAD MICS**
>
> 1. Coles 4038, or a large-diaphragm condenser
> 2. Neumann KM84/184 or other pencil mics from AKG, Schoeps, Audio Technica or DPA
> 3. Condenser mic of choice

KICK DRUM

Kick drums with a resonant head decay longer. Although this results in more tone and low end, it could cause the mix to sound muddy or cover up the bass. As a solution, a hole can be cut, or the resonant head can be removed completely. With blankets or coats, decay can be shortened further. The length

FIGURE 7.2
Multi-mic setup for drum kit. AKG D112 at kick drum, Shure SM57 at the snare, Sennheiser MD421 at floor tom, Neumann KM184 at both hi-hat and crash cymbal. Yamaha NS10 woofer for sub lows.

FIGURE 7.3
Two kick-out mics at kick drum: NS10 woofer and Neumann U47.

of the kick can make all the difference for the mix's low end. Although a short kick leaves space for other instruments, it could cause the mix to lack the low end. The "right" decay time is dependent on the tempo of the song and repetition of the notes. Each song might need custom damping.

A **kick-in** mic can be used for picking up the attack/beater aspect of the kick. Varying its position and angle has a significant effect on tone. Close to the beater, you'll capture more beater and less tone, while angling it toward the resonant head results in more tone. To further reduce the high frequencies of the beater, the mic can be tilted. The most-used position is probably just through the resonant head's hole. As long as the mic is in, the mic picks up little leakage, which is a great advantage. To further reduce spill, a blanket can be used to cover the resonant head side of the kick.

Cardioid dynamic mics are most popular for this application. Some dynamic mics (like the AKG D12 and its successor D112) have a tailored frequency response to emphasize the kick's sweet spots, namely, the bottom end (100 Hz) and the beater frequencies (3 kHz). As an alternative, the click of the beater can be recorded at the batter side of the kick. As long as this (cardioid) mic faces down, leakage from the rest of the kit is minimized, as it arrives at the mic's (insensitive) backside.

PHASE

When multi-miking, it's impossible for all signal to be in phase. How can you approach this, which signals should be inverted? Well, the overheads are the reference. Once these are set up, the phase relationship of all other mics can be aligned with the overheads. Start by comparing the snare with the overheads, then proceed with the other mics. The setting with the best bottom end is generally the right setting.

As low-frequency sound waves need some time to build, a **kick-out** mic is the ultimate for picking up big bottom end. Suitable distances vary from a few inches to 3 ft. At a greater distance, the decay will be slightly longer, but the leakage will increase. To reduce leakage, you can build a tunnel with blankets over mic stands.

As low frequencies are omnidirectional, a kick-out mic can also be positioned next to the kick drum. Microphones with a linear low end, such as large-diaphragm condenser mics or ribbons, are suitable candidates.

TOP 3

KICK-IN MICS

1. AKG D12/D112, Audix D6, Shure Beta 52A
2. Sennheiser MD421
3. Electrovoice RE20

TOP 3

KICK-OUT MICS

1. Neumann U47 Fet-i
2. A large-diaphragm condenser or ribbon
3. NS10 woofer unit (cheap option)

A CHEAP KICK-OUT MICROPHONE

As a transducer, dynamic microphones are equivalent to speakers. By connecting an SM57 to an amplifier, it will produce sound (do not try this at home, though). Similarly, a speaker generates a current once the cone is set into motion by sound waves. This opens possibilities! By accident, engineers found that the woofer unit of a **Yamaha NS10** speaker works particularly well on kick drum (see Figure 7.3). After soldering a cable with an XLR connector to the woofer, it's output is sufficiently high for recording through any standard pre-amp or interface. By Googling "ns10 woofer replacement," you will find either the original or a replica. Non-Yamaha woofers may perform just as well.

FIGURE 7.4
AKG C414 XLS with multiple polar patterns.

SNARE DRUM

Top Mic

As the snare usually needs to be loud in the mix, leakage is more critical than with any other instrument. After adding treble with EQ, spill from hi-hat (and cymbals) increases even more. That's why you'll want to point the (insensitive) backside of the mic toward the hi-hat. Another trick is to stick a piece of **foam** over the mic or mic stand, just between the snare and hi-hat.

As for positioning, there are basically three different techniques that yield a specific timbre. The most popular is the position where the mic points toward the center of the snare, at a height of 2 to 8 in. above the rim (see Figure 7.5). This position captures the best attack. Pointing the mic toward the rim while looking downward emphasizes overtones and will result in a more metallic tone. Swiveling the mic toward a horizontal position yields more tone and decreases attack.

FIGURE 7.5
Shure SM57 pointed at the snare's center.

The "desert-island" microphone for snare must be the Shure SM57. Although the poor bottom end of this mic helps reduce kick bleed, it may, at the same time, cause the snare to sound thin. For a better bottom end, a Sennheiser MD441 is a good alternative. For more "crack" in the sound, that is, better transients and a beautiful top end, a condenser mic such as an AKG C451, a Neumann

KM84/184 or an AKG C414 can be used. Tying multiple mics with tape minimizes phase issues and allows finding a suitable tone in the mix.

Bottom Mic

Equally important for the snare is the rattling metal wires at the bottom head. The resulting top-high frequencies provide for sizzle and help the snare's audibility in a full mix. In order to capture and control this aspect, a second mic can be used under the snare, facing upward (see Figure 7.6). Greater distance yields more tone, but kick bleed may prevent going too far. Popular mics are the Sennheiser MD441 and AKG C414/C451.

ANTIPHASE

Having both a top mic and a bottom mic on the snare (or tom) poses us with a clear case of antiphase. Here's what happens: hitting the snare causes the top head to move downward. This will **pull out** the membrane of the top mic. In the bottom mic, the opposite happens: as the bottom head moves downward, the membrane of the bottom mic (which looks upward) is **pushed**. Therefore, you'll want to switch polarity on one of the mics. As you know, the position that yields the most low end has the least phase issues.

FIGURE 7.6
Sennheiser MD441 pointing at the metal wires of a snare.

TOP 3

SNARE TOP MICS

1. Shure SM57 (cheap option)
2. Sennheiser MD441
3. Neumann KM84/184, AKG C451/C414

TOP 3

SNARE BOTTOM MICS

1. Sennheiser MD441
2. AKG C414
3. A condenser mic

> ***WITH A GREAT PERFORMANCE, IT DOESN'T REALLY MATTER HOW IT SOUNDS.***
>
> **—T-Bone Burnett**
>
> In the context of this book, it may sound contradictory, but it is better to have a fantastic performance with less-than-ideal sound quality than vice versa. A performance communicates with the listener on an emotional level; a good recording **helps.**

TOMS

As toms and snares are similar, most snare techniques can be used on toms too. Although hi-hat spill is less of a problem here, there's always the risk of bleed from the cymbals and snare. Pointing the mic's insensitive side away from those instruments might help. As for mic choice, toms produce quite a bit of low end. That's why the Sennheiser MD421 is a go-to choice for many engineers, although large diaphragm condensers are popular too.

FIGURE 7.7
Sennheiser MD421 on tom. Be aware, this mic has a hidden five position switch in the XLR plug. At the "S" setting ("*Sprache*"/speech), a hi-pass filter causes the mic to produce less bottom end. For toms, the "M" setting ("*Muzik*") will provide better bass response.

Toms produce most of their tone at the bottom head. Using a bottom mic allows capturing and controlling this aspect in the most beautiful way. As many drummers use three toms, adding three mics will cause bleed to increase correspondingly. This can be destructive for the total sound of the kit. On the other hand, recording is often about anticipating and covering all possible situations. So why not sacrifice some extra tracks and mute the bottom mics for the time being? On a featured tom section, the bottom mics could just give that little bit of extra "oomph" that makes the sound outstanding.

TOP 3

TOM MICS

1. Sennheiser MD421 (see Figure 7.7)
2. Neumann U67
3. A dynamic mic (economic choice)

HI-HAT

Many engineers don't mic the hi-hat, as it often arrives sufficiently loud enough in other mics. But an indirect sounding hi-hat could become a problem sooner or later in the production. So what are the options for miking the hi-hat? To reduce snare spill, a hi-hat mic can be angled at the outer edge of the top cymbal. Positioning it toward the middle results in a tighter sound. Never mic the hi-hat in between its cymbals, as the airflow of closing cymbals can cause damage to the membrane. Preferred mics are condensers from Neumann (KM84/184), AKG (C451), Schoeps, Audio Technica or DPA.

THE ULTIMATE FIX FOR LOUD CYMBALS

If cymbal bleed causes so much trouble, why not record them separately? For Queens of the Stone Age's "No One Knows," drummer Dave Grohl played the drums in two takes. In the first take, rubber cymbals were used. In the second take, the drum heads were covered with rubber pads. The result is a very controlled sound, with the cymbals soft in the mix. Way before, the same trick was used for the drum recording of Tom Petty's *Wildflowers* album.

ROOM

A properly recorded room signal may turn out to be the ultimate glue for the multi-miked drum signals. Room mics can also be used for making the drums sound big, if not huge. A good sounding space "begs" to be recorded, as natural acoustics usually sound better than artificial reverb. Even with just one crappy mic left, recording the room is worth the effort. Many positions can be used: on the floor, in far corners, behind gobos or even down the hallway.

For stereo recording, A–Bs are most often used, because of a superior bass performance and wide *soundstage*. In case you have sufficient mics available, multiple A–Bs at different distances can be set up, each with its own *dry/wet* ratio. This allows finding out later which distance works best. The chosen A–B might even change per song section. In the verse, for example, the close A–B may be used, while the chorus features the distant and bigger-sounding array.

TRICKS WITH FIGURE 8s

Although the sensitive backside of bidirectional mics regularly poses us with problems, it can be used to our advantage too.

- Set up two 8s next to the kit, at roughly 1 ft high, with their nulls pointing at the cymbals. That's the position with the least direct sound and most of the acoustics.

(continued)

- Set up two 8s, 2 to 6 ft out in front of the kit. The mics should be angled up–down instead of front–back. At the same height as the cymbals, the figure 8 characteristic rejects direct sound while favoring the room.
- Sticking an 8 halfway between kick and snare could provide you with alternative colors for these instruments. The signal can be added to the other mics. Cymbal bleed is rejected as it arrives at the mic sideways.

LAST BUT NOT LEAST

Context

It's not a drum solo you're recording. A kit that sounds fantastic in itself might sound bad in context. Therefore, every change should be verified **in context** of the song.

Experiment

There are engineers that swear by using an SM57 for the kick drum. Cheap mics at unlikely positions could add useful colors to the kit. Processing such a signal with a compressor or guitar amp opens even more possibilities. On many classic records, these unconventional techniques have turned out crucial for an interesting sound.

Percussion

Recording percussion can, on one hand, be easy and difficult at the same time. It all depends on the genre and sound needed. If a natural sound is what you're after, pencil condensers are the most obvious choice. Having a percussionist perform in front of a stereo array will probably give good results. An X–Y seems obvious, as for its detailed stereo-imaging and good mono-compatibility. Recording levels should be watched closely, as percussion instruments produce loud peaks that easily cause overloads. So far the easy part.

The fierce transients and unforgiving amount of high-frequency energy of many percussion instruments are so different from other instruments that it can be hard for them to blend in the mix. This is often the case with instruments like tambourine, shaker or triangle. Therefore, we have to come up with solutions to "smear" transients and to make the signal warmer. Here are a few tips to tame the energy:

- Use large diaphragm condensers or ribbons.
- Move away from the mic. Room reflections that interfere with direct sound will cause smearing of transients, which in this situation is exactly what you want.
- Other than panning a single mic, recording in stereo causes the instrument to appear in the stereo image naturally. This works even for single instruments like tambourine or shaker.

TOP 3

PERCUSSION MICS

1. Neumann KM84/184 or other pencil mics from AKG, Schoeps, Audio Technica or DPA
2. Condenser mic or ribbon of choice

CHAPTER 8

Recording | Other Instruments

Once the drums are recorded, one of the most challenging tasks is out of the way. Although guitars, bass, piano, organ and vocals generally need fewer microphones, that doesn't mean recording them is easy. Just because there are fewer signals, the quality of a single mic can be crucial. So what's the approach, and which pitfalls can be expected with other instruments? A mic that sounds best on its own may not be the best choice in the mix!

Bass Guitar

Every successful recording session starts by listening to the acoustic sound of an instrument first. This will direct your ear to shortcomings like resonances, noises or any missing frequencies. In case of a bass guitar, listen to the amp first. If the cabinet contains multiple speakers, one of them could sound better than others. A close listen (on a moderate volume) will reveal the best-sounding speaker.

As far as positioning is concerned, pointing the mic at the heart of the speaker (**on-axis**) produces the brightest tone. Moving the mic toward the edge of the speaker (**off-axis**) will dampen the highs. At a distance of 0 to 8 in. from the grille, the mic will usually capture the full spectrum of the bass (see Figure 8.1).

As the low frequencies of a (four-string) bass extend to 41 Hz (low E), mics with a proper low-frequency response are needed. Fortunately, there are many good options in the dynamic department, like the Electrovoice RE20, Sennheiser MD421 or Shure SM7. Large-diaphragm condensers and ribbons will work too, given their linear frequency response in the lows. Don't forget to shield a ribbon's backside, though; otherwise, leakage mixes up with the bass signal. Last, dynamic kick mics may work well too, as of their bass-tailored frequency response.

FIGURE 8.1
Blue Microphones "The Mouse" on bass amp.

Be careful, some amp-mic combinations lack energy in the lowest octave (40–80 Hz). When bottom end lacks, it is impossible for the bass to carry the song. Playing a chromatic scale downward will easily reveal any deficits; this can be verified on a **spectrum analyzer** (see pp. 126–127). In case of doubt, always capture an extra signal of the bass by means of a *DI box* (see Figure 8.2).

Guitar amplifiers may suit the sound of **bass** very well.

TOP 3

BASS GUITAR MICS

1. Electrovoice RE20, Sennheiser MD421, Shure SM7
2. Large diaphragm condensers like Blue "The Mouse," Neumann TLM103 or a ribbon
3. Kick mics: AKG D12, D112, Audix D6, Shure Beta 52A

BORN TO DI

A *Direct Injection box* or "DI," converts the bass's tiny voltage so that it can be recorded on the computer, without an amp/cabinet. DI boxes can either be *passive* or *active*: the latter type needs phantom power (or a 9-Volt battery). Most audio interfaces contain a DI circuit too; this is activated by pressing the "Hi-Z/Instrument"-button.

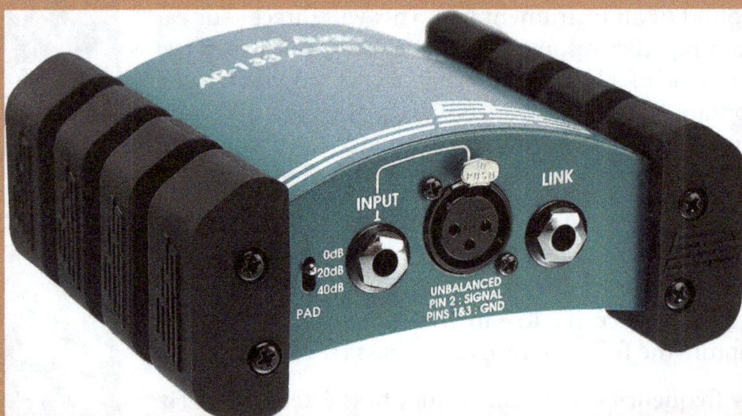

FIGURE 8.2
BSS AR-133 DI box.

Source: Photo courtesy of bssaudio.com.

▶ DI RECORDING

In case there's no amp available, or if you want to prevent bass bleed in other mics, a DI box allows for easy bass recording (see Figure 8.3). DIs have a very linear frequency response and therefore produce a natural and clean bass sound. At the same time, a DI signal will contain all the detail of fret-rattling and string-sliding, and this won't suit all genres.

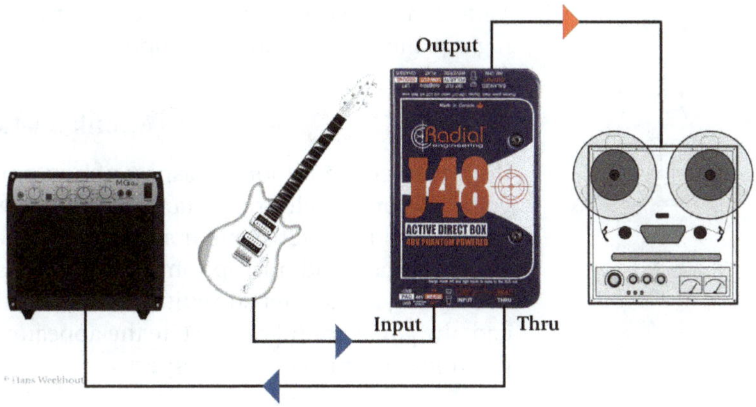

FIGURE 8.3
DI recording
Source: Courtesy of radialeng.com.

If color is what you want, then an amp provides the best results. An amp evens out dynamic differences, which reduces the need for compression in the mix. An amp will also provide more of that "thump in your chest" quality. In practice, most engineers opt for recording **both** signals, as it allows them to balance DI and mic signals afterward. In a way, DI recording can be seen as an insurance policy, as it leaves options open for *re-amping*.

▶ RE-AMPING

Re-amping allows adding the sound of an amplifier **after** recording. This can be done by sending the DI signal back to the recording room and record the mic signal to a new track (see Figure 8.4). As re-amping is done later, tailor-made settings can be found that fit the track precisely, without the noise of the band. It can be even be done in the mix! Re-amping is common for bass and guitar, but there are also creative possibilities for keyboards, drums and vocals. On a technical level, the signal from the computer is not appropriate for feeding an amp. That's why a re-amp box is needed, which, in fact, is a DI box working in reverse (there are, in fact, DI boxes that can be used for re-amping).

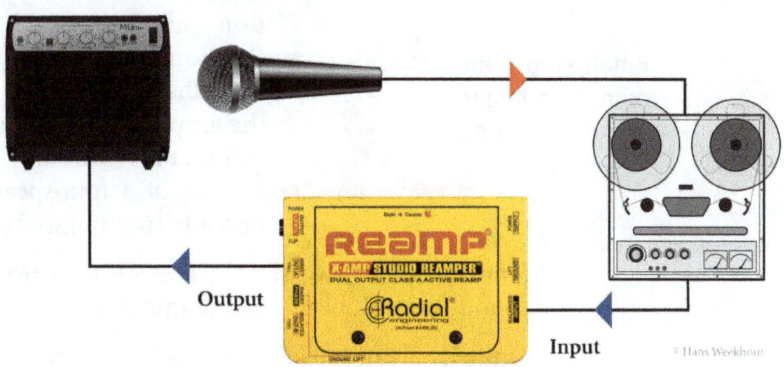

FIGURE 8.4
Re-amping.
Source: Courtesy of radialeng.com.

Popular re-amp boxes are from Radial (Pro RMP), ART (RDB), Little Labs (Redeye) and Eventide (Mixinglink).

Millennia Media has the TD-1: besides a re-amp option, this device includes a high-quality microphone pre-amplifier too, with both tube and transistor options.

FIGURE 8.5
Modern classic: Royer 121 ribbon.
Source: Photo courtesy of royerlabs.com.

Electric Guitar

If **sound** were an important aspect of pop music, then this is especially true for guitar. Recording electric guitar is almost an art. Just ask the famous hard-rock bands from the 1980s: guitar sessions involved endless assessing of dozens of amps, cabinets and microphones. Nowadays, sessions are less extensive usually, but finding a characteristic guitar sound is still considered the holy grail. In the end, the guitar sound can dictate the appearance of a record. The mics and their positions are crucial in this respect.

TOP 3

ELECTRIC GUITAR MICS

1. Shure SM57
2. Sennheiser MD421/441, MD409, E604
3. Royer R121(see Figure 8.5), R122
4. Beyerdynamic M160

▶ MIKING CABINETS

Finding the right sound starts with the best position of the amp. Here are some considerations:

- Positioning an amp in a room's corner might cause low frequencies to build up unevenly.
- To avoid wall reflections, a diagonal and/or slanted position could help.
- Positioning a cabinet on chairs helps reduce reflections from the floor. Always use two chairs left and right under the amp instead of one chair in the middle, as this could obstruct the airflow.
- In case of a four-speaker cabinet, miking top speakers instead of bottom speakers helps to reduce floor reflections.

Small amps can produce a larger sound than **large** amps.

Similar to bass, we should first verify which speaker of a particular cabinet sounds best. How should the mic be positioned?

On-axis, a mic picks up more brilliance while moving it toward the edges produces a warmer sound; the bottom end will slowly take over. **Angling** the mic has a similar effect. A frequently used hot spot is the position where the mic is at a small angle with the cabinet, slightly off-axis, and close to the speaker grille (0–1 in.). Starting off with this position allows experimenting: small changes to the angle and position of the mic will cause significant changes to the sound! After finding the ideal position, this can be marked with painting tape on the speaker grille. To help make decisions, remember that maximum presence (or even aggression) is often required, but the guitar may **never hurt the ear**.

Using **two microphones** on a guitar cabinet allows for many more sound options. Tying a second mic to the first minimizes phase differences (see Figure 8.6). Alternatively, aiming the second mic on a different speaker could add other useful elements to the sound, be it at the expense of phase issues (. . .).

Surprisingly, a cabinet's **side** or **back** is useful for miking too (see Figure 8.7). Although there's obviously not much aggression to capture here, the warm tone with a plentiful low end could support the front mic's signal in an unexpected and beautiful way. This technique allows composing the guitar sound with just two faders: one for the lows and one for the mid-highs.

As far as **phase** is concerned, this situation is similar to the snare. Once the speaker moves forward, the front mic's membrane is pushed while the backside mic's membrane is pulled. This means out-of-phase mic signals. Well, not exactly out of phase, of course, as the spectra on both sides of the speaker are

FIGURE 8.6
Shure SM57 and Sennheiser MD421, off-axis and tied for best phase relationship.

FIGURE 8.7
Beyerdynamic M160 (ribbon) at the back of the cabinet.

different. In order to match the signals as closely as possible, positioning both mics equidistant to the speaker's cone could help. Then, one of the mics can be phase-reversed.

After close miking, the sound may lack width and depth. That's where **ambient mics** can help. Positioning a ribbon or large diaphragm condenser anywhere between 1 and 4 ft, pointing at the exact middle of the speakers, allows for some air in the sound. Although ribbons are popular for this application, faraway omnis catch more low end as they lack the proximity effect. Last, a *true* far mic can be considered, for instance, by positioning a ribbon in a corner.

FIGURE 8.8
For an engineer, walking back and forth between studio and control room in order to change the mic or its position, is confusing. After making an adjustment, it's hard to remember the exact sound of the previous position. That's why working with an assistant is almost a necessity. Producer Eric Valentine came up with an ingenious alternative: a remote-controlled mic robot. Now, the Dynamount company is producing a Wi-Fi-controlled version of it. The included app allows storing positions as presets.

Source: Photo courtesy of dynamount.com.

▶ INTONATION

It's good practice for guitarists to tune their guitars in between takes. But perfect tuning for open strings doesn't mean perfect tuning when played in position. In practice, many guitars lack intonation. As a result, precious studio time is lost because of tuning and discussion between takes. Therefore, guitarists should always meticulously check their guitars' intonation **before** recording. Adjusting the bridge and neck is not too difficult; there are many DIY videos on the internet. Or send your guitar to a guitar store.

Acoustic Guitar

Acoustic guitars produce an extremely wide frequency spectrum, with many overtones. Similar to shakers and cabasas, it can be hard for the high frequencies to blend well in the mix. In order to slightly smear those unforgiving high frequencies, it can be useful to capture some ambience along with the signal; even when the acoustic conditions are less than ideal. Be careful, though; compression during mixing and mastering will cause the volume of the reflections to increase. When in doubt, always choose for the drier option, or track room mics separately.

▶ POSITIONING THE MIC

Many people expect the best tone of an acoustic guitar to come from the sound hole, but actually this position emphasizes the low end. In a sparse arrangement, the low end could prove a useful foundation for the sound, but in a dense arrangement, the bottom end can easily get in the way of other instruments.

Actually, the most common position for miking acoustic guitar is at the join between the guitar's neck and body (see Figure 8.9). This is the spot that often has a pleasant balance throughout the spectrum. Pointing the mic toward the sound hole will pick up more attack of the picking/strumming while suppressing fret noise and finger squeaks. Pointing or moving the mic toward the guitar's head will result in less attack and more brightness of the strings themselves—at the expense of increased fret noise and finger squeaks, though.

Common mic distances are anything between 5 in. and 3 ft. Closer positions yield a detailed and dry sound, while moving the mic farther away produces a rounder tone, with slightly fewer details and more acoustics. Distant mic positions prevent loud finger squeaks.

FIGURE 8.9
Sweet spot for acoustic guitar mic: at the join between the guitar's neck and body.

▶ ALTERNATIVE POSITIONS

For acoustic guitar, there are many good alternative mic positions. One is with the mic looking sideways over the pick hand, pointing at the 12th fret. Another hot spot is the mic looking downwards at the 12th fret, over the fret hand shoulder. This captures more of a warm sound. Mics at different positions can be combined, but of course, phase issues will increase.

Which mics are suitable? The brilliant character of an acoustic guitar is usually best served by a condenser mic. Pencil condensers capture the most natural sound, due to their fast transient behavior and extended frequency response. To soften the tone, pencil mics can be exchanged for either a large diaphragm condenser or ribbon, although the latter could sound too bassy and lack clarity.

FIGURE 8.10
Singer-songwriter setup with two ribbons. Leakage is rejected as it arrives in both mics' nulls (insensitive sides).

Last but not least, never underestimate an SM57. For an alternative, rock 'n' roll type of acoustic guitar sound, it could just provide the appeal that's needed.

TOP 3

ACOUSTIC GUITAR MICS

1. A pencil mic from Neumann, AKG, Schoeps, Audio Technica or DPA
2. A ribbon or large diaphragm condenser
3. Shure SM57

ACOUSTIC GUITAR WITH PICKUP

Some acoustic guitars have built-in pickups. Although a pickup allows convenient DI recording, the quality of these devices varies, and they often lack warmth. Acoustic guitars with an internal microphone generally sound better, but the signal will be bone dry.

▶ RECORDING IN STEREO

If it is appropriate for the track, a beautiful wide stereo guitar is not hard to record, actually many stereo techniques could work. The signal of an X–Y has good mono compatibility and is relatively tolerant for bad acoustics (depending on the distance, see Figure 8.11). An A–B captures good bottom end with more acoustics, while an M/S array produces the widest stereo image, with zero phase issues. M/S also allows adjusting the width **after** recording, without any negative side effects.

Stereo recording is especially suited for guitars in **sparse** arrangements. Then there's space for them to occupy the full width of the stereo image. In **dense** arrangements, it is more common to assign **mono** guitars a specific spot in the stereo image.

FIGURE 8.11
X–Y stereo array for acoustic guitar.

> **TIP**
>
> By listening on their headphones, guitar players can help the engineer by varying their position in front of the microphone. Small movements will cause the guitar sound to change drastically!

Hammond Organ

The Hammond organ has been part of rock 'n' roll's furniture for decades. As the organ itself doesn't produce sound, an amp-speaker combination is needed. Since 1941, the go-to brand for such a device is Leslie (Figure 8.12). The amp and speaker are crucial for the raw edge and width of a Hammond. Before recording Leslie, we'll take a sneak peek in his box.

FIGURE 8.12
Leslie 122.
Source: Photo courtesy of www.hammondorganco.com.

A Leslie box is a large speaker cabinet containing a tube amp, an (optional) spring reverb, a bass speaker and a treble speaker. The bass speaker, which faces downward at the bottom of the box, projects sound onto a rotating drum. Mid-high frequencies are propagated by two horns at the top of the cabinet. The horns are responsible for the famous chorale and tremolo effect and can be controlled from the organ.

Because the low and high frequencies are physically separated, this has consequences for mic placement:

- As **low frequencies** are omnidirectional and carry far, positioning the bottom mic is less critical. A distance of 0 to 12 in. is often a good starting point. Due to its warm tone and proper bass response, a large diaphragm condenser is obvious here, but really any bass guitar mic (see the earlier discussion) could work well. Note: although you could record the bottom end in stereo, our ears can hardly discern any directional information from the low frequencies. Besides, wider bass could cause the low end of the mix to become unstable.
- **Mid-high frequencies**: to capture the widest image of the rotor effect, two close mics can be positioned sideways to the Leslie box, pointing at the sound holes. SM57s are popular because of their 6-kHz edge. An X–Y at a distance of 2 to 8 in. with either SM57s or pencil mics could produce good results too.

Using a **Leslie box** to re-amp instruments like guitar, keyboards or vocals opens a world of **creative** options.

TOP 3

HAMMOND/LESLIE MICS

Top: SM57s, condenser mics
Bottom: most kick or bass guitar mics will work, like an Electrovoice RE20 or a large diaphragm condenser

Grand Piano

With a grand piano, the quality of the acoustics is more critical than it is with acoustic guitar. In the case of bad acoustics, walls and floors can be covered with gobos, curtains or carpets. The ceiling can be covered with acoustic foam over mic stands. Then you should decide whether to position the mics **above** or in **front** of the piano. In the latter case, the lid will cast sound toward the front mics. With mics **above** the piano, it might be better to take **the lid off.** This will result in a slightly more even frequency response as high frequencies arrive in the mics without the lid interfering. When recording a band, the lid can be closed in order to prevent spill.

Stereo recording an instrument this big seems obvious. An X–Y, for example, will result in a beautifully detailed stereo image. By moving the X–Y closer, bottom end and presence increase as of the cardioid mics' proximity effect. In a proper room, M/S can also produce good results, although an A–B at some distance will result in a better bottom end. Depending on the desired dry/wet balance, mics can be positioned further away or closer. Close positions lead to a more detailed, bright and direct sound, while farther away, room reflections and high-frequency loss cause the tone to become warmer and more spacious. Let's look at three different miking techniques for grand piano, ranging from classical (warm) to pop (aggressive):

1. The **classical** tone (Figure 8.13). With the lid fully off, an A–B at a distance of 1 to 2 ft from the casing, at a height of 5 to 6 ft, will produce a warm and relatively spatial tone. Use a mic spacing of approximately 1 to 2 ft. The array can be positioned either in front of the piano or at its foot.

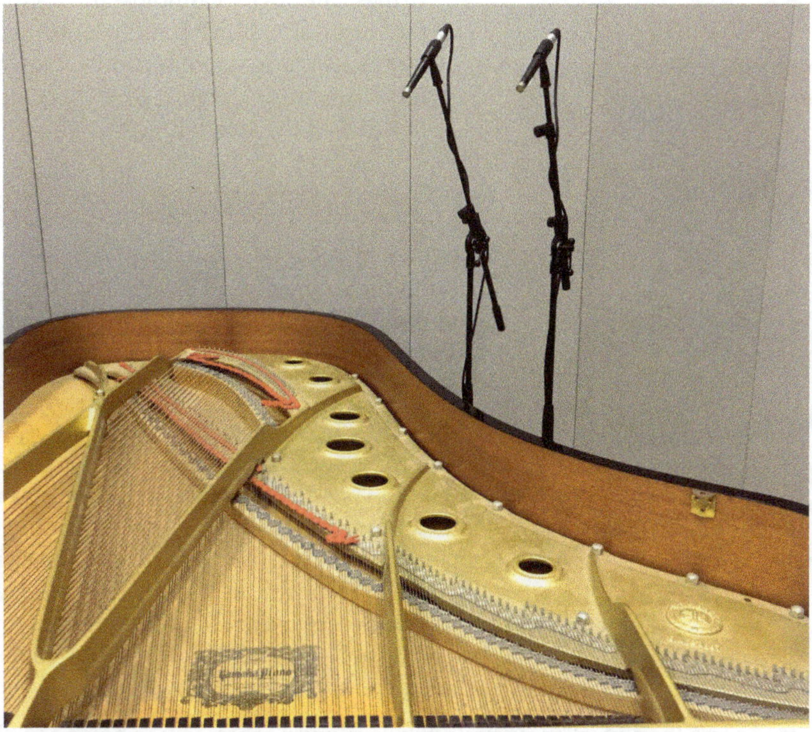

FIGURE 8.13
Grand piano, classical tone: distant A–B at the piano's foot with two DPA 4006 omnis.

> **TOP 3**
>
> **PIANO MICS**
>
> 1. Pencil condensers from Neumann (KM 83–84, KM 183–184), AKG (C451), Schoeps (MK2, MK4), Audio Technica or DPA (4000 series)
> 2. Large diaphragm condensers like Neumann U87 or AKG C414
> 3. Ribbons
> 4. Budget choice: Shure SM57s for a more "rock 'n' roll" type of sound

> When in doubt, always choose for the **dry** option, or record room mics on separate tracks. **Reverb** can always be added in the mix.

FIGURE 8.14
Grand piano, brightest tone: X–Y with Neumann KM184 cardioids close to the hammers.

2. For more bite and less ambience, the array can be positioned 1 to 2 ft above the soundboard, approximately 2 ft behind the music stand. Pointing the mics at the hammers yields more attack.
3. Finally, an X–Y close to the hammers over the high strings of the piano results in the most aggressive tone (see Figure 8.14). Despite being far away from the low strings, the X–Y's cardioids are bound to pick up quite some bottom end, as low frequencies are omnidirectional. Note that too close an X–Y results in a "hole in the middle" of the stereo image.

Upright Piano

Before recording an upright piano, its position in the room should be assessed. Positioning the instrument off the wall (and corners) will prevent any low frequencies to build up. Because of the size of an upright, stereo recording seems obvious.

In a proper room, an A–B at a distance of 2 to 4 ft at the back of the piano will capture a natural and warm tone. For a brighter timbre, the piano's panels can be taken off, and the (wide) A–B can be positioned close to the player's shoulders (see Figure 8.15). With the lid open, an A–B (or X–Y) can also be positioned above the piano.

In case of less than ideal acoustics, such as a living room, you're bound to close miking, probably with an X–Y because of the cardioids capturing less room reflections.

Keyboards

Keyboards are commonly recorded with DI. DI recording is convenient, and DIs capture the most natural sound. In a band setting, however, the direct character of electronic sounds may not match the organic sound of the rest of the instruments. Recording keyboards through an amp (or re-amping afterward) will help in such a case. Guitar amps cause the spectrum to shrink, and they'll add character and distortion. Electromechanical keyboards, such as Fender Rhodes, Wurlitzer, Yamaha CP70/CP80 or Hohner Clavinet often benefit from re-amping too. Don't forget the Leslie option here!

FIGURE 8.15
Piano A–B: DPA 4006 omnis close to the strings.

FIGURE 8.16

Vocals

Although you could use multiple mics for recording a vocal, even the smallest phase cancellation would cause the human voice to sound unnatural. Apart from this, comping and processing extra tracks isn't really an activity to look forward to. Therefore, vocals are usually recorded with one mic. This doesn't mean that recording vocals is easy. On the contrary, the vocal is **the** element that communicates the song. And just because there's only one signal, its quality is decisive. So how do you get the most out of a vocal recording?

1. Match the Microphone With the Voice

The best microphone for vocalist A isn't necessarily the best mic for vocalist B. A female falsetto calls for a different mic quality than a low male voice. Last, the best sound for the lead vocal in solo isn't necessarily the best sound in the mix. Therefore, experimentation is the motto here.

Almost without exception, classic vocal mics are large diaphragm cardioid condensers. Why? Well, cardioids feature the proximity effect and capture less of the room. Large diaphragms produce a slightly warmer tone than pencil mics. Although they lack the proximity effect, don't rule out omnis completely. With an omni, voice quality will hardly change upon movement of the vocalist. Besides, omnis are (way) less sensitive for plosives, while their sound is slightly richer than cardioids. Legendary vocal mics are the (very) expensive AKG C12 (and its siblings AKG C414, Telefunken ELA-M250/M251), Neumann U47/U67 and Sony C800G (Figure 8.17).

FIGURE 8.17
The AKG C12 is considered legendary, it's been the favorite mic for Frank Sinatra, Prince, Jamiroquai, Tom Petty and Alanis Morisette. As phantom power had yet to be invented, this 1953 (tube) condenser was companioned with a dedicated power supply. A secondhand C12 in good condition could easily set you back US$15,000.

Source: Photo courtesy of akg.com.

The good news: good quality clones are flooding the market by brands like Warm Audio, Peluso, Bock and Sony itself. Virtual mics can closely mimic the sound of the classics. On the internet, there are manuals to build vintage mics yourself. And the best news: a cheap mic may turn out superior to an expensive model! An illustrious example is the Shure SM7, as it was used on Michael Jackson's album *Thriller*. The Shure is a favorite for hard rock and metal vocals too. And Tom Yorke (Radiohead) is fond of the Electrovoice RE20. Good results can also be achieved with the (cheap) Shure SM58. Although ribbons may appear dull at first sight, their signal could reveal a beautiful top end after applying EQ.

Here are a few characteristics that may help you decide for a specific microphone:

- How natural does it sound?
- How loud and natural are the *s*'s?
- How close, or forward sounding, is it in the mix?
- How much of the proximity effect is exhibited?
- How directional is it (i.e., how much of the room is recorded)?
- How's the dynamic reproduction? Compare both soft and loud sections of two microphones.

FIGURE 8.18
Microphone shoot-out with Coles 4038, Shure SM57, Sennheiser MD441 and Electrovoice RE320.

TOP 3

VOCAL MICS

1. Neumann U47, U67, U87, AKG C12, Sony C800G, C414
2. Shure SM7B, SM58, Electrovoice RE20
3. A ribbon, such as the Coles 4038

Budget tip: Audio Technica 2020

2. Microphone Technique

By changing their distance to the mic, vocalists can radically change the sound. When singing softly, it usually works well to come **close to the microphone** (1–10 in.) so that the proximity effect can cater for a big, warm bottom end. The vocalist will appear **upfront** in the mix, while the performance will sound **intimate**. Close to the source, the mic will pick up more high frequencies, allowing details to be preserved. All the little crackles, noises and breathing add up for a vocal that sounds **specific**.

When singing loud, on the other hand, it's usually better to step back from the mike (up to 3 ft or more). This will make the vocal sound warmer, as high-frequency waves lose energy when traveling a long distance. Extra distance will also cause room reflections to enter the signal; they will provide for the grease that rounds off any harsh elements in the sound.

FIGURE 8.19
The giant RCA R44, released in 1936, is 12 in. high and weighs no less than 9 lb. This legend is famous for its huge bass response and ultra-warm sound. True-to-the-original replicas are still manufactured by AEA microphones.

3. Record Dry

Reflections that weren't audible during recording may become an unpleasant surprise once the mix and the vocal are compressed. Therefore, it's better to capture soft and close vocals dry. Bone dry.

A dry sound can be achieved by damping the room with absorbent materials, like curtains, blankets or mattresses. The wall **behind** the singer is the most important. This is because any reflections bouncing off the back wall travel over the vocalist's shoulders before arriving at the **sensitive** side of the mic! Another solution for damping reflections is the "**Reflection Filter**" (see Figure 8.20): its circular construction will reduce reflections entering the mic quite effectively.

FIGURE 8.20
sE-Electronics Reflection Filter.

▶ POP SCREEN

Microphones don't like air blasts on their membrane. When singing *p*'s or *b*'s (*plosives*), the gust of wind can easily lead to severe distortion, especially with condensers and ribbons. As a solution, a *pop screen* (see Figure 8.21) can be positioned at 1 to 8 in. in front of the mic. A pop screen is a circular nylon-mesh screen that breaks down any blast of air. Because of their ultralight and fragile membrane, ribbons may need two pop shields. Contrary to what some people think, a pop screen does not reduce *s* sounds.

FIGURE 8.21

FIGURE 8.22
Hi-pass filter on Neumann U87.
Source: Photo courtesy of Neumann Berlin.

▶ HI-PASSING

With vocal recording, it's good practice to use a hi-pass filter of the mic (see Figure 8.22), pre-amp, mixer or audio interface. Not only will this prevent any unwanted sub-low rumble from passing trains or tapping feet; it will also reduce the effect of plosives. Setting this no higher than 100 to 120 Hz can be considered safe for vocals.

▶ SIBILANCE

Although *s*'s and *t*'s (*sibilance*) are essential for the vocal's articulation and aggression, there are many stages during production that may cause them to become overly loud, colored or distorted. Exaggerated sibilance can hurt the ear, disturb the listening experience and cause distortion on both vinyl and radio. In order to prevent *s*'s and *t*'s to become a problem, it's important to prevent excessive sibilance already during recording. When can you expect sibilance to become a problem?

- Certain vocalists produce louder *s*'s than others. With the same singer, sibilance can vary per day.
- Some lyrics contain more *s*'s than others.
- Some mics emphasize *s* sounds more than others (in combination with a certain vocalist).

To combat excessive sibilance, you could try using another condenser mic, or you could replace a condenser with a ribbon. A more affordable trick is to vertically fix a pencil to the mic, straight across its membrane (see Figure 8.23). In case you lose too much of the (precious) high frequencies, your last resort is filling the gap of the vocalist's front teeth with a piece of chewing gum. Although not every vocalist will be excited by the very idea of it (. . .).

FIGURE 8.23
Pencil strapped to the mic to prevent loud *s*'s.

▶ EXPERIMENT

If it suits the style, the vocal can be recorded in the bathroom, with a cheap mic or through a Leslie box. Organic treatments usually sound better than computer plugins. Bold decisions help shape the production and inspire everyone involved. That being said, a lo-fi sound can never be changed into hi-fi, while a clean and natural recording can always be transformed into something vibey.

CHAPTER 9

Digital Audio Workstation and MIDI

A *digital audio workstation* (DAW) is a computer program that records, edits and mixes both audio and MIDI. Such a large-studio-in-a–box has great advantages over the real thing, as it is powerful, mobile and cheap. Now that the last 10 years new DAWs have flooded the market, it is important to know about their differences. Why do the pros work on Pro Tools? What can you do with MIDI, how does it work and how different is the workflow from analog recording?

▶ FAIRLIGHT

The mother of all DAWs must be the **Fairlight CMI**, released in 1979 (see Figure 9.1). It could record audio digitally and playback recordings via a keyboard, and its sequencer allowed *quantizing* (position audio samples rhythmically in time). These features were unheard of in those days. The instrument

FIGURE 9.1
Fairlight CMI (Computer Musical Instrument). A light pen allowed activating functions and draw waveforms.

was quite expensive: only well-to-do artists like Herbie Hancock, Kate Bush, Trevor Horn, Peter Gabriel, Thomas Dolby and Stevie Wonder could afford it. Regarding its limited sound quality, it is hard to believe that the British musicians' union had serious worries if its members would survive competition of the Fairlight (. . .). In hindsight, the gritty sound character caused by early digital technology (8 bit, 28 kHz) has been crucial for the Fairlight's classic status. Songs like "Owner of a Lonely Heart" (Yes), "Relax" (Franky Goes to Hollywood) and "Rockit" (Herbie Hancock) would sound quite different without the use of the instrument.

▶ PRO TOOLS 1.0

With more powerful and affordable computers in the latter end of the 80s, it became possible to use them for audio recording. In 1989, an early version of "**Pro Tools**" came to the market, costing US$6,000. The system allowed recording four (mono) audio tracks on an Apple personal computer. Revolutionary was the possibility to view, cut, copy and paste audio by means of a screen and mouse. And to **undo** actions. Considering the US$150,000 price tag of an analog 24-track recorder, Pro Tools was a bargain for professional studios.

Since then, many DAWs have come to the market, running on more powerful computers. Modern DAWs are capable of replacing a complete studio. Even a medium-priced laptop can handle projects that contain a hundred tracks or more. In practice, the size of a project is limited by the computer's CPU (*central processing unit*) only.

Benefits of the DAW

How does working on the computer compare to working on tape?

1. **Total recall.** In an analog studio, the next day's client will reset the mixer and other gear. This means that, when leaving the studio, you must be absolutely sure about the mix to be final, as there's hardly possibilities for recall. In the computer, simply opening the project allows proceeding exactly from the point where you left off.
2. **Undo.** The last action(s) can be cancelled. Saving the project with a new name allows reverting to versions of days or months ago.
3. **Nondestructive editing.** Editing and copying can be done endlessly and up to the finest detail—without any loss of audio quality. In case the edits are no good, the original, unchanged audio file can be dragged into the project. By the way, DAWs can also process audio **destructively**. This can be useful for sharing audio files. Before executing a destructive function, the computer will display a warning.
4. **Automation.** In an analog studio, multiple hands are needed to "perform" a mix. In the computer, a mix can be built fader by fader, by means of automation. Automation is not limited to faders only; actually all buttons can be automated. Movements can be performed live, or curves can be drawn

with the mouse. Zooming in on the project allows the volume of individual syllables of a vocal to be fixed. After finishing, it's time to sit back, relax and enjoy the mix.

5. **Mixing in-the-box.** In an analog studio, music is stored on tape and mixed on a console. DAWs offer both storage and a mixer in a single device. It is therefore obvious to stay in the digital domain and mix without the help of external devices. This is called mixing *in-the-box*. During the *bounce* (mix), the computer sums all channels into a single audio file. The bounce file can then be used for CD-manufacturing or for digital distribution. Ricky Martin's "Livin' La Vida Loca" (1999) is known to be the first in-the-box mix ever.
6. **Overview.** The overview of a DAW is unrivaled. Zooming in allows viewing and editing individual notes or automation curves. Zooming out gives a nice overview of the project while it's easy to see which instrument plays at a specific position. Individual parts or complete song sections can be moved or copied easily.
7. **Sound quality.** What goes in comes out exactly the same.
8. **Sharing projects.** In order to send a project or collaborate with other people, only an internet connection is needed. With a fast-enough connection, this can be done even in real time.

Disadvantages of the DAW

1. **Latency** (delay). When recording a track on the computer, audio has to be converted to digital in the *audio interface*. The resulting digital stream of zeros and ones travels through the computer's OS (*operating system*) and DAW and is converted back to analog in the audio interface in order to hear it on speakers or headphones. Each individual stage adds up to the total amount of latency, causing the microphone signal to arrive later than the rest of the music. Although a little latency poses no problems, larger latencies prevent a musician from playing in time. With faster computers and interfaces, latency can be minimized down to a few milliseconds. Although that may seem negligible, it can easily hinder an experienced musician.
2. **Making music with your eyes.** Although the overview on a screen is convenient, it may also distract, and lead to decisions based on what you **see** rather than what you **hear**.

> What we can **see** is limited. Much more limited than what we can **hear.**
> —Karlheinz Stockhausen

Most Popular DAWs

Although the looks, terminology and philosophy of different DAWs may vary, most programs offer similar functions. This means that any type of music can be recorded with any DAW. Although exceptional, there are electronic artists who use Pro Tools, and bands who use Live. Every DAW has its strengths and weaknesses and a particular workflow. That's why certain programs fit certain music styles better than others.

- **Avid Pro Tools** (see Figure 9.2). Since the 1990s, Pro Tools has become the de facto standard for music recording and audio for picture. Almost every professional can work with the program, while projects can be mixed in every professional studio around the world. Pro Tools is a reliable and no-frills, straightforward program. Although it initially focused on recording, editing and mixing audio, it now includes features for creating music. Since then, Pro Tools is also a viable option for musicians and composers that work at home or in small studios. The program runs on both Mac and PC.

FIGURE 9.2
Avid Pro Tools.

ULTIMATE VERUS NATIVE

Pro Tools comes in two different versions. "**Ultimate**" is the professional, more expensive variant, under the hood it works differently compared to other DAWs. Pro Tools Ultimate uses dedicated hardware for executing audio calculations (like plugins). That's why the processing of audio is almost instantaneous, resulting in near-zero latency. By adding extra hardware, more tracks and plugins can be run. To give you an indication, a fully expanded Pro Tools Ultimate system can record 256 tracks simultaneously, with near-zero latency (as of 2019). The options for scaling and the absence of latency are important requirements for working in a professional environment. That's why Pro Tools has become the standard for professionals. As the cheaper "**Native**" version of Pro Tools uses the computer's internal CPU for audio processing, project size and latency are similar to other DAWs. As far as features are concerned, Pro Tools Native is almost identical to its professional brother and is fully compatible with it as well.

- **Apple Logic Pro** (see Figure 9.3). In 1986, German company C-Lab released an innovative MIDI sequencer for Atari computers under the name "Notator" (see Figure 9.9). In 2002, Apple bought the company and christened the program Logic. Through the years, Apple has added many instruments and audio content to the program in order to **create** music. Viewing and editing MIDI data with it are considered superior, while functions for editing and mixing audio match those of Pro Tools. It has an interface that is highly

FIGURE 9.3
Apple Logic Pro.

customizable. Since Apple's profit comes from selling hardware, the software is relatively cheap or even comes for free. That's why other manufacturers have a hard time competing with Logic. Logic is Mac-only.

- **Ableton Live** (see Figure 9.4) is quite different from other DAWs; it has a workflow that especially suits the repetitive character of electronic music. Functions such as looping musical phrases, switching song sections and playing with tempo have all been made easily accessible. As its name implies, Live is also suited for use on stage, and all operations can be applied in real time, without the audio dropping out. Actually, **Bitwig Studio** is a relatively new DAW that has a similar workflow. Live and Bitwig run on Mac and PC.

FIGURE 9.4
Ableton Live.

- **Steinberg Cubase.** Together with Pro Tools and Logic Pro, this is one of the oldest DAWs, having originated in the 1980s. Cubase is a solid and all-around program for both Mac and PC. It is known for its rapid implementation of the latest technologies and functions. Its bigger brother **Nuendo** specializes in mixing and editing audio for games and postproduction (A/V).

WHICH DAW SOUNDS BETTER?

As far as sound quality is concerned, there is no difference between DAWs like Pro Tools, Logic or Cubase. Although the various programs may perform calculations differently, it's the **audio interface** that's transducing. So that's where artifacts may enter the signal.

Other DAWs

Since 2010, the DAW market has exploded. Many hardware and software firms have released their own DAW. Some of them offer similar functionality as the "premier league" DAWs:

- MOTU **Performer,** Acoustica **Mixcraft,** and Presonus **Studio One** might be slightly less popular, but they do have a complete and professional functionality. Harrison **Mixbus** is the first DAW with *sound.*
- **FL Studio.** In the 1980s, this DAW was known as "Fruity Loops." Since 2018, it is available for Mac too, and it has been upgraded, offering many unique functions. FL Studio is popular in electronic music and in hip-hop.
- **Apple GarageBand.** Music production doesn't get easier than with GarageBand. It shares its looks, audio quality and sound library with bigger brother Logic but lacks certain functions and editing possibilities for professional use. Logic loads GarageBand projects, which allows for easier switching to Logic.
- **Propellerheads Reason.** Introduced in 2000, this DAW has many instruments and samples on board. It is a beloved tool for people that need to quickly construct songs. It is efficient with CPU, which allows working on older computers.

DAWs can even be run on mobile devices, such as an **iPad.** Mobile CPUs are powerful enough to run larger projects at a high quality. A good example of such a DAW is **Auria** (by WaveMachine Labs). Who knows, maybe professional studios will be run on a telephone somewhere in the future. At least computing power and sound quality are hardly an issue anymore.

DAWs are not compatible. Projects started in one program cannot be opened in another. Although manufacturers have agreed on the "OMF" and "AAF" standards, those formats do not include any mix information or plugin settings. If necessary, projects can be exchanged by exporting individual tracks as audio files. These files can then be imported in the DAW of choice.

▶ PLUGINS

Plugins are little software programs that can be used inside a DAW for manipulating sound. There are reverb, delay, compression, EQ, analog tape simulations, distortion plugins or virtual instruments like synthesizers and samplers. All DAW manufacturers supply a standard set, plugins from other brands can be added. Just like physical studio equipment, each virtual device can be used to add a specific character to an individual instrument or to the mix. A complex project may contain dozens, if not a hundred or more, different plugins. Since the computer has to calculate all plugin processes in *real time*, there is a limit to how many plugins can be used. Once this limit is reached, audio may crackle or stutter, or the DAW displays warnings. There are a few different plugin standards, as shown in Figures 9.5 through 9.7.

FIGURE 9.5
VST (Virtual Studio Technology): this is the oldest plugin standard, patented by Steinberg. Besides Cubase and Nuendo, many DAWs are compatible with the VST format.

FIGURE 9.6
AU (Audio Units): this format is patented by Apple, and can be used in Logic, GarageBand, Ableton Live and FL-Studio.

FIGURE 9.7
AAX (Avid Audio eXtension): this is the Pro Tools plugin format.

Generally, DAWs work with a single, specific plugin format. Therefore, AAX plugins cannot be used in Logic for example, while Audio Units plugins cannot be used in Pro Tools. In case you buy a third-party plugin, the manufacturer generally offers an installer for all popular plugin formats. People that use two or more DAWs can use the plugin in any program, with identical sound quality across all formats.

When buying a plugin, it is authorized for use on one computer only and is protected with a software key or a USB dongle. Anyone is allowed to install the software, but the plugin will only run on a computer that is authorized. When saving a project, it's not the plugin that is saved but,

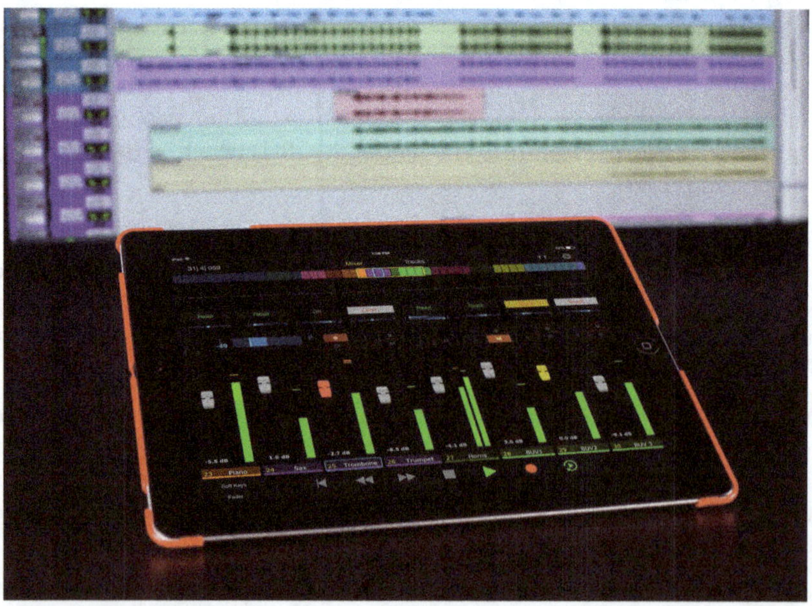

FIGURE 9.8
iPad remote control for Pro Tools.

Source: Photo courtesy of avid.com.

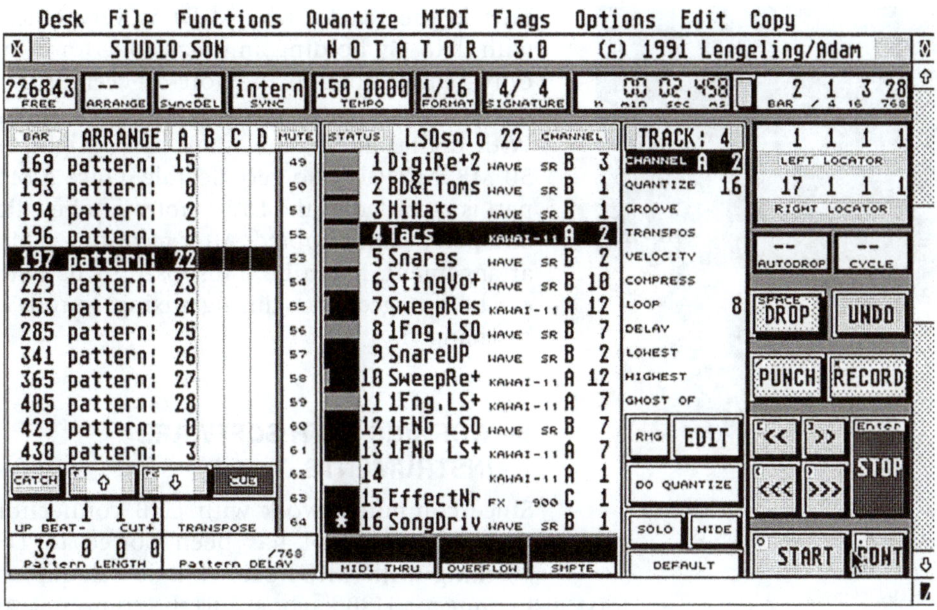

FIGURE 9.9
Logic Pro in its infancy: C-Lab Notator for Atari ST computer s (1986).

rather, its **settings**. This means exchanging a project requires the other party to own the same plugin(s). In case a plugin is not (or no longer) present on the computer, the DAW displays a warning. The audio of the relevant track will play, but the sound will be different.

MIDI

Analog synthesizers in the 1960s and 1970s sounded fantastic, but making them play simple parts was cumbersome and required many cables. The introduction of "MIDI" (*Musical Instruments Digital Interface*) in 1982, has dramatically changed the workflow in the studio. It allows computers, keyboards, software instruments, mixers, effects and even stage lights to communicate.

MIDI is a digital language that allows communication through either a 5-pin DIN cable (see Figure 9.10) or a USB cable (see Figure 9.12). With MIDI, **real-time** information can be sent, for example from a master keyboard to a synth. When a key is pressed, not only its **note number** (1–127, see Figure 9.15) is transmitted but also its **velocity** (1–127). Velocity indicates how hard the key was struck. After choosing the desired preset on the receiving synth, the corresponding note will sound. Upon release of the key, a **note-off** command is transmitted, and the corresponding sound will stop playing. So MIDI sends note information only, not the sound itself.

FIGURE 9.10
MIDI In, Out and Thru by means of 5-pin DIN ports.

This is the reason why MIDI is very efficient with data. By holding an organ tone for 3 min, only a single note-on and note-off event is transmitted, which take up just a few bytes in total. With digital audio, that same tone might take 50 MB. MIDI is also very flexible. Once a MIDI part is recorded in the DAW, notes can be edited and parts can be copied. After recording a MIDI arrangement, it can be transposed or its tempo can be changed. Even during mixing, sounds can be changed.

Via MIDI, only **note information** is sent, not the sound itself.

▶ WORKING WITH SOFTWARE INSTRUMENTS

Since computers work with USB connections, the MIDI protocol has been ported to USB. Working with MIDI-over-USB is self-explanatory: just connect a USB keyboard to the computer, and it will usually be recognized by the DAW automatically. Then, software instruments can be played and recorded instantly.

▶ ALTERING SOUND VIA MIDI

Apart from note information, MIDI can transmit other data as well. Like the movement of a sustain pedal, sostenuto pedal (damping), the modulation wheel (vibrato), pitch wheel or a fader on a master keyboard or mixer. This information is called **MIDI controllers,** and their **values** range from 1 to 127. In total, 127 MIDI controllers are defined in the MIDI specification (see Table 9.1). As all manufacturers adhere to this standard, devices of multiple brands can be combined in a single setup.

Although being a relatively old standard, MIDI is more relevant than ever. There are numerous devices that can transmit MIDI, like joysticks, light beams, touch pads, wind controllers, breath controllers, Roli keyboards, MIDI guitars, MIDI drum kits and so forth. In order to control the mix in a DAW with hardware faders, there are MIDI control surfaces that send fader values over MIDI. MIDI-controllers are very suitable for use on stage too. Not only can they be used to manipulate sound, but also for controlling stage lights (DMX). In case you're interested in this area, you should definitely look at Max for Ableton Live.

▶ CONNECTING OUTBOARD SYNTHS

To extend their sound palette, many people combine hardware synths with a computer setup. In case the synth has a 5-pole MIDI connector only, you'll need a USB-to-MIDI interface. This translates the MIDI-over-USB signal from the computer into a MIDI signal. In order for the synth to receive MIDI from the computer, the MIDI-out of the MIDI interface must be connected to the MIDI-in of the synth. Subsequently, additional synths can be connected by connecting their MIDI-in with the "**MIDI-thru**" of the preceding synth.

Table 9.1 Often-Used MIDI Controllers

MIDI controller	Controller number
Modulation	1
Volume	7
Panning	10
Sustain Pedal	64 (0–63 = off, 64–127 = on)
Sostenuto Pedal	65 (0–63 = off, 64–127 = on)

In case you want both synthesizers to play unique parts, synth 1 can be set to MIDI channel 1 and synth 2 to MIDI channel 2. This can be done in the system/utility menu of the synth. Then, each track in the project should be assigned the corresponding (external) MIDI channel. As one MIDI cable can transmit up to 16 channels, 16 different synths can be made to play their own unique part. This is called *daisy-chaining* (see Figure 9.11), every next synth is called a *slave*. By setting several synths to the **same** MIDI channel, it is possible to **layer** sounds.

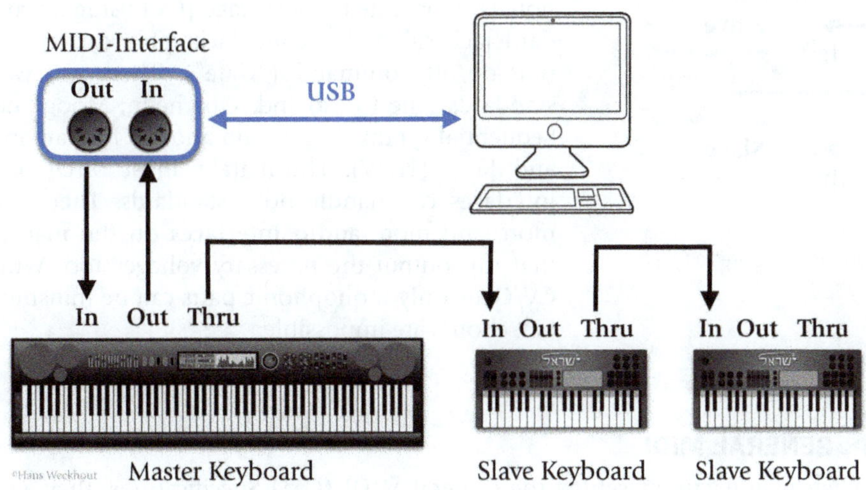

FIGURE 9.11
MIDI daisy-chain setup.

FIGURE 9.12
Roland USB–MIDI interface.

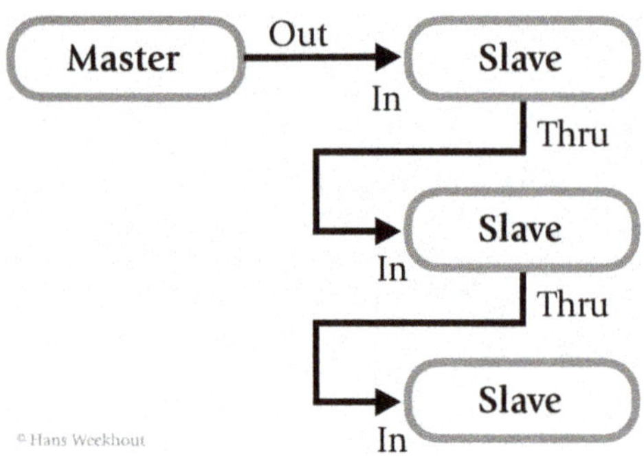

FIGURE 9.13
In a master–slave setup, the MIDI-out port is used only for the master keyboard.

▶ CONNECTING PRE-MIDI SYNTHS

Old analog synthesizers and Eurorack equipment can be connected to the computer by using their **CV/Gate** jacks. A "MIDI->CV"-interface is needed for this. With the CV/Gate system, the pitch of a note is transmitted as a voltage (CV) using a standard jack cable. A second jack cable carries the note-on/off command ("Gate"). There are two standards: one for Roland, Oberheim, Moog and Sequential synths (V/oct) and another for Yamaha and Korg (Hz/V). Fortunately, most MIDI->CV interfaces can handle both standards. There are more and more **audio interfaces** on the market that can output the necessary voltages too. With CV/Gate, only monophonic parts can be transmitted; chords are impossible.

▶ GENERAL MIDI

When manufacturers settled the General MIDI (GM) Specifications, they also agreed for preset numbers to correspond with specific sounds. Therefore, GM keyboards always have an electric piano on program number 12 and an acoustic guitar on number 31. GM drum kits always have a kick drum on C1 (note number 36) and a snare drum on D1 (note number 38; see Figure 9.14). This way, GM allows for easy sharing.

▶ DISADVANTAGES OF MIDI

Although it can be called a great advantage that MIDI allows parts or sounds to be changed up to the last minute, this can also be a disadvantage. For instance, when sharing MIDI arrangements. If the other person doesn't own the same physical synthesizer or software instrument, the arrangement will sound (very) different. To prevent this from happening, MIDI parts can be recorded on audio tracks. But actually, even if you're not planning to share projects, it makes sense to capture MIDI parts as audio. With analog synths, for instance, recalling a sound will be almost impossible, as only the slightest movement of a knob or a different temperature will cause the sound to change. Software instruments suffer from another problem: due to updates, they might refuse to load. Only when MIDI is captured as audio files, you can rest assured that a project can be recalled reliably.

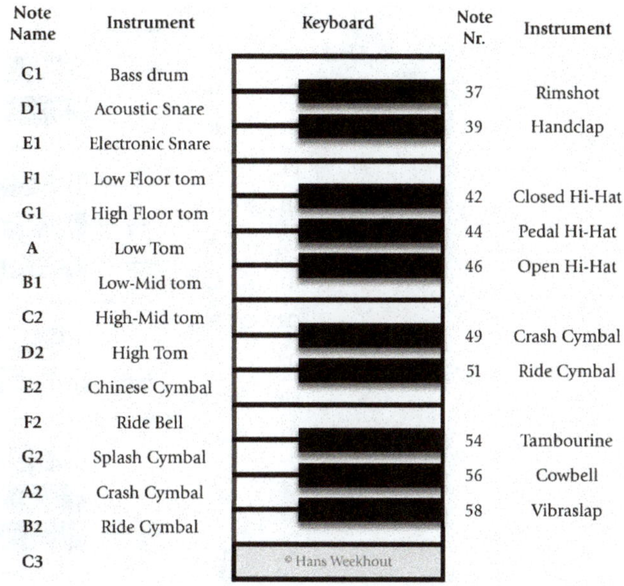

FIGURE 9.14
GM standard drum format.

Note Nr.	Note Name	Freq. (Hz)
36	C1	32.7
38	D1	36.7
40	E1	41.2
41	F1	43.7
43	G1	49
45	A	55
47	B1	61.7
48	C2	65.4
50	D2	73.4
52	E2	82.4
53	F2	87.3
55	G2	98
57	A2	110
59	B2	123.5
60	C3	130.8
62	D3	146.8
64	E3	164.8
65	F3	174.6
67	G3	196
69	A3	220
71	B3	247
72	C4	261.6
74	D4	293.7
76	E4	329.6
77	F4	349.3
79	G4	392
81	A4	440
83	B4	493.9
84	C5	523.3
86	D5	587.3
88	E5	659.3
89	F5	698.5
91	G5	784
93	A5	880
95	B5	987.8
96	C6	1046.5

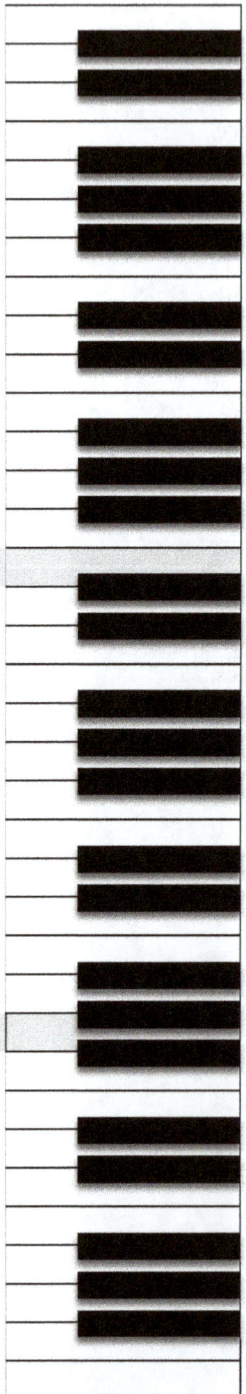

Note Nr.	Freq. (Hz)
37	34.7
39	38.9
42	46.3
44	51.9
46	58.3
49	69.3
51	77.8
54	92.5
56	103.8
58	116.5
61	138.6
63	155.6
66	185
68	207.7
70	233.1
73	277.2
75	311.1
78	370
80	415.3
82	466.2
85	554.4
87	622.3
90	740
92	830.6
94	932.3

FIGURE 9.15
MIDI note numbers, with corresponding frequencies.

CHAPTER 10

Recording on the Computer

In order to get audio in and out of the computer, an audio interface is needed. The audio interface contains an *A/D converter* (analog-to-digital) that converts analog sound into zeros and ones. When the computer plays, the D/A converter converts the digital stream back to analog. Although digital audio is robust and reliable, it is not foolproof. Only the right handling and the right settings will result in good sound. When is analog better, and when is it better to go digital? How can you preserve sound quality throughout the production process? And which format should be used when delivering a project?

▶ PCM

Analog signals can be digitized by means of Pulse Code Modulation (PCM, see Figure 10.1). Invented in 1937, PCM has become the standard for both Compact Disc and digital audio on the computer. With PCM, a signal's amplitude (volume) is measured at regular intervals. Each sample (measurement) is rounded ("quantized") to the nearest value. The resulting stream of zeros and ones can be stored on the computer. Upon playback, the shape of the original signal is reconstructed by reading the numbers.

The quality of a digital signal is determined by

- **sample rate:** this represents the number of times the signal is analyzed. With CD, this is 44,100 times per second. Most used rates in music production are 44.1 kHz and 48 kHz, and
- **bit depth:** this represents the number of steps available for measuring a signal's amplitude. Bit depth for CD is 16 bit, every measurement can take on 65,536 different values (2 to the power of 16). Nowadays, 24 bit has become the standard. This means that the signal's volume can be measured at more than 16 million steps (2 to the power of 24).

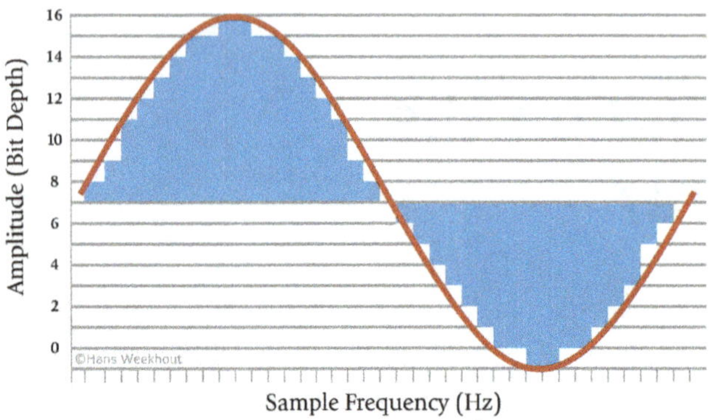

FIGURE 10.1
Pulse Code Modulation with a 4-bit converter: 16 steps are available for measuring the signal's volume (2 to the power of 4). Actually, by zooming in sufficiently on the computer, the stepped digital response can be made visible, even with 16 bit or 24 bit files.

The higher the sample rate and bit depth, the better the approximation of the signal, and the better the sound quality. However, this comes at the cost of larger files and a higher load on the computer's CPU. For a small project on a modern computer this may not be a problem, and sample rates of 88.2 kHz, 96 kHz, 192 kHz or even 384 kHz can be used. But in larger sessions, the CPU must perform extra calculations for every track, synthesizer or effect added. As pop projects often use many tracks, this means that we're often restricted to using lower sample rates. Fortunately, computers become more and more powerful, so high sample rates and 32-bit recording will eventually become the standard.

▶ HOW GOOD IS DIGITAL?

During conversion, rounding errors occur as the signal is quantized. At higher settings, the steps may become smaller and smaller, but in the end, A/D and D/A converters are transducers, so information **will** get lost eventually.

Therefore, it is better to keep audio in the digital domain once it has been converted. For example, when processing a digital recording of drums with an analog device and then recording the signal back into the DAW will result in losses due to the extra D/A and A/D steps. With modern, good-quality interfaces, the artifacts may not quickly become apparent, but it is not a good idea to do this many times.

▶ IS DIGITAL BETTER OR WORSE THAN ANALOG?

It depends. Digital is in any case cheaper. When storing or transferring audio, the information is not transduced, so audio quality will always remain the same. By listening to a CD, sound quality will be exactly the same as it was in the studio.

Over the years, digital technology has advanced considerably. Modern audio interfaces have an extremely linear frequency response, almost immeasurable distortion, and produce hardly any noise. What goes in, comes out exactly the same. Where tape "smears" transients, digital reproduces them exactly as they are. That's why digital may be perceived as "tighter." All in all, the **absence** of sound can be a good reason to choose for digital.

In practice, many professionals make use of both analog and digital. After recording drums analog, they might "dump" individual tracks onto the computer so that each digital file carries an analog imprint. Then, they'll record the rest of the

FIGURE 10.2
Avid HD I/O and PreSonus Quantum audio interfaces.
Source: Photos courtesy of avid.com and presonus.com, respectively.

instruments digitally on the computer. Others may record all instruments digitally and then use tape as a destination format when mixing or mastering. With a hybrid approach, each instrument has "seen" an analog circuit at least once.

The choice for either analog or digital depends on taste and personal preference. How audio is treated, is more important than how it is stored. To illustrate this, let's compare Black Dub's untitled debut album with Foo Fighters' *Wasting Light*. Although the latter was recorded on tape, its sound can hardly be called warm. The opposite applies to Black Dub; the album has a warm, thick sound "**despite**" being a digital recording! It's the people in the driver's seat that determine the sound of a production, not the storage medium.

With analog versus digital, it is more important how audio is **treated** than how it is **stored**.

Audio Interface

Many audio interfaces are on the market today. How do they differ, and which features are important when choosing an interface?

- **Firewire** has long been the professional standard, but it is no longer supported. Old Firewire audio interfaces can often be connected to a modern computer by using a dedicated Firewire-to-Thunderbolt cable.
- **Thunderbolt/USB-C** is the fast, professional standard. It can handle hundreds of audio channels over a single cable. Thunderbolt has enabled manufacturers to further decrease latency. Due to its high specifications, Thunderbolt interfaces and cables are slightly more expensive.

- **USB (2.0 and 3.0)** is the oldest and most compatible standard. It is widespread and therefore cheap. As long as you mix in-the-box and record only a limited number of channels at the same time, even a USB 2.0 interface could do fine.
- **AVB ("Audio Video Bridging")** is a relatively new format that uses ethernet cables (network, CAT5/CAT6). This allows sending data over distances up to 300 ft., which is handy in live situations or when recording on location.

FIGURE 10.3
The Universal Audio Apollo Twin MKII thunderbolt audio interface is also a *monitor controller*. It can record both mic and line signals, and it has a talk-back option, while the speaker/headphone volume can be controlled conveniently with a large knob.

Source: Photo courtesy of uaudio.com.

▶ SOUND QUALITY

Due to a large home-recording market and technology advancing, interfaces have become better and cheaper at the same time. But quality comes at a price. More expensive interfaces use better converters, which will result in better sound quality, with more depth, detailing and a more stable stereo image. Then there is another consideration: in case you only record single tracks of vocals, guitars or bass, it might be better to stay away from multi-input/output interfaces. Because each additional input/output requires an extra converter, it is more likely for a 2-in, 2-out interface to contain better converters, with a better sound quality.

SYSTEM AUDIO

A/D and D/A converters can also be found behind the mini-jack audio connectors of a computer. In order to keep computer prices down, manufacturers often use low-cost chips for this application. This results in an inferior sound quality when compared to dedicated audio interfaces. Apart from this, mini-jack connections are unreliable and often produce a hum. All in all, computer jacks are not an option for professional use.

Digital Formats

▶ LOSSLESS FILES

When recording a signal in the DAW, the A/D converter's digital stream is written to disk in a so-called *lossless* file. This is an exact representation of the signal, and it offers the best possible sound quality. That's why lossless files are the de facto, professional standard for storing and exchanging audio. Most popular formats are WAV (patented by Microsoft) and AIFF (Apple Interchange File Format); both deliver the same sound quality. A newer version of WAV is BWV (Broadcast WaVe). The bit depth of WAV and AIFF ranges from 8 bit to 32 bit, while sample rates can be anything between 1 to 384 kHz. Their channel format can either be mono or stereo.

Mono/Stereo

Stereo signals, such as the overheads of a drum kit, should not be recorded on two mono channels but, rather, on a stereo channel (see Figure 10.4). Only then can their phase relationship be preserved. Stereo files have other advantages too: they only need one plugin for altering sound, and they take up less space on the screen and in the mixer.

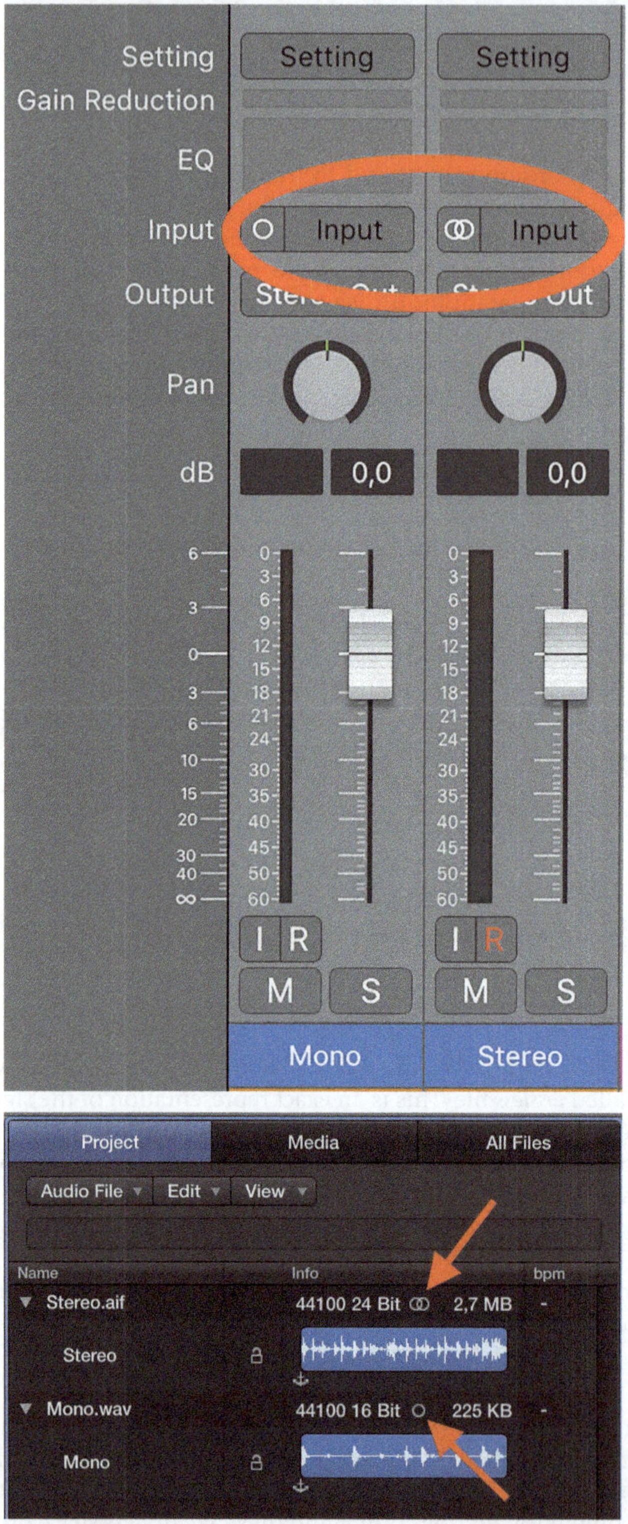

FIGURE 10.4
(a) Logic channels can be switched from mono to stereo, in Pro Tools you'll select mono/stereo when creating tracks. (b) Stereo files and mono files can be distinguished by their symbols.

▶ LOSSY FILES

Ultimately, music should reach consumers. They want to store music on their telephones, or listen to music streams on the internet, without spending too much data. As lossless files are too bulky for this purpose, various *lossy* formats have been developed, such as MP3, AAC (by Apple) and Ogg Vorbis (as used by Spotify).

Lossy files take advantage of limitations in our hearing. When creating a lossy file, an advanced computer algorithm discards information that is considered beyond human perception. The resulting **compressed** files are easily a tenth of the size of their high-quality brothers. But no process exists that could magically re-create any discarded information. That's why lossy files are unsuitable for use in the studio. Throughout the production process, audio stays in a full-quality format; lossy files are relevant for **distribution** only. Streaming platforms use compressed formats too.

Note that **compressing** digital files has nothing to do with the audio compression effect that reduces the dynamic range!

▶ MP3

Released in 1993, MP3 was the first popular compressed audio format. MP3s have a bad reputation. This originates from the days when people started sharing corrupt files illegally on the internet. Bad MP3s can be recognized by their grainy, bit-crushed quality. Although "good" MP3s suffer from subtler by-effects, the signal will always lack a certain amount of depth and detail, while transients will be smeared. That being said, with a proper encoding program (like Pro Tools, Logic or iTunes) and higher settings (like 256 Kbps or 320 Kbps, see Table 10.1), the mix can be converted into an MP3 that's good enough to serve as a listening copy for band members, record company or management.

▶ AAC

Ten years after mp3, Apple came up with AAC (Advanced Audio Coding) as a compressed audio format for iTunes. At the same bitrate, AAC files have a slightly better sound quality than MP3s. AAC files are protected with DRM (Digital Rights Management), which is a great advantage for record companies and composers. AAC files can be identified by the ".m4a" extension in their file name.

TRUTH OR MYTH

FIGURE 10.5
"As long as people listen to streaming audio and MP3s, why bother for a good recording?"

In order to find an answer, we'll use an analogy with photography. Let's compare a full-quality picture (a), with its low-resolution sibling (b). Although the compressed version clearly lacks details, its colors are unscathed and it shows that the picture originates from a high-quality camera. The same is true for audio: although playback quality may suffer from compression artifacts, the frequency spectrum is complete, and the quality of the recording is clearly noticeable.

Table 10.1 File size for three minutes of stereo audio at 24 bit, 44.1 kHz (WAV/AIFF, MP3 and AAC)

File type	Bitrate (Kbps)	File size (MB)
WAV, AIFF	2116	47.7
MP3, low quality	128	3.1
MP3, high quality	256	6.3
AAC	256	6

The quality of an MP3 is specified in "Kbps" (kilobytes per second). This rate can be anything between 32 Kbps and 320 Kbps, while the sampling frequency can be either 44.1 kHz or 48 kHz. Note that AAC files are slightly smaller than MP3s.

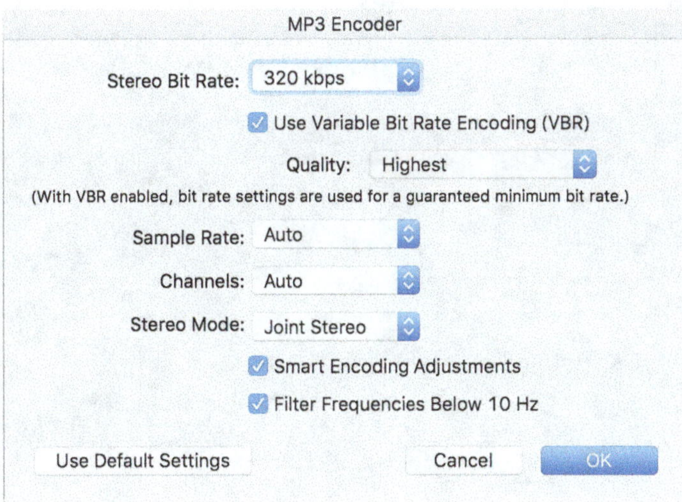

FIGURE 10.6
iTunes-> Preferences-> Import Settings: settings for importing.

METADATA

An MP3 file not only contains audio; other information can be added too, like song title, artist, tempo, composer, lyrics, URL or e-mail address. We call this metadata. Before sending an MP3 to a record company or booker, it is a good idea to fill out all the relevant fields. Once the recipient loads the MP3 in a program such as iTunes, its metadata comes into view. Not filling out metadata may therefore result in missed opportunities!

Recording Settings

▶ BIT DEPTH AND SAMPLE RATE

Before starting a project, you'll need to choose settings both bit depth and sample rate. Then, recorded files will conform to those settings, while imported files will be converted.

Although 16 bit is the standard for CD, 24 bit has now become the professional standard. Of course, 24-bit files are bigger by 50%, but sound quality is substantially better than 16-bit files. With current hard disk prices in mind, larger files can hardly be called a problem.

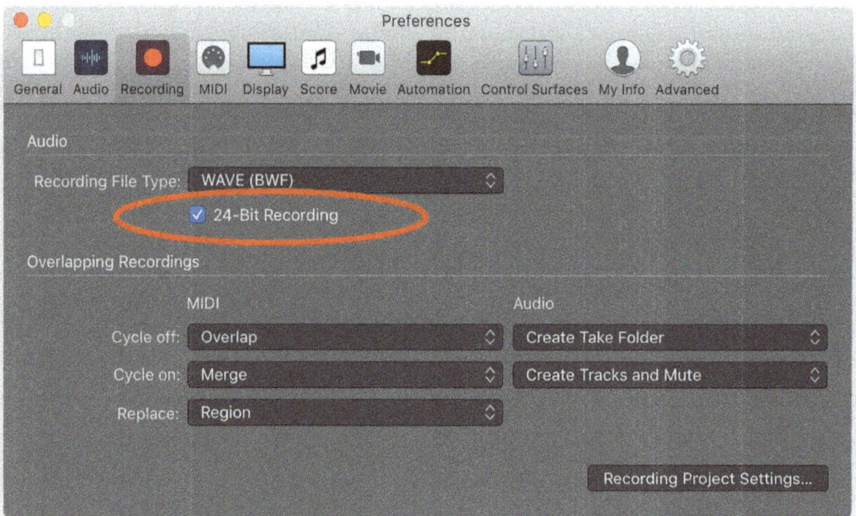

FIGURE 10.7
24 bit in Logic: Logic Pro X/Preferences/Recording.

When choosing **sample rate,** things are a bit more nuanced. Historically, 44.1 kHz and 48 kHz have always been the default sample rates for DAW projects. The advantage of 44.1 kHz is that files don't need to be converted in case you produce music for CD, whereas 48 kHz offers better sound quality, taking up only 10% more disk space. Plus, 48 kHz is the standard in broadcast and postproduction. This allows for easy exchanging in case you produce music for picture. But the CD is almost dying, and the benchmark for audio quality increases. Therefore, higher sample rates, like 88.2 kHz, 96 kHz, 76.4 kHz, 192 kHz and even 384 kHz, are getting more and more popular.

> **WHAT'S THE AUDIBLE EFFECT OF HIGH SAMPLE RATES?**
>
> As you know, A/D and D/A converters produce rounding errors. Simply stated, both higher bit depths and sample rates result in more data available, thereby decreasing the effect of rounding errors. Apart from this, high sample rates allow for a frequency response beyond 20 kHz. Although we strictly can't hear any higher, frequencies beyond 20 kHz are believed to influence our perception. All in all, higher sample rates translate in more depth, better stereo imaging and a sound that's perceived as less straining for the ear.

So how feasible are high sample rates in practice? Well, your computer's CPU dictates the limits, best thing is to find this out by trial and error. High sample rates tax the computer's CPU, as all plugins must perform their calculations at this higher speed too. A smaller project, with a limited number of tracks, might run perfectly fine at a higher sample rate. But in case of a project with many tracks and plugins you may be restricted to use 48 kHz (or 44.1 kHz).

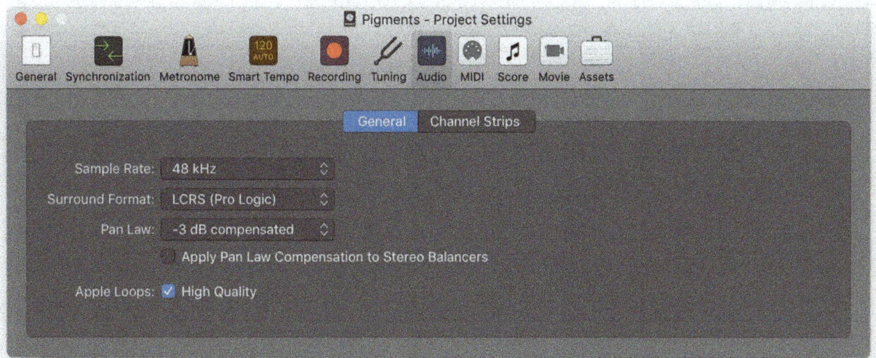

FIGURE 10.8
Choosing sample rate in Logic: File/Project Settings/Audio.

▶ IS IT ALLOWED TO CONVERT SAMPLE RATES?

The short answer is no. During calculation, information gets lost. That being said, some calculations lead to better results than others. Let's look at an example to make things clear: by converting a 44.1-kHz file to 48 kHz, the software has to come up with new values, about every 10 samples. This will lead to compromises in sound quality. Instead, converting a 48-kHz file to 96 kHz is painless, as the computer simply duplicates every sample, although this will not add any sound quality.

Before conversion, it is a good idea to always overthink the process beforehand. Conversion from 88.2 kHz to 44.1 kHz is less harmful than converting from 88.2 kHz to 48 kHz. So understanding the process of conversion is important, but in practice you may have no choice. In case a music project at 44.1 kHz will be used for a movie, for instance, the 44.1-kHz mix must be converted to 48 kHz.

Files with different sample rates cannot be combined in the same project. When starting a new song, the chosen settings apply to all files. When importing audio files from other sessions, these will be converted into the project's format.

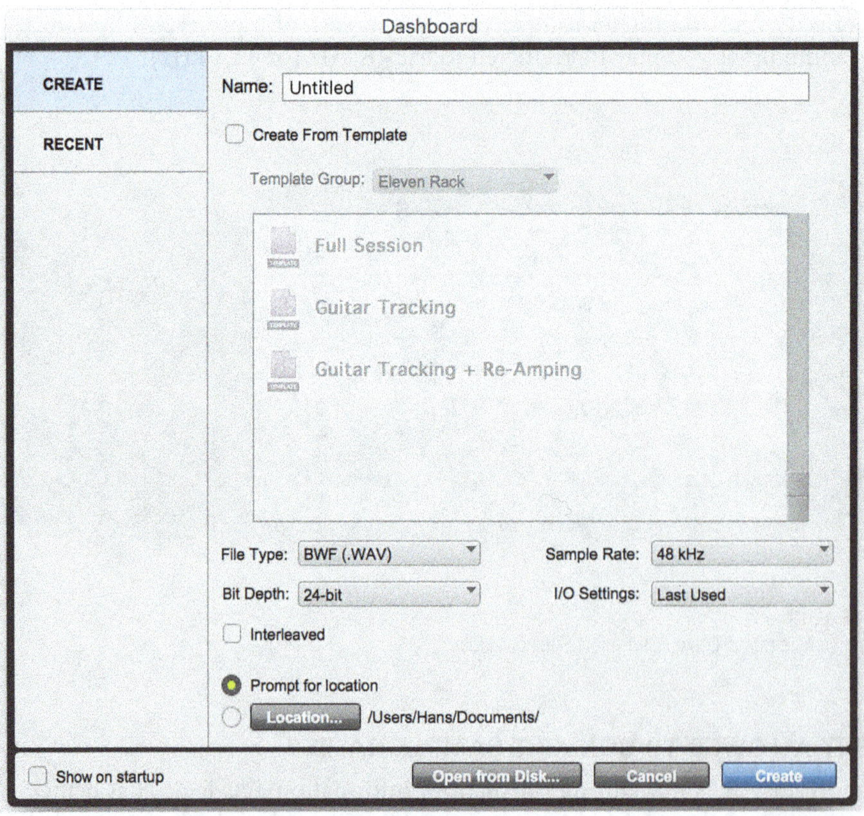

FIGURE 10.9
Before starting a new project in Pro Tools, bit depth and sample rate must be chosen.

▶ HOW LOUD SHOULD YOU RECORD?

In theory, the correct answer is **as loud as possible, but without overloads**. Why "as loud as possible"? As you know, the amplitude of a signal is measured along the Y-axis (bit depth). A 16-bit converter has 2 to the power of 16, or 65,536, steps available. Only by fully modulating the system are all 65,536 values available to describe the signal. Softer signals are described with fewer data, which will result in a lower sound quality.

Why should you record "**without overloads**"? Well, in the analog world, character can be added by overloading a device. Distortion gradually increases with the input level. Although the color of analogue distortion is generally perceived as pleasing, this is not the case with digital. In certain situations, a signal can be perfectly clean, while increasing that level by only 1 dB results in an ugly digital distortion. This is easy to understand if you look at the picture of the digital staircase of Figure 10.1: signals can never get louder than the highest digital step. Every signal that's louder is decapitated. Once digital distortion is contained within the signal, it can never be removed. Peaks that were lost because of **clipping** will forever be gone.

So far the theory; it is important to keep this in mind when working in the computer. But in practice, we're working with musicians, so it will be hard to predict the level. Therefore, the best practice is to go for the safe option, namely, a substantial lower level that prevents overloads. By leaving a **6- to 12-dB headroom** above the highest peaks, you can rest assured that the system will never overload.

Decreasing the recording level in a 24-bit project is less critical than it is in a 16-bit project. That has to do with **dynamic range**. A system's dynamic range expresses the difference between the **softest** possible signal (before disappearing in the system's noise) and the **loudest** possible signal (before distortion occurs). A 16-bit system has a dynamic range of "only" 96 dB, while a 24-bit system has 144 dB. With these numbers in mind, no harm is done by disobeying the last 12 dB in a 24-bit recording.

▶ GAIN

Adjusting the recording level is done with the **(pre-amp) gain knob** on the audio interface. After leaving the A/D converter in the interface, the digital stream's volume cannot be changed anymore. So this is exactly the level shown on the DAW's channel meter and what gets recorded in the audio file. Provided there are no plugins in the channel of course.

Leaving 6 to 12 dB headroom **above** the **loudest** signals is a safe recording level.

Latency

Depending on both the power of your computer and the complexity of the project, latency (see Figure 10.10) may be high—so high that it prevents a musician from playing in time. Are there any workarounds?

Solution 1: Pro Tools Ultimate

- The professional version of Pro Tools uses dedicated hardware for audio processing, with (near-)zero latency as a result. Although this is a must for professional studios, your budget may not allow. Fortunately, there's also Solution 2.

Solution 2: Optimize DAW Settings

- Every DAW has an "I/O-buffer size" setting (see Figure 10.11). This setting tells the computer how much power it is allowed for processing audio from input to output. After setting the I/O-buffer size to a low value (64 or lower), latency could become workable (or even barely audible), depending on the sensitivity of the musician. But the computer has to work hard in this scenario. Audio could start to glitch, crackle or stutter, or "system overload" messages pop into view. This will not harm the computer or the audio, so in practice, an appropriate buffer size can be found by trial and error. As high latency is of no concern when **mixing**, the buffer can be set to high values (like 1024 or even 2048).

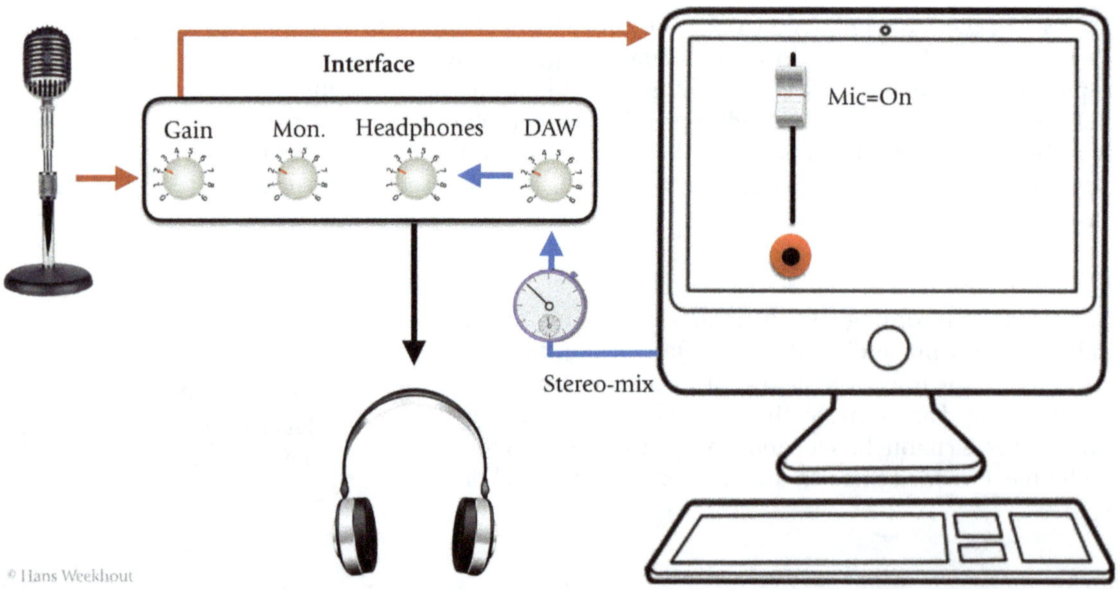

FIGURE 10.10
Monitoring with latency.

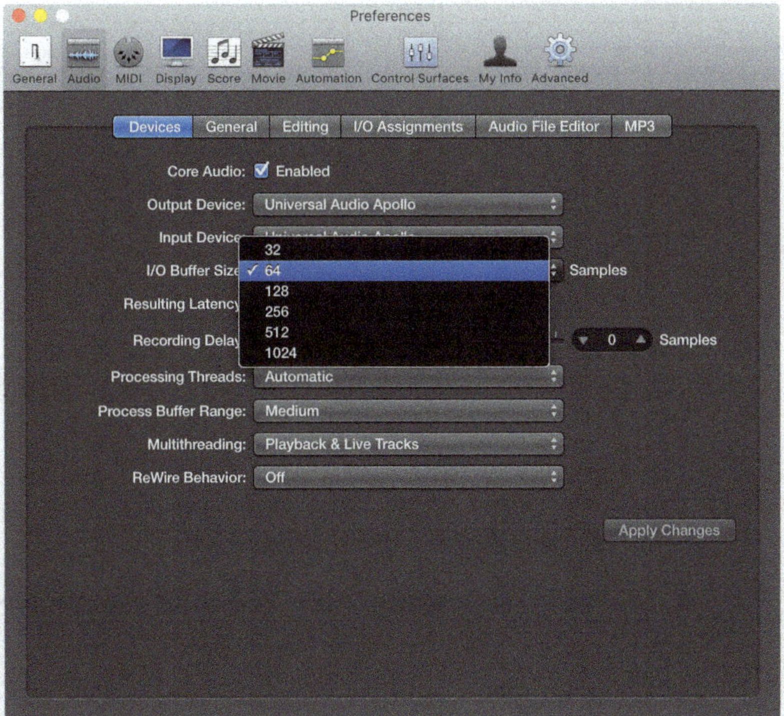

FIGURE 10.11
I/O-buffer size in Logic Pro X (Logic Pro->Preferences->Audio->Devices). In Pro Tools, I/O-buffersize can be set in Setup->Playback Engine.

- Don't use plugins on the track to be recorded.
- Enable "**Low Latency Mode.**" In this mode, plugins and sends that tax the system most, are disabled. In **Logic** you can find Low Latency Mode in the "Record" menu; in **Pro Tools,** it is in the "Options" menu.

Solution 3: Zero Latency Monitoring

- Many audio interfaces have an option for mixing the mic signal directly with the output of the DAW. Before starting a "latency-free" recording, the recording channel should be muted. Otherwise, both the direct signal (from the interface) and the (delayed) signal from the DAW become audible. After recording, the channel must be unmuted in order to review the take. When recording entire takes, zero latency monitoring (see Figure 10.12) is a great solution. But for intensive punch-in/punch-out sessions, it will be hard (if not impossible) to constantly push the mute button *and* the record button both at the same time.

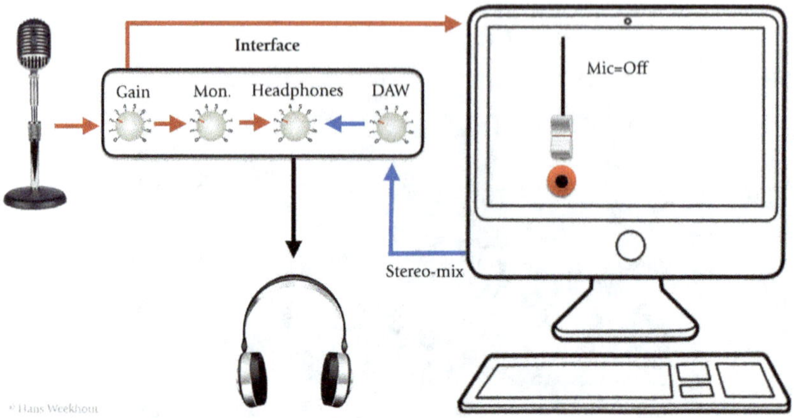

FIGURE 10.12
Zero latency solution.

At **higher** sample rates, latency **decreases.**

Solution 4: Use a Higher Sample Rate

- Latency **decreases** with a **higher** sample rate. At 96 kHz, for instance, latency is only half as long as it is at 48 kHz.

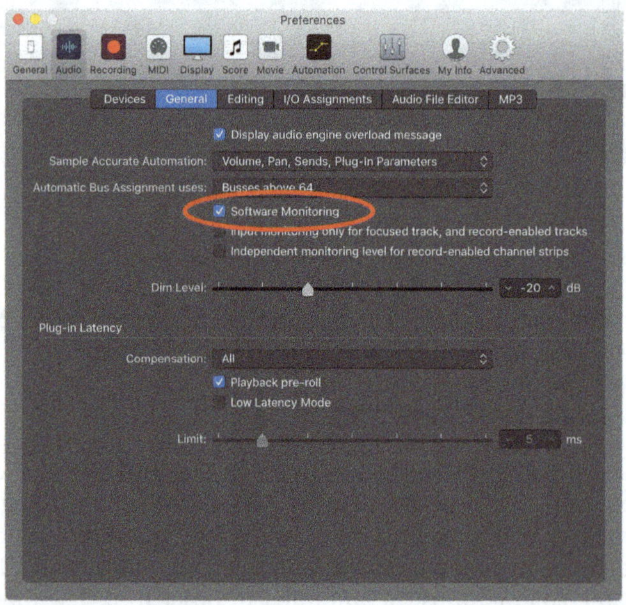

FIGURE 10.13
Logic Pro: by disabling "Software monitoring" (in Logic Pro X/Preferences/Audio/General), zero latency monitoring is a viable option for intensive punch-in/punch-out sessions too. Don't forget to turn this setting back on when recording is finished, as it will prevent playing software instruments too.

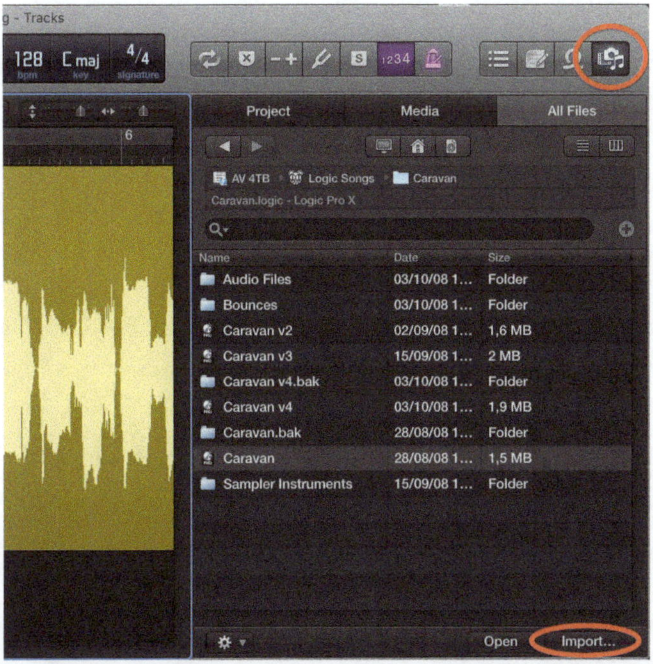

FIGURE 10.14
Import tracks from another project in Logic: click "Browsers," "All Files." After double-clicking the desired project, select the relevant tracks, and click "Add" (bottom right). In Pro Tools you can use "Import Session Data" (in the "File" menu).

SLAVE

Chances are, that in an almost finished DAW project, you may have set the I/O buffer at a high value to prevent audio from glitching. What if you still need to record a big choir? As a solution, the audio file of an instrumental bounce, or **slave mix,** can be dragged into a new/empty project. After recording, you can either drag the choir mix into the original project or import all the individual tracks separately (see Figure 10.14).

CHAPTER 11

The Recording Session

For a proper recording, a lot is involved. Not only must you create the right atmosphere for the musicians to feel comfortable; you also need to guide them during their performance, and take care of the technical side of the session, which can be quite demanding. This chapter shows you various strategies to collect takes from an artist. It also explains how headphone mixes can be adjusted to improve performance. Is it allowable to use effects like EQ, compression or reverb while recording? Last, we dive into the creative possibilities of varispeed.

Headphones Mix

▶ INDIVIDUAL HEADPHONE MIXES FOR THE MUSICIANS

When recording a **band**, musicians need individual mixes on their headphones. A drummer, for instance, will need sufficient volume of the rest of the band in his cans, as he can already hear his kit acoustically. A vocalist in the vocal booth, instead, cannot hear the drums acoustically, so the drums must appear relatively loud in her headphone mix. Chapter 16, "Organizing a Project," explains how to create individual mixes on headphones.

▶ MAIN MIX TO HEADSET

Although individual headphone mixes are essential for recording a **band**, they could work adversely when **overdubbing**. As the producer and engineer listen to the control room mix, it may never come to their attention that the mix in the headset fails. Not only can the balance be wrong; there's also the risk of unwanted signals in the cans, like the click track, a loud guitar solo or any previous takes that interfere with the current take. As musicians focus on performance, not on technique, they might never notice these shortcomings, while the delivery has suffered in the meantime. To prevent this from happening, it's usually better to send the control room mix to the cans instead. Using the "main-mix-to-headset" method, people in the control room listen to the same signal as the musician and can adjust the mix accordingly. This will prevent miscommunication and improve the chance of a good performance.

▶ IMPROVING PERFORMANCE WITH THE RIGHT HEADPHONE MIX

Generally, musicians want their instrument loud in the cans. Although you might have set their instrument loud already, it is not unlikely for the musician to ask for more. This is normal, musicians need control. But as soon as a performance turns out either untuned or untimed, it could well be caused by the musician being too loud in the mix.

Technically, no stone should be left unturned for getting the best possible **performance**.

But there's more: the intensity of a musician's performance can be controlled by adjusting their volume. For example, if you want a vocalist to sing a verse softly, you'll raise her volume in the cans. After she hears herself louder, singing will turn out **softer**. In the chorus, where a loud intention is needed, decreasing volume in the cans will result in **louder** singing. Be very precise with the level, changes of only 1 to 2 dB can make all the difference!

▶ BACKING VOCALS

When double tracking **backing vocals**, there's another thing to be aware of. Vocalists get tired during double tracking. Especially in long sessions, you may see them fade away. The result is sloppy timing and loss of intensity. In case the current take is masked by any previously recorded tracks, this might not come to your attention. To prevent this, try muting the choir's already-recorded tracks when doubling. Now you'll be able to properly assess the current performance and motivate and guide the singers if necessary. Of course, this method can only work if every individual choir member is totally sure about the notes and their phrasing. During recording, the performance should be scrutinized for intensity, timing and tuning. After recording, the result can be verified in context by unmuting the earlier vocal tracks.

TIP

CLEAR-CUT HEADPHONE MIXES

On headphones, it can be hard to oversee complex arrangements. It will help a musician to simplify a complex arrangement by taking out tracks that are less significant. Always be sure to supply sufficient rhythm in the cans!

▶ TIMING AND TUNING ISSUES WITH HEADPHONES

In case a vocalist has trouble finding the right intonation when singing with headphones, taking off one side of the cans could help. It allows a musician to hear herself acoustically. But with one can off, it can be hard to keep track of the rhythm. That's when you might need to raise the volume of drums and percussion in the cans. Not unimportantly, the sound of the can that's off will arrive in the mic as **crosstalk**. Therefore, always ask vocalists to slide the can to the back of their heads or pull out the can's cable (in case of a studio or DJ headset).

▶ HEADPHONE VOLUME CHANGES PITCH

Once you start recording in the studio, try getting used to a **safe headphone volume**. As noted in Appendix 2, loud sound not only causes hearing loss but will also cause listening fatigue. And there's another problem with high listening levels: they change the perception of pitch. At high headphone levels, our brain will interpret note pitch as **sharp**. As a reaction to this, you'll compensate by singing **flat**. So in case a vocal take turns out flat in its entirety, this could well be due to headphone volume being too high!

A **flat** vocal performance could be due to a **high** headphone volume.

▶ WHICH HEADPHONES?

Basically, there are two types of headphones, **open** and **closed**. Although the sound of open headphones (see Figure 11.1) is a little more hi-fi, open shells could cause feedback or at least bleed in the mic. Therefore, closed headphones (see Figure 11.2) are the preferred type for recording. Not only do they reduce headphone bleed; they also provide for better isolation from sound sources outside, a necessity when recording with a band. But even closed headphones cause bleed. This can turn out to be problematic when a vocalist sings a soft outro with a loud click track in the cans for example. As it is impossible to remove spill from a recording, always be alert when capturing soft performances. Closed headphones also provide better isolation for noise from outside, which can be called an advantage in case the drummer seems to be determined to demolish his drumkit. . .

FIGURE 11.1
Beyerdynamic DT990-Pro open headphones.

Source: Photo courtesy of beyerdynamic.com.

FIGURE 11.2
Sennheiser HD25-1 closed headphones, for studio or DJ use. One of the cans can be detached.

Photo courtesy of sennheiser.com.

Recording Strategies

▶ STRATEGY 1: RECORDING ENTIRE TAKES

Now that you know about making the best settings and working with headphones, it's time for recording. What's the considerations and strategies for a recording session?

When recording a single musician, collecting multiple takes on individual tracks has many advantages. This way, he or she can stay in the same musical flow and is least hindered by the technical side of the recording. After recording, you'll *"comp"* (compile) a final take by selecting the best bits of the individual takes (see Figure 11.3). Comping afterward can take considerable time and energy, however. Finding the best fourth line of the third verse can turn out quite a job if you have nine takes to choose from (. . .). Therefore, it is better to compile the best bits in between takes as good as you can. This is when momentum and concentration are peaking. After recording, the artist can be presented a finished result. This will speed up the process and keep everybody inspired. If you can't keep up with comping during the session, then at least make notes during the different takes.

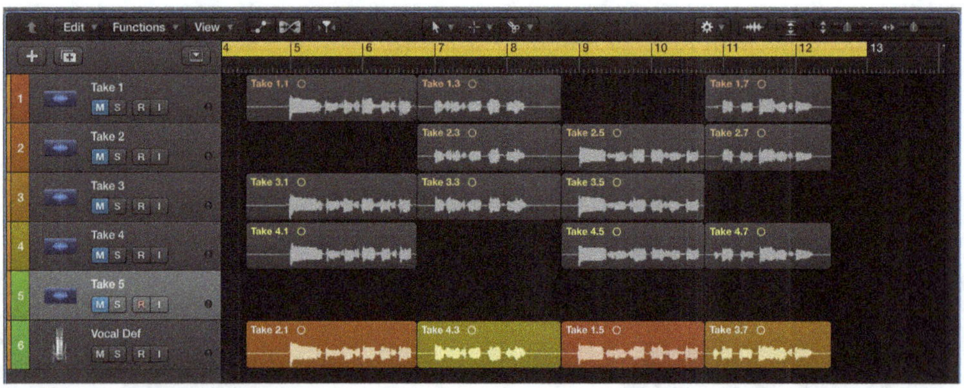

FIGURE 11.3
Tetris advanced: comping the vocal.

▶ STRATEGY 2: CONSECUTIVE RECORDING OF SIMILAR SECTIONS

Recording an entire take is often what's necessary to preserve a natural flow of the song. But with multiple takes, this can be straining for the artist, as it requires her to go through intensity and mood changes many times. On the technical side, different intensities require custom settings for mic gain, EQ, compression, effects and distance to the mic. As an alternative, consecutive recording of sections with a similar mood could work better. By first recording all the verses and then all the choruses, both artist and engineer can fully concentrate on a single intensity. Keep in mind that it may work better to first record the less straining parts while leaving the difficult sections for later. In a vocal session, voice quality can be preserved by recording the higher parts **after** the lower sections. Once the lead vocal is done, the vocalist can squeeze out all remaining energy with the *ad-libs* (improvisations).

▶ STRATEGY 3: OVERDUBBING WITH PUNCH IN/PUNCH OUT

After recording a longer section, a single phrase (or word) can be rerecorded by using the *punch-in/punch-out* method (see Figures 11.5, 11.6 and 11.7). This means that the previous take plays until the moment the record button is pressed (punch-in). Then, the old take is erased and the microphone signal becomes audible. At the moment of punch-out, recording stops, and the original take becomes audible again. Automatic switching between recorded audio and the mic signal is called "**Auto Input Monitoring**" (see Figure 11.4) After some training, words or even syllables can be punched. In case you want to revert to the original take, you can simply click and drag the right border of the original clip/region to the right.

FIGURE 11.4
Logic Pro: "Auto Input Monitoring."

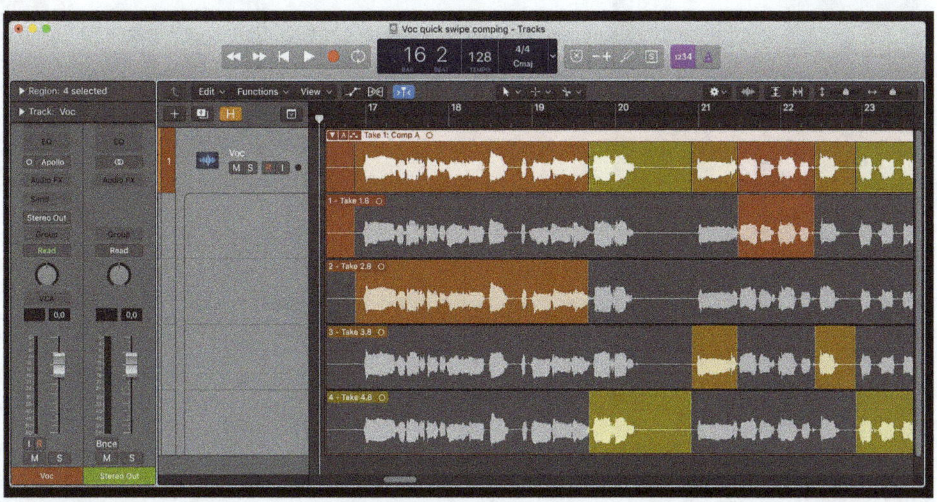

FIGURE 11.5
When punching in/punching out in Logic, a "take folder" is created by default. The best bits can then be selected with "Quick Swipe Comping." (Click–hold and drag the mouse).

To prevent changes in mood or intensity, it helps if the musician plays along, both before and after the punch section. Although the punch-in/punch-out method takes a lot of everyone involved, it can be called a great advantage that comping is done on the fly. Once decisions are taken, the project can move forward.

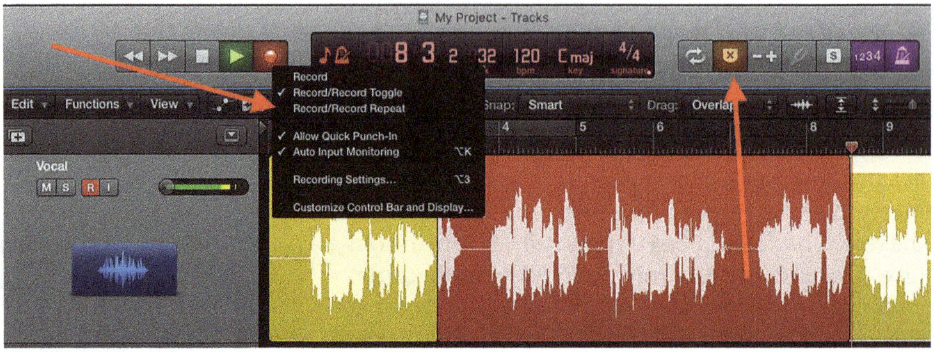

FIGURE 11.6
In case you don't want Logic to create a take folder but, rather, want to record regular regions, the "X-button" ("Replace recording") should be activated. Recording is (as always) nondestructive.

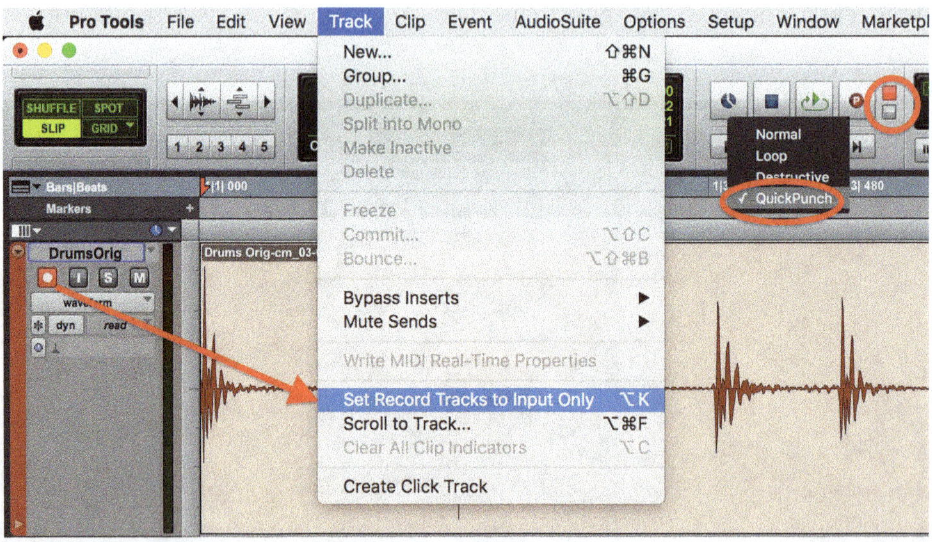

FIGURE 11.7
Punch in/punch out in Pro Tools: right-click the record button to select "Quickpunch." After arming a track, "Auto-Input Monitoring" can be enabled in the "Tracks" menu. The green LED in the transport bar (under the red LED) will reflect your choice.

▶ PRERECORD/POSTRECORD

Both **Pro Tools** and **Logic** record audio, *before* the actual press of the record button. It is called "**prerecord,**" and this is how it works: by arming a track, a new audio file is created on the hard disk, named after that track. Although this file is empty as long as the DAW is idle, it gets filled with data (i.e., the mic signal) upon pressing "play." Only after punching, a new clip becomes visible in the screen. Now, any good lick that was played before the punch-in point can be recovered by simply dragging the left border of the clip/region to the left. "**Postrecord**" works in a similar fashion: playing beyond the punch-out point causes data to be added to the audio file. To bring back this audio after recording, just drag the right border of the clip/region to the right. Prerecord and postrecord are not only great to recover "unrecorded" material but are also priceless for fixing bad punches, like incomplete breathing or chopped-off note endings. In **Logic**, postrecord is limited to one beat.

▶ GETTING THE ULTIMATE TAKE

After all sections have been recorded properly, it's always worth the trouble to collect a few ultimate takes of the entire song. Knowing that the job is basically done, musicians ease off and feel free to take risks. Although such a take can rarely be used in its entirety, it could eventually contain some fantastic notes that propel the song to the next level.

For both the engineer and the musician, overdubbing is often demanding, especially with vocal sessions. The musician is keen on nailing the best take, the engineer puts everything to work in order to accommodate this—and to come up with a sublime sounding end result. This requires section-by-section tweaking of the monitor mix, microphone pre-amp gain, compression, EQ, effects and, last but not least, accurate locating and punching. In case you hear the vocalist's intonation lacks in the verse, you increase the volume of the piano chords in her cans. When timing lacks while doing the ad-libs, you increase the level of the kick and the snare in the cans. Once the rhythm has ceased playing in the outro, you unmute the click track in the cans in order to rhythmically support the musician. When punching in bar 8 of the chorus, you'll play her the whole chorus before punching in. But with the seventh take of that same section, waiting will get tedious, so you'll give her a *pre-roll* of four bars or even two bars. Only when the engineer responds instantly to the artist's delivery, results can be optimal. This requires anticipation, a positive attitude and empathy toward the musician and her performance. During recording, you keep asking yourself: How can we do it better? What is technically needed for capturing the ultimate take?

▶ DOUBLING

Doubling is an often-used production technique that has been key to the sound of artists such as Brian Wilson, John Lennon, Kurt Cobain, Billie Corgan and

Dave Grohl. By rerecording the vocal on a new track, small differences in timing and tuning cause a thickening effect. Doubling smooths out any differences, resulting in better overall pitch. That's why this technique is also a great solution for vocalists that lack stability. Apart from this, doubling can be used to reinforce the song's musical flow: a single vocal in the verse could work well as a contrast to a doubled vocal in the chorus.

Commonly, a doubled **lead vocal** consists of either two or three tracks. The choice is personal and depends on the sound wanted. Although with two tracks, the individual performances can still be distinguished, this will be impossible with **three** vocals. Three is a crowd!

▶ BACKING VOCALS

With backing vocals, the size of the choir is determined by the number of vocalists and their doubles. For an intimate choir, a single voice per harmony could work well. For a medium-sized choir, three tracks per harmony can be considered a minimum, taking up nine tracks in total in case of a three-part harmony. For a large choir, it is not uncommon to double a single harmony four, eight, or more times. In case of doubt, always opt for more doubles, as you can always decide to leave them out later.

For soul or rhythm-and-blues (R&B)-style choirs, it may work well if the vocalists perform multiple harmonies in one take, as this will add to the vibe. But you'll not be able to change the volume of a single harmony afterward. This can be called a disadvantage of recording multiple harmonies in one take. That's why in most pop sessions, backing vocals are recorded per harmony. With multiple backing vocalists, they will all sing the same note per take. With harmonies on separate tracks, balance (and processing) can be determined afterward.

Don't forget the bidirectional mic option when recording backing vocals, as it allows vocalists to stay in touch visually!

ADVANCED GAIN

In case the intensity of the performance changes *within* a take, it can help the production if you adjust the mic's pre-amp gain correspondingly. It works like this: first, ensure that loud phrases (maybe the choruses) do not exceed the 6- to 12-dB headroom requirement. Then, set the recording channel's fader at the appropriate monitoring level. Now, when recording the softer phrases (maybe the verses), increase the mic's pre-amp gain until the instrument is sufficiently loud in the monitors *while leaving the recording channel's fader untouched.* After remembering the exact pre-amp settings for each song section, the level can be adjusted live, along with the performance. Of course, this is advanced; don't do this during your first recording session or with unpredictable performances. It is also unsuited for "organic" recordings. That being said, the longer the overdub session, the better you get acquainted with the musician's performance, which allows for adjusting levels precisely. Instruments that sit stable in the mix will help the production.

Using Effects

Adding effects during recording can improve the delivery dramatically. As long as you'll insert them in the DAW's recording track, they will not be printed, so you can always change things afterward. Note that effects in the recording track could increase latency.

Reverb

Musicians generally like a little reverb on their instrument. Although the effect blurs details in the recording, no stone must be left unturned for making the artist feel comfortable. Having said that, too much reverb causes too much blur and can "spoil" them. Therefore, always use as little as possible. Do not **insert** reverb on the recording track, as this will increase latency. Although adding reverb with an "**aux**" (Chapter 16, "Organizing a Project") causes the **reverb** to be late by a few milliseconds, this is always better than increased latency.

EQ

Even the best mic signal can often be improved with EQ. For instance, cutting a vocal's sub lows, under 100 Hz, will make space for other instruments. Or, boosting a little bit of mid-highs on a ribbon mic may help the vocal cut through the mix. Any EQ correction could help the mic signal to sit better in the mix and add to its appeal.

Compression

The dynamic range of an instrument can be so wide that the artist loses track of the band when performing loud parts while during softer parts her instrument may almost disappear—especially with vocals. In order for the artist to stay in contact with both her performance and the band, a compressor (Chapter 15, "Effects | Compression and Limiting") can be inserted in the recording channel. Apart from this technical reason, compression can also add to a certain sound that **inspires** the artist. Some producers may exaggerate the amount of compression during recording, causing the finest details to become audible in the cans. This allows the artist to play with every nuance.

Unity Gain

Every device has an optimum level: it's the level that causes the least noise and distortion. In the manuals of EQs, compressors, mixers or other devices, the manufacturer lists the optimum position for input and output knobs. This is called "**unity gain**." At unity gain, the device **neither amplifies nor attenuates**. With some devices, unity gain is when the knobs are at 12 o'clock, others are unity gain at 3 o'clock. With faders, it's at the 0 dB mark.

In case there are multiple devices in a chain, for instance, mic-pre->compressor->EQ-> mixer channel->audio interface, you try to maintain the same level throughout the chain. With every device at unity gain, no losses can occur. Unity gain is a requirement for proper A–Bing too: only when the level stays the same, an objective comparison can be made between the clean and processed version of the signal. When recording, all devices should be set to unity gain, while the mic's pre-amp level is adjusted according to the required record level and the volume of the musician.

Spectrum Analyzer

A spectrum analyzer (see Figure 11.8) shows a visual overview of the signal's frequency content. By inserting an analyzer in the mix bus, it's possible to see not only the spectrum of the mix but also any instrument in **solo**. When recording, a spectrum analyzer helps reveal unwanted frequencies in the signal. And, it could prevent making plain **mistakes**. Here are a few examples:

FIGURE 11.8
Voxengo "Span" spectrum analyzer (free).

- Microphones can produce large amounts of sub-low energy. That may be due to a train passing by, the musician tapping his feet or any stumbling against the mic stand. This sub-low energy does not contain any musical value but will tax amps, speakers and compressors. Eventually, it will get in the way of other instruments.

- While listening on small speakers during recording, the accidental activation of a hi-pass filter in the recording chain may go by unnoticed. For a kick drum or bass guitar, this can turn out disastrous.
- Software instruments can produce unnatural amounts of sub lows, even down to 0 Hz. That's earthquake territory! Some synthetic sounds produce energy at 20 kHz and beyond (local dogs will be yawning at the studio's door).

Many of these problems may never appear on small speakers. Only when the mix is played on a proper system may they become apparent. By keeping an eye on a spectrum analyzer, you'll minimize the chance for mistakes.

That being said, the visual representation of music should not be overestimated. Never counteract a specific peak in the spectrum with EQ by its looks. The peak might contribute to the character of the instrument or help it cut through the mix. Ears are always more important; a spectrum analyzer "just" helps.

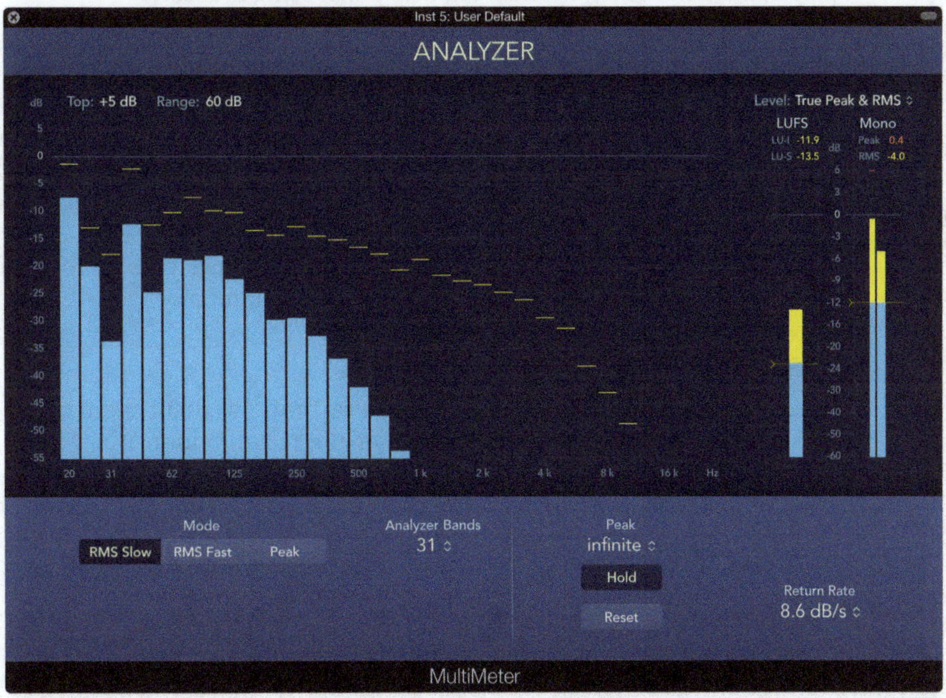

FIGURE 11.9
Synth bass with energy up to 0 Hz on Logic Pro's MultiMeter.

Creative Recording With Varispeed

Varispeed is a standard option on tape machines. Increasing or decreasing the speed of tape causes both tempo and pitch to change. The Beatles' "In My Life" (piano solo), Led Zeppelin's "Four Sticks" (vocals) and "When the Levee Breaks" (drums) and Prince's "If I Was Your Girlfriend" would sound quite different without the effect. Varispeed has many creative applications, and can also be used to solve problems. Here are a few examples:

- A virtuoso guitar solo (that's actually too complicated) can be recorded one or two semitones lower. Tempo will decrease by either 6% or 12%.
- In case a vocalist doesn't get to the top note, varispeed can be set to −1 semitone. By punching the top notes only, you might get away with the sound being slightly different.
- Recording male backing vocals a few semitones higher results in a more masculine sound when played back at normal speed. This trick may work well for drums too;
- Double tracking multiple vocals at various speeds (like ±1/4 tone) results in a thicker sounding choir.
- Doubling the horn section at −1 semitone will result in an extra aggressive sound when played back at normal speed. As it is not easy for brass players to play in a different key, this trick requires transposing the score.
- Recording at half speed allows singing an octave on top of the lead vocal.

Slow speed recordings result in better musical timing when played back at **normal** speed.

When playing at a slower tempo, musicians have more time to position individual notes. That's why slow speed recordings result in **better musical timing** when played back at normal speed. High-speed recordings, on the other hand, require more precise timing from a musician. In case you want an instrument to sound natural, varispeed is usually limited to one semitone. Greater values cause a **formant shift**, aka the Mickey Mouse or Chipmunk effect.

▶ VARISPEED IN DAWS

There aren't many DAWs that offer varispeed. **Pro Tools** has a half-speed option: by right-clicking on either the play-button or the record-button, "half-speed" can be checked. **Logic** has a full implementation of varispeed (see Figures 11.10–11.13). It is a playback preference for the entire project. It cannot be automated but will be reflected in the bounce.

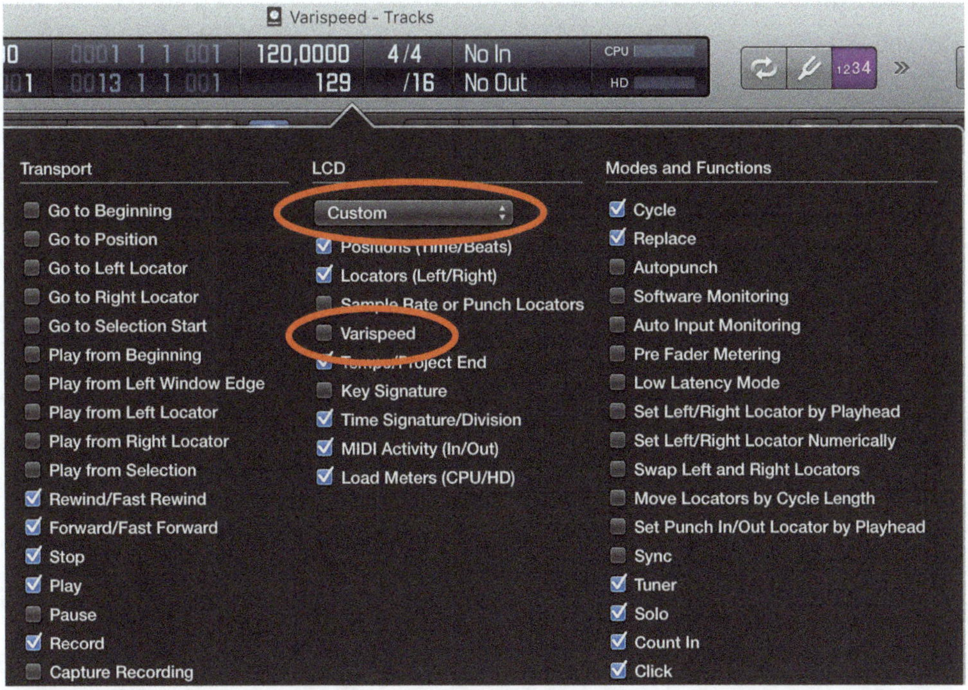

FIGURE 11.10
Varispeed in Logic is not visible by default. By Ctrl+clicking in the "Control bar" (between the knobs), the "Customize Control bar window" opens, and "Varispeed" can be checked.

FIGURE 11.11
Click–holding "Varispeed" allows choosing the varispeed mode. "Speed Only" causes the tempo to change while keeping the pitch. Although this mode is definitely handy for learning the licks of your favorite artist, *audio quality will suffer*. For recording purposes, it is better to choose one of the other modes.

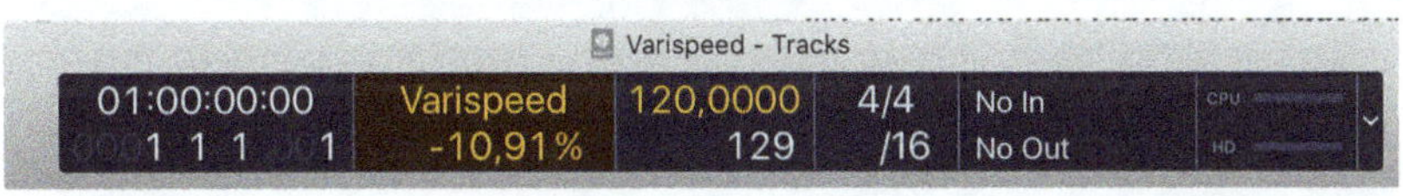

FIGURE 11.12
Clicking the "%" field allows changing varispeed units. Options are "Percentage," "Resulting Tempo," "Semitones," and "Tuning" (if A = 440 Hz).

FIGURE 11.13
Musically skilled people may find the "Detune (Semitone Cents)" setting most useful.

PART II
Mixing

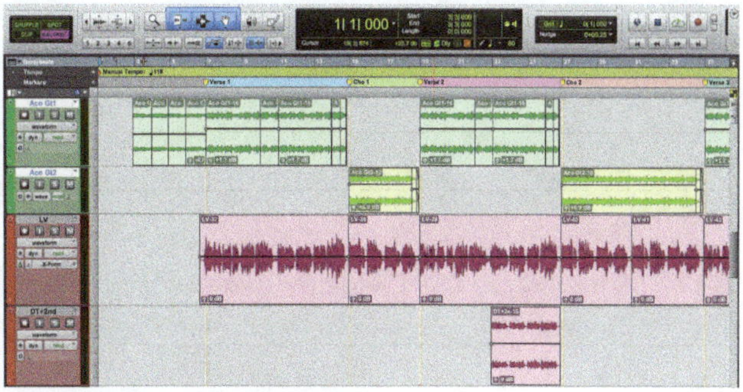

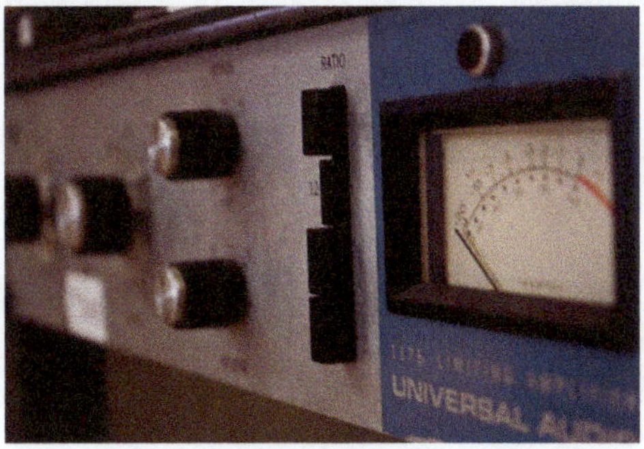

CHAPTER 12

Effects | Equalizers

Already in the 1950s, the frequency spectrum of instruments could be altered by means of simple equalizers. Since the invention of the *parametric EQ* in 1967, it is possible to apply corrections surgically. Although each instrument has its own unique space in the spectrum by nature, EQ can help emphasize differences in the mix. Once you start mixing, you'll notice that even well-recorded instruments can often be improved with EQ, causing separation and contrast to improve. Not only can this essential production tool help shape and focus an instrument's spectrum; it can also be used as a creative tool to rightfully mangle a sound. Which type of EQ should you use? And how do you know at which frequency to work?

Common EQ Types in Music Production

1. A **shelf equalizer** can be found on the shiny silver stereo of your parents and on the average guitar amp that has "bass" and "treble" controls. With shelf EQ, you can gradually boost or attenuate frequencies up to a certain level. This level is maintained until the end of the spectrum, hence the name "shelf" (see Figure 12.1). A shelf EQ never operates on the mid frequencies only but, rather, on either **end** of the spectrum. Due to its straight curve, shelf EQ changes the spectrum more globally and musically than other EQs.
2. The **parametric equalizer** or "**peak EQ**"/"**bell EQ**" was invented in 1967 by George Massenburg (Earth Wind and Fire, Toto, Weather Report). The term *peak* or *bell* refers to the shape of the curve when boosting or attenuating frequencies. A parametric EQ has three parameters: "**frequency**" (the frequency to boost or cut), "**gain**" (how much boost or cut) and ***Q-factor*** ("Quality"; see Figure 12.3). Quality determines how much of the surrounding frequencies will be included in the boost or cut. A higher Q equals a narrower bandwidth. When attenuating a certain frequency with a high Q, we call that a "**notch filter**." In music production, a notch filter can be used to **surgically** remove ugly resonances or frequency peaks from a signal.

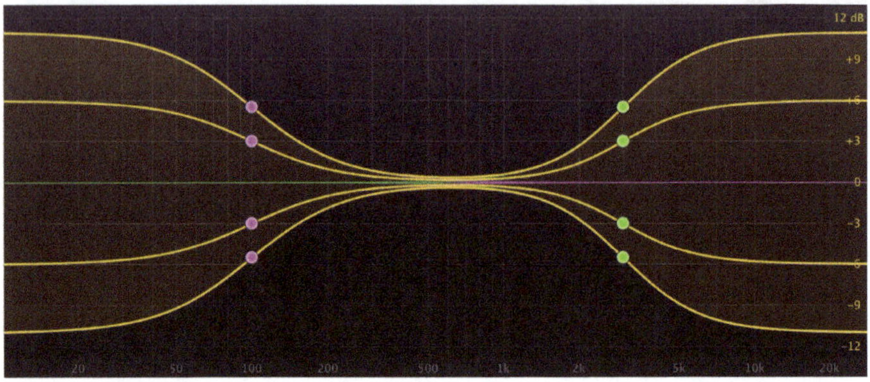

FIGURE 12.1
Shelf-EQ curves when boosting/attenuating 100 Hz and 3 kHz by 6 dB or 11 dB. Note that the spectrum above or below the displayed frequency is affected as well.

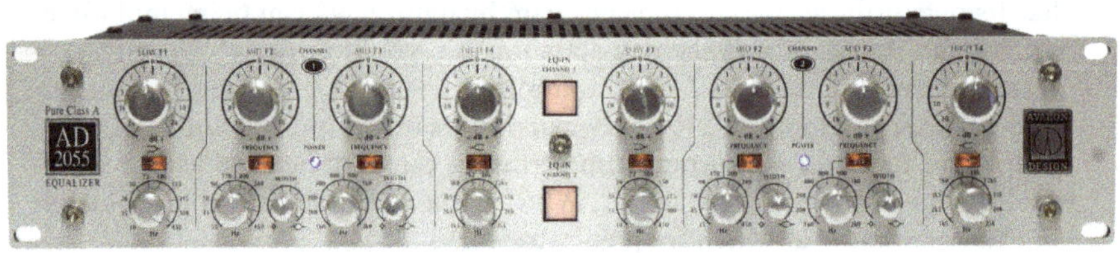

FIGURE 12.2
Avalon AD2055 stereo parametric equalizer, with two shelf bands and two parametric bands per side.
Source: Photo courtesy of avalondesign.com.

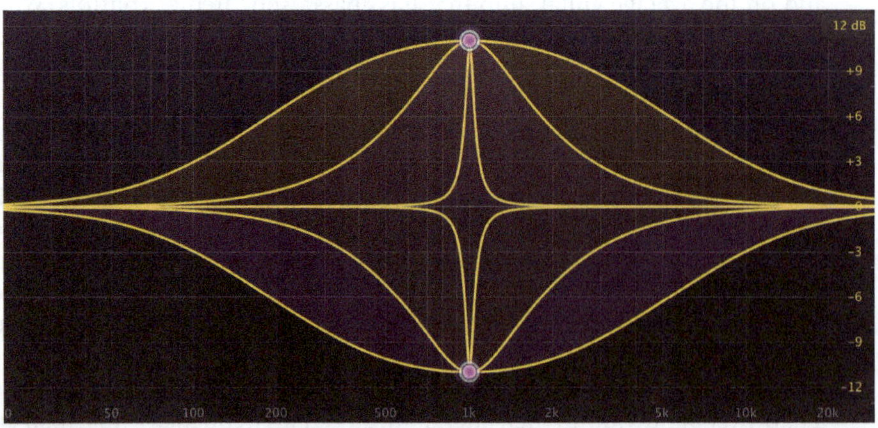

FIGURE 12.3
Bell-EQ curves when boosting/attenuating 11 dB at 1 kHz. The Q-factor in this example varies from 15 (narrow) to 1 (medium) to 0.25 (wide).

3. **Hi-pass** and **lo-pass filters**, which are also known as lo-cut/hi-cut filters, are integrated with many equalizers. They remove either the lows or the highs from a signal with just two parameters: **cutoff frequency** and **slope**. "Frequency" indicates the frequency at which the filter works, while "slope" determines how drastic filtering occurs. Slope is indicated as "dB-per-octave"; common figures are 6 dB/oct, 12 dB/oct, and 24 dB/oct (see Figure 12.4). The cutoff frequency is the point at which the signal is attenuated by 3 dB. In speaking language, hi-pass and lo-pass filters are often referred to as "EQ," but they are in fact filters. Both hi-pass and lo-pass filters are powerful and often-used tools to remove unwanted energy from signals.

Manufacturers have agreed for the cutoff frequency to be the –3dB point. Let's say, a 6 dB/oct hi-pass filter is set at 400 Hz. Compared to higher frequencies, 400 Hz will be attenuated by 3 dB, while 200 Hz (exactly one octave below 400 Hz) will be attenuated by 9 dB. In short, even though a hi-pass filter **indicates** "400 Hz," the area above 400 Hz is affected too. Therefore, always keep a safety margin in order to prevent losses.

In the mix, the **color** of an instrument is just as important as its **volume.**

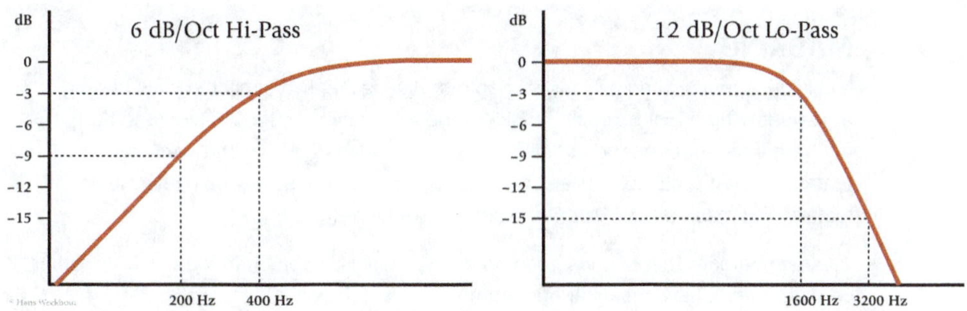

FIGURE 12.4
Left: a 6 dB/oct hi-pass filter at 400 Hz. Right: a 12 dB/oct lo-pass filter at 1.6 kHz.

▶ EQ IN THE MIX

With more instruments in the mix, it gets harder to distinguish instruments individually. The more different their spectrum, the easier it is for our ears to tell instruments apart. By nature, cymbals focus in a different part of the spectrum than the kick (although you might be amazed by the amount of bottom end in cymbals). Or a steel string acoustic guitar has more energy in the higher part of the spectrum than a distorted electric guitar. Emphasizing these natural differences with EQ allows for better separation between instruments. Boosting certain frequencies reinforces a specific quality of an instrument and may help it cut thru the mix. By attenuating less important frequencies, space is created for others. **Focusing** each individual instrument's spectrum helps create **contrast** in the mix.

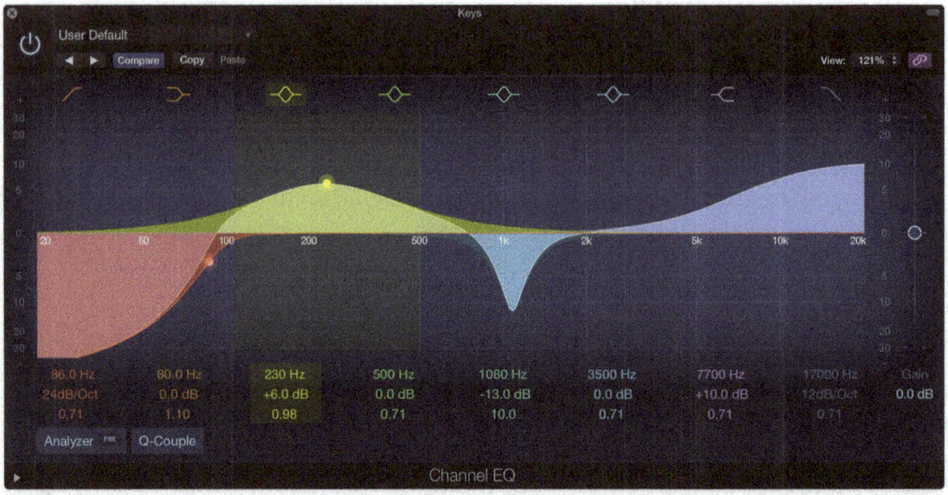

FIGURE 12.5
Channel EQ in Logic Pro: two shelf EQs, four bell EQs, a lo-cut filter and a hi-cut filter.

MASKING

Masking happens when our ear is unable to distinguish two similar signals playing at the same time. For example, a 3-kHz tone will mask a softer 2.5-kHz tone but will have little effect on the audibility of a soft tone at 1200 Hz. The **harmonics** of a mask tone may cause masking too. A 2-kHz tone with a strong 4-kHz harmonic (octave) will mask a 3.8-kHz tone. What can we learn from this?

- An instrument that sounds good on its own might disappear in the mix when other instruments have a similar spectrum.
- When the spectrum of two instruments **partially** overlap, they will **appear** to sound differently.

With more instruments in the mix, more masking happens. Masking can be prevented by altering the microphone and its position, panning, EQ or the sound of the instrument.

Here are a few examples:

- **A vocal has trouble cutting through the mix.** Boosting a bit of the highs (4 kHz and beyond) with a shelf EQ will cause the focus of the spectrum to be shifted to an area where the vocal experiences less competition from typical mid-instruments such as guitars, snare and keyboards.
- **The piano is interfering with the vocals.** By cutting the piano at 1 kHz with a bell EQ, space is created for the vocal, while the piano doesn't suffer too much. This is called "**carving.**" Executed at the right frequency, 1 to 2 dB may make all the difference.

- **The bottom end of the mix feels cluttered or muddy.** By cutting the (often) unnecessary low end of cymbals, guitars and other instruments, space is created for kick and bass.

Hi-pass filters are widely used in pop music. Removing low end is relatively harmless, as our brain will automatically make up for any low frequencies missing (see Appendix 1: "Characteristics of Sound").

Once you start EQing, you'll notice that each instrument has its own unique **sweet spots** (see Figure 12.6). For example,

- the "body" of a snare drum can be found around 100 to 200 Hz,
- the aggressive mid-range of a guitar is in the 800- to 1.5-kHz band and
- the click of the bass drum (which makes the instrument audible in a dense mix or on small speakers) is at 2 kHz and beyond.

After cutting an instrument's **bottom end**, our brain will make up for any missing information **automatically**.

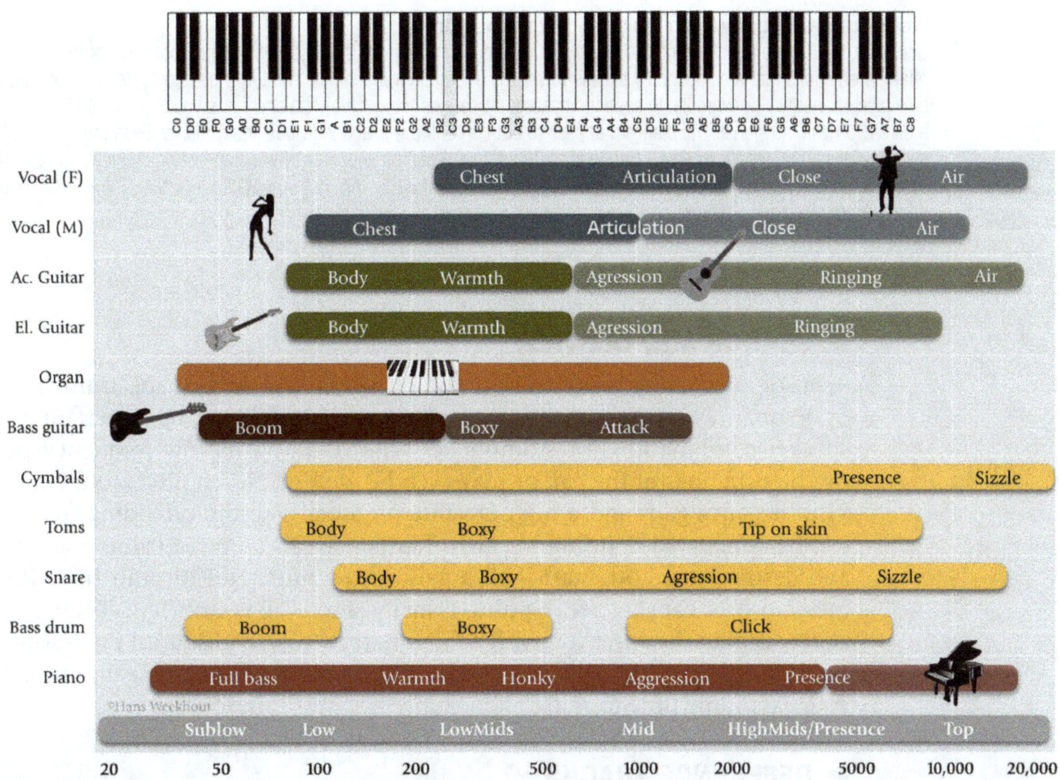

FIGURE 12.6
Frequency range and sweet spots of common pop instruments. Dark bars indicate root notes; light bars indicate overtones.

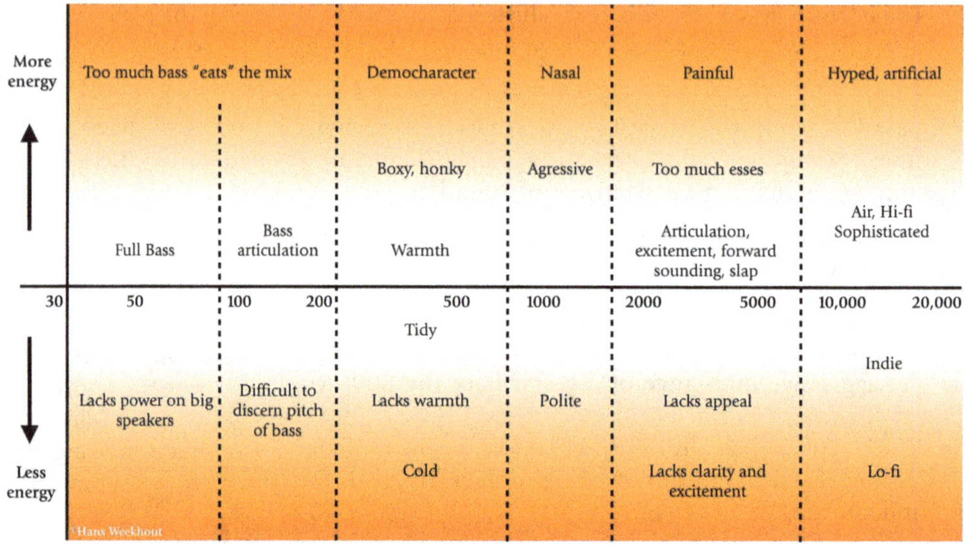

FIGURE 12.7

How do you know which frequency to work on?

Generally, we like the spectrum of our mix to end up around the *x*-axis. The adjectives in the upper half of the diagram relate to emotions that are caused by excessive energy of certain frequencies. The lower half contains emotions caused by a lack of energy. For example, in case the bass drum sounds "boxy," then the chart tells you there might be too much 400 Hz in the signal. The diagram applies to both an entire mix and individual instruments. Even though certain instruments cannot produce certain frequencies, the diagram can be accurate by imaginary shrinking it horizontally. For example, a violin hardly produces energy below 200 Hz. Nevertheless, this instrument has a certain "fatness," or bassy quality at approximately 200 Hz.

▶ SEARCH AND DESTROY

Acoustic recordings often suffer from irregularities in the spectrum, caused by resonances in the instrument, bad acoustics, nonlinear mics, certain amp–speaker combinations or stompboxes. The resulting narrow frequency peaks can be straining for the ear, or may even be painful. Notch filters to the rescue! With negative gain and a high Q-value on a bell EQ, the offending frequency can be attenuated. Finding the right frequency can be done using the "search and destroy" method. Start with a gain in the **plus**. Not too much, as it will mess up your ear's reference point. Usually, 4 to 8 dB is sufficient. Then, slowly start scanning the spectrum with a fairly narrow bandwidth until the resonance reacts most violently. This is the offending frequency. Null the gain and allow your ears some time to reset. Then, apply negative gain to taste.

▶ REFERENCE TRACKS

Comparing your own mix to a good sounding reference track allows for easier finding the right EQ settings. With a specific goal, the purpose of your actions becomes clear. It's important for the reference track to be in the same genre and (approximately) the same intensity, tempo and instrumentation. Ask yourself questions such as How bright is the snare in the reference track? and How much low end does the bass have? Make sure to compare tracks at the **same volume**.

▶ FREQUENCY BANDS ARE RELATED

As soon as you alter a specific part of the spectrum with EQ, perception of another area changes. For example, by adding treble to a mix, the bottom end will appear to be less impressive. Or, by reducing the low mids of a vocal, "presence" seems to increase. So EQing is not about absolute amounts of energy but, rather, about **balancing** frequency bands.

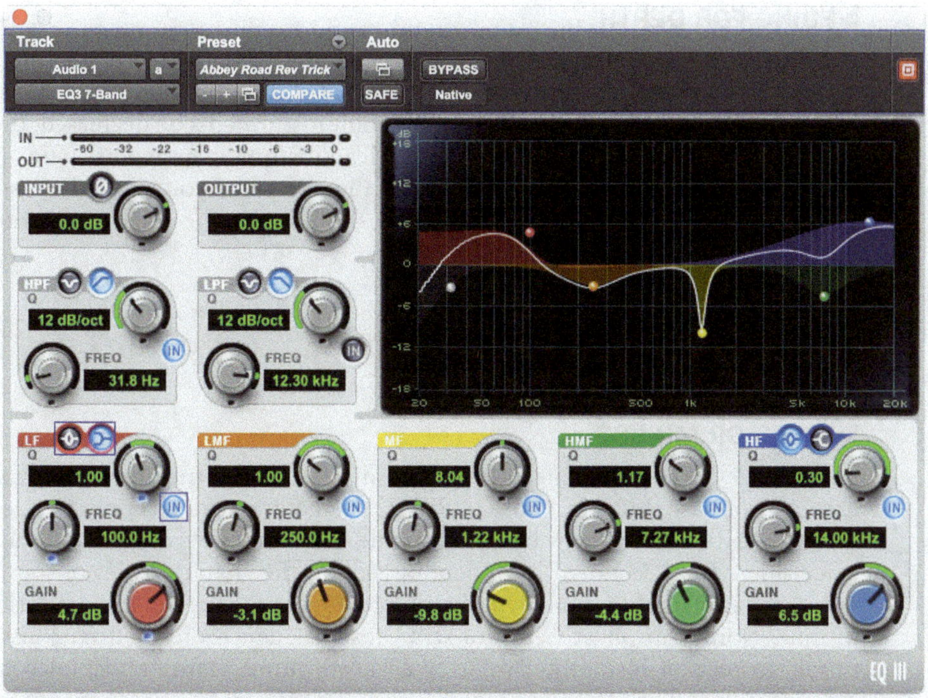

FIGURE 12.8

Pro Tools: EQ3–7 has both hi-pass- and lo-pass filters, three parametric bell EQs and two bells that can be switched to shelf.

▶ MORE TREBLE?

In the mix, you aim for sufficient energy in each frequency band so that every cilium in your ear is triggered. With sufficient energy in the high-frequency band, for example, the mix's appeal will increase, with added excitement, definition, "slap" and overtones. These are no small feats! That's why the highs can easily become addictive, resulting in overuse. With too much treble, the mix, or any of its instruments, will lack depth, sound artificial and be tiring to listen to. Be aware of this trap and try to achieve a proper balance between the highs and the lows. Comparing your work to other productions will help greatly.

▶ MORE BASS?

Similar to the highs being necessary for sparkle and slap, the lows are necessary for power, punch and energy and may invite listeners to dance. Unfortunately, "more is better" does not apply here either. Too much low end will eat up the mix, as it will divert our ear from the highs. It will prevent the mix from being exciting.

▶EQING FOR DEPTH

For our ear, the amount of high frequencies determines how close (or far away) we perceive an instrument. This knowledge comes in handy when mixing. Not only can the volume of less important instruments be decreased; they can also be made less bright with EQ. For lead instruments such as snare and vocal, it's easier to occupy the forefront of the mix when you allow them sufficient energy in the highs. Backing vocals should never be brighter than the lead vocal.

Advanced EQ

▶ PLUS OR MINUS?

As if phase wasn't an issue during recording, EQ will introduce a certain amount of phase shift too. Close to the EQ's target frequency, certain frequencies will be delayed. Narrow bandwidths and larger amounts of gain will cause more side effects. As noted in Appendix 1 about the characteristics of sound, it is in the **transient** that many different frequencies are fired in a short period of time. Nonlinear phase behavior will cause frequencies to be fired consecutively instead of instantly. Due to attacks' smearing, clarity and punch are reduced. It is not an effect you look forward to when trying to rescue a poorly recorded drumkit with large amounts of EQ (. . .).

Conventional EQ causes **smearing**, especially with narrow bandwidths and larger amounts of gain.

Cutting frequencies causes smearing effects to be less noticeable than boosting. This is why many professionals prefer **minus curves** instead of **plus curves**. For example, with an instrument that lacks both lows and highs, **cutting** the mids will result in a similar sound as **adding** bass and treble. The phase response of an EQ differs per brand and type. In fact, it is partly responsible for the "character" of an EQ.

▶ LINEAR-PHASE EQ

Things that aren't possible in the analog world, can be done in the digital domain: EQ without phase shift. Linear-phase EQ is more transparent than conventional EQ, it will preserve a better musical balance and can colorlessly treat the harmonic spectrum of an instrument or mix. Surgical EQ can also work well with this type of EQ. Unfortunately, there are also downsides. For one, linear-phase EQ is CPU-hungry. It may cause latency to increase to over a second, so it is not suitable for recording, only for mixing and mastering. Second, whereas conventional EQ suffers from post-ringing, linear-phase EQ can suffer from **pre-ringing**. This effectively causes you to hear sound before the actual transient starts. The effect is more likely to happen with steeper curves and with extreme settings of a linear-phase hi-pass filter. **Logic** has linear-phase EQ built in.

▶ PASSIVE EQ

Passive EQ is the oldest type of EQ. Although not completely free from phase issues, passive EQ exhibits fewer smearing effects than conventional EQ. Rather than being surgical, passive EQ is suitable for **global tone shaping** only. Often, large amounts of gain can be applied, without too many artifacts. The most famous passive EQ must be the Pultec EQP1-A. **Logic** has a virtual model of this device, "Tube EQ" (see Figure 12.9) while **Pro Tools** has "Pultec EQP-1A."

FIGURE 12.9
Logic passive EQ: Tube EQ

CHAPTER 13

Effects | Echo/Delay

Everybody knows the sound of natural echo: a scream in the mountains causes sound to arrive a little later. Since Les Paul came up with the idea to produce echoes with tape machines in the 1950s, the effect can be heard on many pop classics, like "Apache" by The Shadows, "Great Balls of Fire" by Jerry Lee Lewis and "Imagine" by John Lennon. Genres like reggae and dub would sound entirely different without the use of tape echo. In today's music, echo/delay is just as popular, although the effect is usually applied in more refined ways. Delay has at least one big advantage over reverb: it takes up less space in the mix!

With tape echo, the recording head of a tape machine records sound on tape. After the tape rolls, the signal is read by the replay head a little later. The length of the delay can be adjusted by changing the position of the head, and/or the speed of the tape (see Figure 13.1).

Tape echoes are mechanically fragile; they distort, produce noise, and limit the signal's frequency response. To improve on this, manufacturers came up with new techniques. In the 1970s, analog echoes from Electro Harmonix and MXR entered the market, while in the 1980s, digital units from Lexicon, Korg and Roland became popular (see Figure 13.2). Although designed to sound "better," these devices all had their own imperfections, resulting in a reduced but interesting sound quality. Nowadays, it's easy to design a software delay with zero distortion and a flat frequency response. But what's more boring than the exact copy of a sound? That's why the old tape delays, analog echoes and early digital gear have gained renewed interest. Although some people still prefer the original boxes, the majority of today's delay effects is applied with software that offers close imitations of the old units.

FIGURE 13.1
Maestro Echoplex EP-2 tape echo.

FIGURE 13.2
Iconic early digital delays with a characteristic sound: Lexicon Prime Time, Marshall Time Modulator, Bel BD80, Roland SDE1000.

▶ FEEDBACK

The feedback knob on a delay unit sends the output back to the input. The signal will then be delayed again—and again (see Figure 13.3). Feedback on vintage devices causes every subsequent echo to degrade progressively. This leads to a very natural decay, as darker signals seem farther away for our ear.

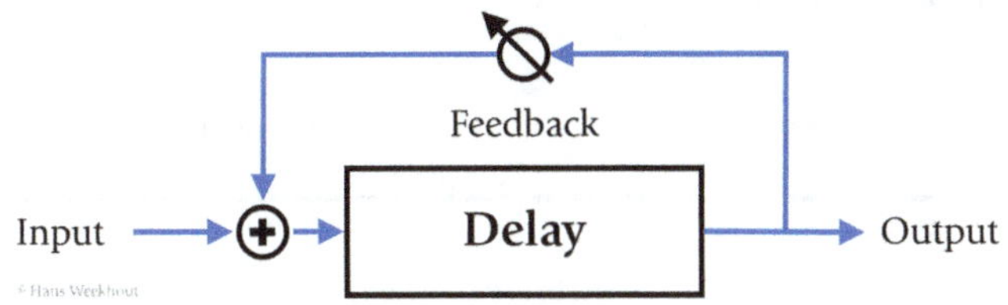

FIGURE 13.3
Delay schematically: the feedback knob controls how much signal is sent back to the input.

Delay Effects

- **Doubling** (30–90 ms) occurs when the delay signal seems to "stick" to the source, causing a **thickening** of the sound. For our ears, delay times shorter than 40 ms are harder to detect.
- **Slapback/Echo** (90–170 ms) will forever be connected to the sound of **rock 'n' roll**. Adding slapback echo to a guitar or vocal instantly causes that rockabilly vibe. Slapback echo sometimes seems to add punch to an instrument.
- **Long Delay** (170–750 ms) adds **depth** to a production, while an instrument can be made to sound bigger. The effect doesn't need to be obvious; only small amounts may subtly detach an instrument from its dry background. Settings of 750 ms and beyond cause the dry and delayed signals to gradually disconnect.
- **Ping-pong delay** is created by panning two different delays left/right in the stereo image. In case of musical values for the delay times, interesting rhythmic patterns can be the result. Ping-pong delays are partly responsible for the sound of U2's "The Edge."

Modulation Effects

Many delay units offer an "LFO" (Low-Frequency Oscillator) that continuously modulates delay time. This results in a **pitch change** of the signal. The strength of the effect can be set with *depth*, while the speed of the LFO can be changed with *rate*. With short delay times, flanger and chorus effects can be created.

- **Flanger** (3–20 ms): the well-known spacey/sci-fi effect. Increased feedback causes the effect to become more intense.
- **Chorus** (20–40 ms): the "watery" effect that causes an instrument to sound doubled or appear wider in the stereo image.

- **Phaser:** may sound similar to a flanger but technically it's different. A phaser works with so-called all-pass filters that change the phase of a signal. By adding the effected signal to the original, certain frequencies are cancelled. By **modulating** the phase shift with an LFO, other frequencies get affected, which results in the well-known up/down jet effect. The number of "stages" of a phaser indicates the number of all-pass filters. At last, phase cancellation is fun in music production!

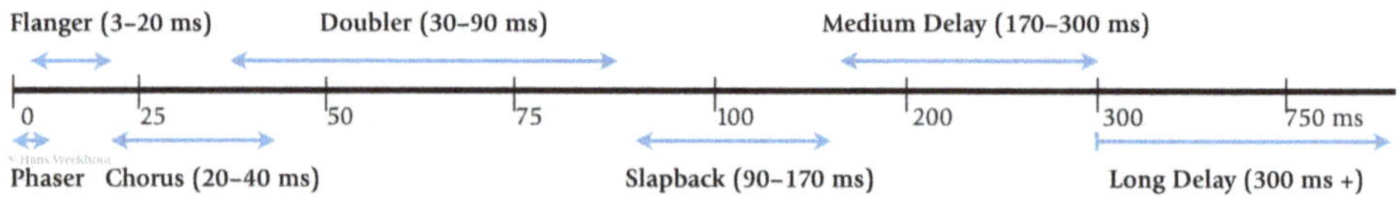

FIGURE 13.4
Different delay times yield different effects.

Both **Logic** and **Pro Tools** have a wide range of modulation effects on board, both as regular plugin-effects and as stompboxes.

▶ ADT AND THE BEATLES

Sometime during the recording of *Revolver* in 1966, Abbey Road engineer Ken Townsend came up with the idea to double John Lennon's voice mechanically. A mechanical double, Townsend thought, could only sound realistic if its timing alternates between earlier and later, just like a real singer's performance. Fortunately, the Abbey Road tape machines contained an additional playback head that read the signal **before** it arrived at the regular playback head. Townsend sent this early signal to a second tape machine, that processed the signal like a regular tape delay. By continuously varying this second machine's tape speed, the vocal arrived either before or after the original vocal. The process is called "Automatic Double Tracking" (ADT) and it can be heard in the first verses of "Tomorrow Never Knows."

Software manufacturer Waves has a plugin version of the effect under the name "Reel ADT."

ADVANCED FEEDBACK EFFECTS

The concept of a delay with feedback can be taken one step further by engaging a **filter** in the feedback circuit (see Figure 13.5). Dial in either a hi-pass or lo-pass filter with high Q-setting. Then, by slowly increasing feedback, the delay will start ringing at the boosted frequency. Just before ringing, altering the frequency causes feedback to move to a different tone. By using rhythmic delay times, the effect can add intricate long tails to vocals or drums.

FIGURE 13.5
Filtered feedback circuit in Logic Pro: create an *aux* (Chapter 16), insert a delay (with zero feedback), followed by a filter of choice. The *send* causes the filtered delay signal to return to the input of the delay. Be careful with feedback: *signal levels can easily get out of hand; this can damage speakers or harm your ears!*

Legendary Tape Echoes

Many different tape echoes have been manufactured over the years. Just because of their unique sound, legendary oldies like the Watkins Copicat, Maestro Echoplex and Klemt Echolette, Roland Space Echo (see Figure 13.6) and Binson Echorec are still in use by professionals that favor the real thing over software.

Binson Echorec

Although Italian manufacturer Binson brought the Echorec to the market already in the 1950s, the device became popular, only when Hank Marvin of The Shadows started using it in the sixties. The Echorec is not tape-based but

FIGURE 13.6
The most famous tape delay ever must be the Roland RE-201 Space Echo. Released in 1973, this device has three different replay heads, as well as a built-in spring reverb (see Chapter 14, "Effects I Reverb"). It has twelve different settings: four echo-only, seven that combine echo and reverb, and one that's reverb only. Universal Audio ("Galaxy tape-echo") and Audiothing ("Outer Space") have released plug-in versions of this classic.

works with steel wires wrapped around a rotating aluminum drum (memory disc; see Figure 13.7). After the wires are magnetized by the recording head, the magnetic energy is read by one of the (4, 6, 8 or even 10) replay heads. Later models could produce reverb effects too. Because of its working principle, the Echorec is more durable and robust than tape echoes. This was reflected in the price: due to the complicated manufacturing process, the Echorec's original price was similar to that of a Fender Stratocaster or Vox AC30!

Famous users of the Echorec include Jimmy Page, Delia Derbyshire (BBC Radiophonic Workshop), The Chemical Brothers and David Gilmour (Pink Floyd). Last but not least, the Echorec has contributed to the unique drum sound of Led Zeppelin's "When the Levee Breaks."

BINSON ECHOREC 2°

« L'Echorec » è il frutto degli sforzi creativi dei tecnici della BINSON per migliorare la riproduzione degli impianti sonori ed offrire ai musicisti la possibilità di ottenere nuovi effetti acustici.

Più volte imitato ma mai uguagliato, produce effetti di eco semplice o multiplo, ripetizioni singole o composte, alone cattedrale e riverbero.

Il pregio dell'« Echorec » è costituito dalla memoria magnetica a disco rotante di costruzione semplice, solida e di alta precisione. Offre la massima garanzia di durata al funzionamento continuo.

Effetti: Echo-Ripetizione-Alone-Riverberazione naturale - 3 canali con tastiera selezionatrice.
Dimensioni: mm. 430 x 280 x 235 Peso: Kg. 12.

FIGURE 13.7
Original flyer of the Binson Echorec.

CHAPTER 14
Effects | Reverb

Even if you're not a professional, you've still experienced reverb; it is the effect that makes singing in the shower or church sound good. Reverb is a mighty tool to glue instruments in the mix or to detach them from their dry background. But reverb could at the same time cause a production to sound undefined or outdated. In this chapter you'll learn how to apply the effect in a refined fashion and how to get creative with it.

In the open air, without any hard surfaces, there can be no reverb. Reverberation occurs when sound is reflected from the floor, walls and ceiling. Natural reverb (see Figure 14.1) consists of two components:

1. **Early reflections**, or **"E/R"**: these are the first reflections to reach our ear after being reflected by **close** surfaces, for example, the floor or a desk. Early reflections give us an indication of the **size** of a room: in a small space, there are many of them, arriving relatively early. In a large space, there are fewer, and it takes a while before they reach our ear. Because of the short path and limited bouncing, the E/R signal's high frequencies are relatively **intact**.
2. **Reverb:** the actual reverb is caused by **later** reflections. They bounce from surface to surface, and it takes longer before they arrive in our ear. Just because the reflections come in big quantities, it's impossible to hear them individually. They *smear*, causing the reverb to sound like a continuous signal. As a result of a longer travel distance and increased bouncing, reverb contains **fewer high frequencies.**

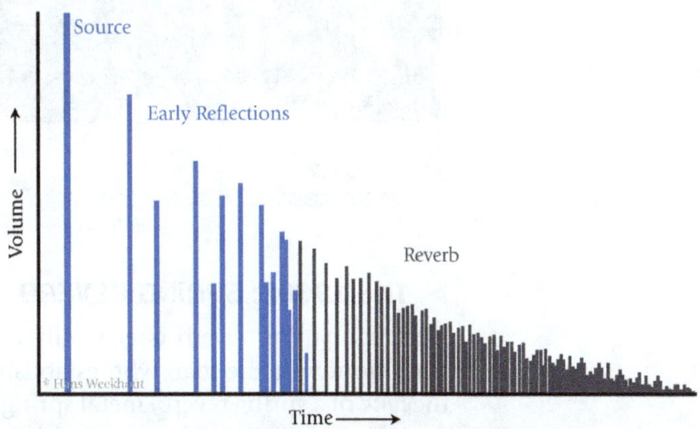

FIGURE 14.1
Reverb consists of both early reflections and the actual reverb.

Reverb in the Studio

Until the 1950s, only the natural reverb of a room was used in music recording. Soon after that, engineers found ways to generate reverb artificially. During the last 50 years, studio reverb has gone through a huge evolution. Despite all modern techniques, the sound of the old methods and devices is still very relevant.

▶ THE 1950S: ECHO CHAMBERS

By playing sound over speakers in a dedicated room (*echo chamber*; see Figure 14.2) and recording the signal with microphones, engineers could selectively add reverb to individual instruments. By moving the microphone(s), color and length of the reverb could be altered. Especially in the US, there are still studios that use their original echo chambers.

FIGURE 14.2
Modern echo chamber (University of Dresden).

▶ THE 1960S: SPRING REVERB

Although spring reverb is generally associated with guitar amplifiers, Laurens Hammond was the man who established a patent on spring reverb in 1939. In the *tank* of a spring reverb, metal springs are set into motion by means of a driver (transducer). The resulting sound is recorded by a pickup. Because of its low price, spring reverb quickly gained popularity in US studios in the early 1960s. The lo-fi sound of spring reverb is very specific and easy to recognize. Of all reverb devices, spring reverb sounds least natural, which is probably part of its success. Best-known studio spring reverb must be the AKG BX20 (see Figure 14.3); it is the favorite reverb of Dave Fridmann (Tame Impala and MGMT), Bloodshy & Avant (Miike Snow and Britney Spears), Adrian Utley (Portishead) and Lana Del Rey. The sound of surf (punk) music would be quite different without spring reverb. In "Take Me to Church" by Hozier, the production largely depends on the sound of—no, not a church—but a spring.

Pro Tools has "AIR Spring Reverb," **Logic** has spring presets in "Space Designer."

FIGURE 14.3
AKG BX20 spring reverb.

FIGURE 14.4
Doepfer A-199 spring reverb with three springs.

▶ THE 1960S: PLATE REVERB

A giant step for *mechanical reverb* was EMT's invention of the 140 *plate reverb* by the end of the 1950s (see Figure 14.5). This device consists of a big housing case by the size of a double bed, containing a metal sheet that's suspended by spring tensioners. The plate is set into motion by a transducer, while contact microphones pick up the sound farther down the plate (see Figure 14.5(b). A wheel is used to press damping material against the plate; this changes reverb time. Studios that formerly had to reserve costly space for an echo chamber could

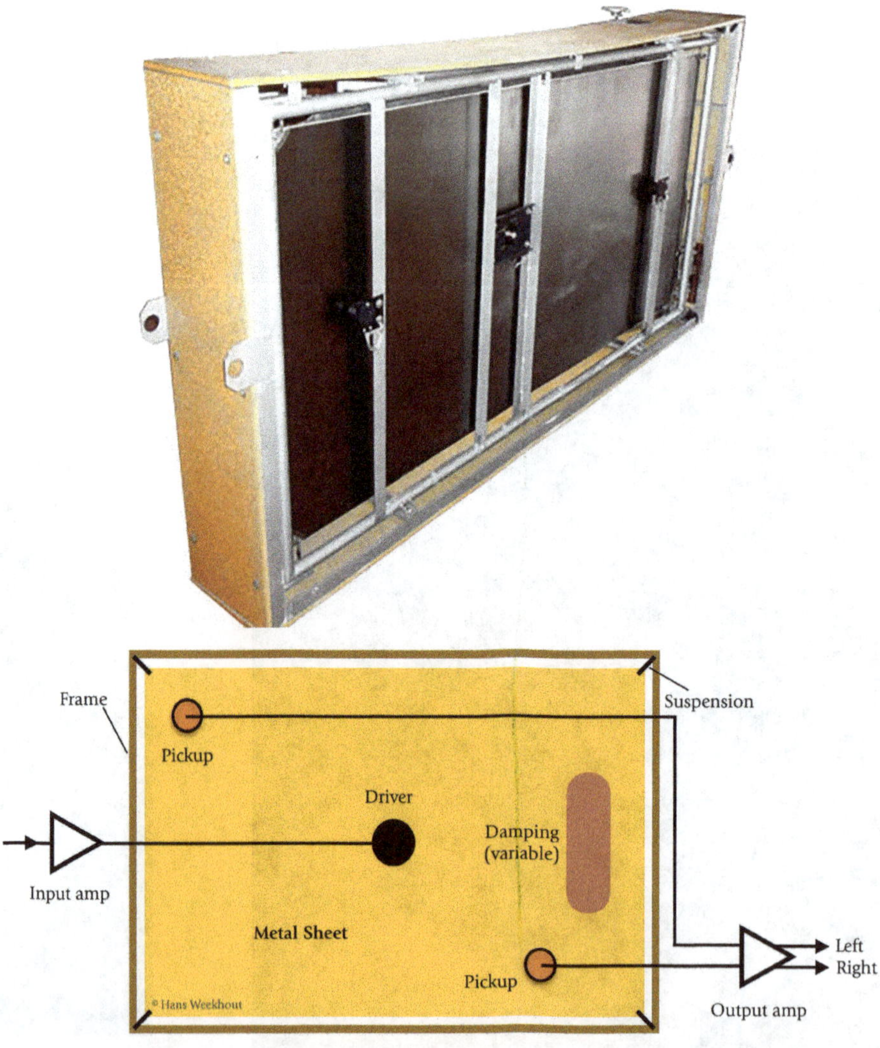

FIGURE 14.5
(a) EMT (Elektro Mess Technik) 140: probably the most used reverb in pop music ever. This baby comes in at the size of a double bed, weighing almost 600 pounds.

now easily house one or more plate verbs. Plate reverb was crucial for the sound of Motown and the American disco records of the 1970s. London's Abbey Road had several plate reverbs too.

Plate reverb doesn't produce early reflections and has a very natural decay. It has a metallic sheen to it, which in itself may not sound "beautiful." Only after adding it in the mix does it often seem to blend well, in a natural and pleasant way. Plate reverb is probably the most-used reverb ever; its sound might be nested in our collective consciousness, similar to the sound of a Stratocaster, Linn Drum, Fender Rhodes or tape echo. Passionate users of the effect are Joe Chiccarelli (The White Stripes, Alanis Morisette), Mick Guzauski (Daft Punk, Mariah Carey) and Mike Shipley (Alison Krauss, Def Leppard). Although some professionals favor the original devices, the majority of plate reverb is nowadays applied with software.

Logic has plate presets in "Space Designer," **Pro Tools** has plate presets in "Space."

▶ 1976: DIGITAL REVERB

The introduction of the personal computer in the mid-1970s made it possible to generate reverb by means of an algorithm in a computer chip. In early units, computer power was limited, resulting in a reverb that sounded more like a delay with feedback. For the first time, however, detailed settings for the reverb's color and length could be made, while spaces could be created that couldn't exist in the real world. Also new was the ability to save settings as presets.

DIGITAL REVERB COMMON PARAMETERS

- **Type:** Hall, Room (wood or tiled), Plate, Chamber or Non-Linear (fantasy). Due to individual algorithms, each space has a unique reflection pattern.
- **Size:** size of the room
- **Decay:** length of the reverb
- **Pre-delay:** delays the total reverb signal (including the early reflections)
- **Diffusion/Density:** increasing diffusion/density adds more reflections, causing a smoother, more "beautiful" envelope of the reverb. It can make a source sound lonelier, while the space becomes hollower and emptier. Less diffusion/density yields more character/"sound" and takes up less space in the mix.
- **Damping:** shortens the reverb time for the high frequencies. This has the added advantage of suppressing s sounds in the vocal's reverb.
- **Low Cut:** prevents the mix from getting muddy or undefined
- **Modulation:** varies the individual reflection's delay times. This can help maintain the listener's attention for the reverb.

Despite its 12-bit technology, the first digital reverb that sounded less grainy and digital was the Lexicon 224 (see Figure 14.6). Upon introduction in 1978, it cost US$7,500. For digital (or *algorithmic*) reverb, Lexicon would become the de facto standard. The 224 was followed by the 224X, the 480L and the 960L, and they all became standards of their own.

For algorithmic reverb, **Logic** has Silver Verb, Platinum Verb, and Chromaverb. **Pro Tools** has Reverb One, D-Verb, AIR Nonlinear Reverb, ReVibe II and Air Reverb.

FIGURE 14.6
Lexicon 224 digital reverb: parameters can be adjusted with faders.

▶ 2000: CONVOLUTION REVERB

Nowadays, computing power has increased, and reverb can be generated by using dedicated audio samples. Such a sample, or "**Impulse Response**" (IR), can be recorded after firing an alarm gun in a real space. Convolution reverb allows instruments to be positioned in the most beautiful sounding spaces, like the Royal Albert Hall, the Taj Mahal, the Grand Canyon or a Volkswagen Beetle's interior.

Most popular convolution reverb must be Audio Ease's "Altiverb." In **Logic** there's "Space Designer" (see Figure 14.7), while **Pro Tools** has "Space."

SAMPLING SPACES IN LOGIC

With Logic, you don't need an alarm pistol in order to sample your own space. After opening Logic's Space Designer plugin, the "Open IR Utility" option can be found under the "IR Sample" menu. IR Utility creates an impulse response by playing back a sine sweep. This opens up possibilities not only for acoustic spaces but also for effects devices. As the sine wave is an electrical signal, it can be sent through your favorite hardware reverb, guitar amp or stompbox. Doing so gets you the sound of countless devices in your computer. Lots of IRs can be found on the internet, often free of charge.

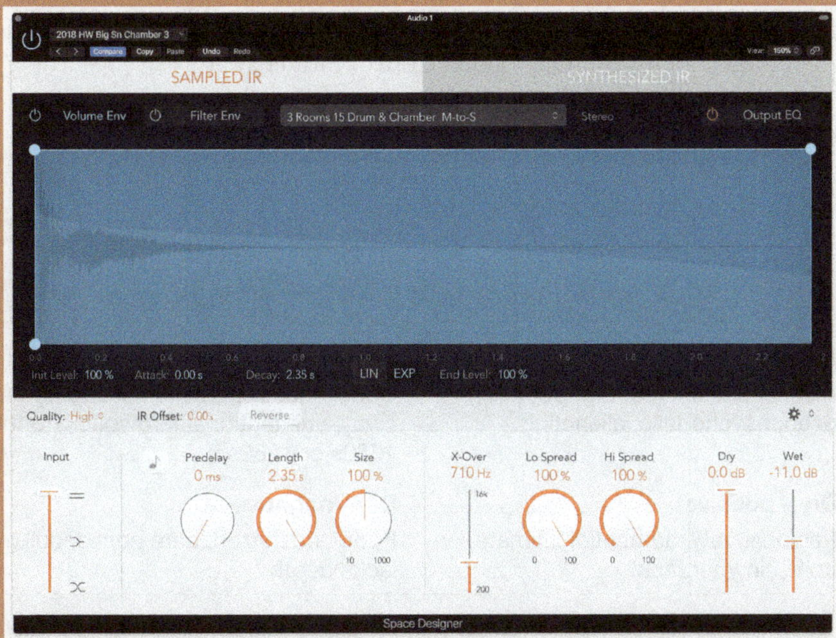

FIGURE 14.7

Reverb in the Mix

Why Would You Use Reverb in the Mix?

1. To **imitate** an existing space (e.g., church, garage or club)
2. To improve **contrast** with dry instruments
3. To **glue** instruments
4. To **blur** an instrument, for instance, to cover up the out-of-tune notes of a guitar solo

5. To make an instrument **bigger**, for instance, by adding room reverb to a snare
6. To create a certain **atmosphere** or as a reference to a certain style (like the 1980s or 1960s)
7. To emphasize a **scene change** between song sections, for example, a dry vocal in the verse and a wet vocal in the chorus
8. To **widen** a sound: a mono source becomes stereo
9. To add **sustain** to an instrument
10. To add **creative effects**, such as gated reverb (see the following discussion), reverse reverb or by using irregular impulse responses or NonLin programs

Commonly, reverb effects cause the mix to become **3D**. The **amount** of reverb determines if an instrument is either close or further away.

▶ HOW MUCH REVERB?

Table 14.1 shows subjective terms when using reverb. In the mix, this can help you decide on the exact amount of the effect.

Table 14.1 Positive and negative associations for reverb

Reverb—positive	**Reverb—negative**
Rich, sophisticated, big, 3D, juicy, deep, psychedelic, alienating	Undefined, impersonal, cheap, too far away, unintentional throwback (to the 1980s or 1960s)
Dry—positive	**Dry—negative**
Personal, raw, authentic, intimate, up front, "in your face"	Poor, undersized, bare bones, dull, lacks depth

Completely dry mixes are not very common in pop music. Maybe apart from a handful of records by The Strokes, Phoenix, The Red Hot Chili Peppers and the early work of Prince. On those records, the personal, authentic and intimate sound is caused by the **absence** of reverb. In case omitting the effect is the wrong choice, mixes or instruments sound boring, demo-ish, poor, or two-dimensional (2D) instead of three-dimensional (3D). Apart from this, leaving out reverb requires the **performance** to be top notch, as every single note will show up completely naked and exposed.

With too much reverb, definition will suffer. The effect might sound too comfortable, as if you are begging the listener for admiration. With the right color and amount, however, reverb can make a production sound rich, juicy, deep, big or even psychedelic. You'll present the listener a picture he can believe in. Good examples that come to mind are *1989* by Ryan Adams, *Morning Phase* by Beck, "Owner of a Lonely Heart" by Yes or "Let It Happen" by Tame Impala.

Which Instruments?

Vocals, guitars, keyboards, snare and toms are common instruments to use reverb on in the mix. A short room on the kick, snare and toms can make the kit sound thicker and wider. Less common instruments for reverb are hi-hat, overheads and bass, as the effect can cause smearing and loss of definition.

▶ LESS IS MORE

Specific reverbs such as spring reverb, echo chamber and reverse reverb can be the eye catchers of a production; they'll add character and personality. Adding exaggerated amounts of reverb to a single instrument can make a great statement. But more often than not, reverb is used only to **detach** an instrument from its dry background. Professionals often apply reverb just for this reason. The effect may hardly be audible as a separate ingredient, but once muted, the mix loses a bit of its magic. For this approach, it helps if you:

- Use **less reverb**. Find the minimum, then switch reverb off and on. You may find that just a little is enough.
- Use **more pre-delay** (50–150 ms). This will separate reverb from the original signal. Reverb will float like a mystical layer in the background of the mix, while the instrument in question no longer feels dry. As an extra advantage, pre-delay preserves definition of the instrument.
- **Filter off the highs** (or use a less bright reverb)

Advanced reverb: you'll **never** know that it's **there** until you mute it.

▶ SHOULD YOU USE ONE REVERB OR MULTIPLE?

It depends. When **gluing** is what you want, for example, with the individual mics of a drum kit; the kit will sound more like a single instrument when using just one room reverb. Or, when you want a studio recording to sound like a live concert (or a garage for that matter), it may work well to use just one reverb for all instruments. This will create more blend. But be careful; too much reverb results in too much blend, which causes the mix to lose definition.

More often than not, reverb is used to **increase contrast** between instruments. Each individual instrument's character is emphasized by using several, unique reverbs. With a downtempo song, for example, a small room for the drums, a plate for the vocals and a hall for the strings will improve separation. You can create even more contrast by keeping other instruments dry.

Using **multiple** reverbs will increase **contrast** in the mix.

▶ HOW LONG SHOULD REVERB BE?

For the length of reverb, you should consider the song's tempo, as well as rhythm and length of the notes. Although a long reverb can make an instrument sound **big**, this could blur following notes. That's why drums are often treated with short, room-like reverbs. In case you want to use a longer reverb for the snare, decay should be short enough for the reverb to end right before the next afterbeat.

▶ AS AN AUX OR AS AN INSERT?

The **amount** of reverb in a mix is more important than its **length**. That's why it's not necessary to use dedicated reverbs for each and every instrument. This is good news, as our computer power is limited. By starting off the mix with both a long reverb (hall or plate), and a short reverb (room or chamber) added as an **aux** effect (Chapter 16, "Organizing a Project"), the send levels can be used to adjust the amount of reverb per instrument. Doing so keeps the project efficient and manageable, as the computer has to calculate a limited number of reverbs only. Special effects, such as the occasional spring reverb on the guitar solo, can be applied on the desired track as an **insert**.

Getting Creative With Reverb

- **Gated reverb.** At the beginning of the 1980s, producer Steve Lillywhite and engineer Hugh Padgham came up with the idea to send the reverb signal into a noise gate. A noise gate is a device that passes audio only when the signal is sufficiently loud. Before the reverb has ended naturally, the gate closes. Just because gated reverb is so short, the effect can be applied in serious quantities. It can blow up a signal dramatically; that's why it's a popular tool to increase the size of drums and percussion. In electronic music you can hear it on lead synths. The exemplary use of gated reverb can be enjoyed on the drums of Phil Collins' *In The Air Tonight* (at 3'41").
- **Logic** has a dedicated plugin for gated reverb: "Enverb." It allows free drawing of the reverb's envelope. As there is no gate involved, even soft sounds can benefit from the effect. In a conventional setup, the gate would stay closed!
- **Reverb follows delay.** In case a delay effect sounds too "choppy," its signal can be sent to a reverb. This will cause each individual echo to obtain a small reverb tail.
- **Reverse reverb.** Playing a tape in reverse while adding reverb to the desired instrument was already possible in the 1960s. By recording the reverb on a spare track and then playing the tape in the normal direction, the now-reversed reverb could be used in the mix. This process took quite some time and effort though: besides changing reels on the tape machine, you had to keep a close eye on the new track order: track 1 turned into 24, track 2 became 23, and so on. Nowadays, the effect can be easily constructed on the computer (see Figures 14.8 and 14.9). Reverse reverb can have an intriguing effect on all sorts of sources, from drums, vocals, piano to guitar.

Just because gated reverb is so **short**, it can be applied in **serious** quantities.

FIGURE 14.8
Although some reverb units offer "reverse reverb" presets, a reverb plugin can never produce audio before the actual sound starts. The "real" thing can be easily constructed in Pro Tools: copy the desired section to a new audio track. Reverse this by selecting Audiosuite->"Reverse" (in other) and clicking "Render." Now, select a reverb plugin from the "Audio Suite" menu, set it at 100% "wet" and click "Render." Finally, "Reverse" the resulting file and drag it to the desired position.

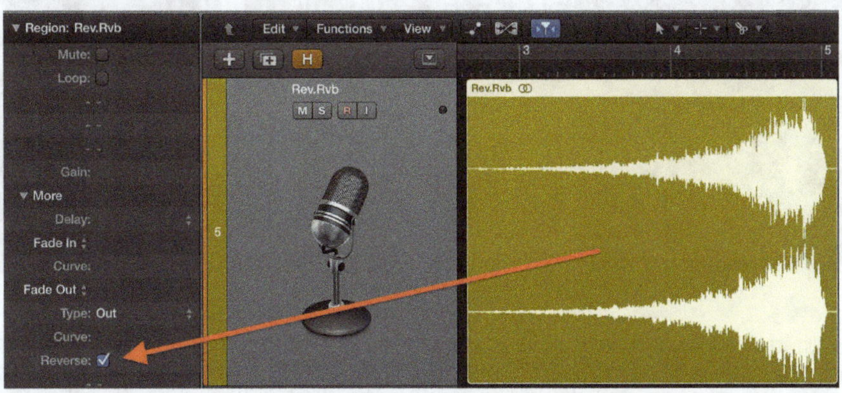

FIGURE 14.9
Reverse reverb in Logic: copy the desired section to a new audio track, and reverse it by checking "Reverse" in the "Inspector." Now, add reverb to this track (100% wet), and choose "Bounce in Place" (under "File"). Finally, "Reverse" the resulting file and drag it to the desired position.

- **Extra-wide reverb.** Two (slightly) different mono reverbs left/right in the stereo image results in even wider reverb.
- **E/R only.** Many digital reverbs allow setting the amount of early reflections and reverb independently. By choosing E/R only, "ambience" can be added to an instrument without any reverb tail. Small amounts detach an instrument from its background, while large amounts can make an instrument sound big, weird and unnatural, not unlike gated reverb.
- **E/R as a rich slapback delay.** Try adding pre-delay (100–150 ms) to an E/R signal; it will result in a more intriguing slapback effect.

FIGURE 14.10
The sound of natural reverb is unsurpassed, and usually sounds better than artificial reverb. Why is that? Let's say, we have a snare recorded with an SM57. After applying reverb to it, the reverb signal will take on the colored response of that specific mic. If you record the acoustic space with dedicated microphones instead, the reverb signal will sound richer and more natural. Provided you use good-quality mics and record in a proper acoustic space of course.

CHAPTER 15

Effects | Compression and Limiting

Of all studio effects, compression is probably the most difficult to grasp. But the effect has great advantages. Compression will allow for instruments to sit stable in the mix so that the listener can stay in contact with the performance. The effect can also be used to add aggression to an instrument. When used on the mix, compression provides excitement, urgency and glue. Understanding the process is one thing, but finding the right settings is another. Even professionals regularly gain new insights when using compression. One thing's for sure: without compression, pop music would sound completely different!

In the early days of pop music, compression was used to reduce the noise of tape machines. Taming the mic signal's peaks with a compressor allowed raising the recording level on tape. In the mix, the corresponding fader level could be lowered, due to the louder signal coming from tape. This resulted in less tape noise flowing into the mix. As a positive by-effect, compression prevented unexpected peaks to overload the tape.

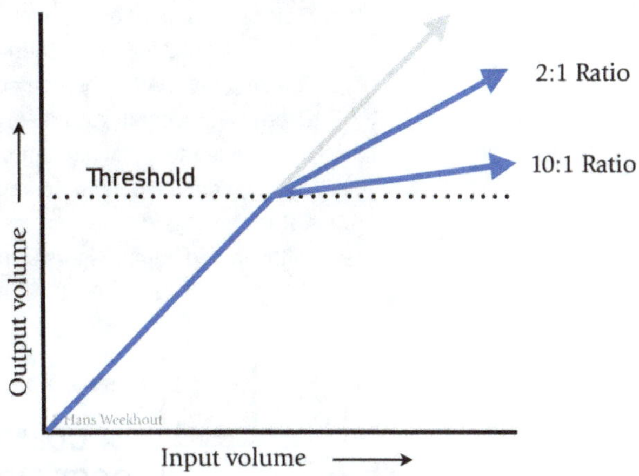

FIGURE 15.1
Compression graphically. For signals that are below threshold, the output signal increases proportionally with the input. However when crossing the threshold, this linearity ends: the compressor starts attenuating the signal. The higher the ratio, the greater the attenuation.

Although we hardly experience noise issues on the computer anymore, there are still good reasons for applying compression. One is for **reducing dynamic range**; the other is to **add character.**

Why would you want to decrease the dynamic range? To explain this, we'll compare the softest possible signals in music with the loudest. The **softest** acoustic signal of a band's performance may be around 40 dB SPL. This is roughly similar to the ever-present background noise in many of our listening environments (living room, car or train). The **loudest** signal of that band could easily exceed 100 dB SPL. Provided that music can be played that loud in a living room, it will cause the listener to turn down the volume. But now, the soft notes will disappear in background noise.

Here you see the need for compression: although the large dynamic range of a band may be suited for a concert, it is not in everyday listening situations. Current pop music commonly has a dynamic range of 3 to 15 dB. So, if the original dynamic range of such a band was 60 dB, this means dynamics are reduced by at least (60 dB – 15 dB =) 45 dB. That's no less than enormous! Such a reduction can only be achieved when using the right tools and techniques. Along the way, quality of the music must be preserved and technical artifacts minimized. So let's dive in!

COMPRESSOR CONTROLS

Essentially, a compressor is an automatic fader that starts working when the input signal exceeds a certain **threshold**. Then, the compressor reduces the signal's volume (see Figure 15.1). How this happens can be controlled with the following settings:

- **Threshold.** Every signal below the threshold remains unaffected. Signals that exceed the threshold qualify for compression. (Some compressors have an **input level** knob instead of a threshold knob: increasing the input level will cause more signal peaks to trigger the compressor.)
- **Ratio.** At a ratio of 1:1, no compression takes place. At a ratio of 3:1, a 9-dB increase at the input will cause the output to increase by only 3 dB. A peak of 8 dB will be reduced to 2 dB if the ratio is set to 4:1.
- **Attack.** Determines how quickly the compressor responds to a signal that exceeds the threshold. Longer attack times allow more transients to pass.
- **Release.** As soon as the signal drops below threshold, the compressor returns to its normal state. Release time determines how quickly that happens.
- **Makeup gain.** After the compressor has reduced peaks, the signal's volume has decreased accordingly. Makeup gain can be used to compensate for this. After increasing makeup gain, not only the peaks return to their original level, but **the soft signals have become louder too**.
- **GR meter** ("gain reduction meter"): indicates the attenuation of the signal (in dB). With vintage compressors, it moves from the center to the left.

Truth Myth

"A COMPRESSOR BOOSTS SOFT SIGNALS."

Although it is true that soft signals become louder after using makeup gain, a compressor actually reduces the peaks of a signal.

Compression to Reduce Dynamics

Let's say, a vocalist alternates soft notes with loud notes or varies her distance to the mic. In this case, mild compression settings can be used to reduce the dynamic range. For example, with a ratio of 3:1 to 4:1, and a threshold that's high enough to prevent the compressor from grabbing every single word. As long as the GR meter indicates a peak reduction of 2 to 6 dB on average, negative side effects are minimal. For this application, medium settings of attack and release are fine.

Mild settings cause the vocal to sound more **compact** (see Figure 15.2). Not only can every word of the lyrics be understood; it will also be easier for the listener to **stay in contact** with the vocal. On a technical level, the signal becomes manageable, which allows for easier finding a stable level in the mix.

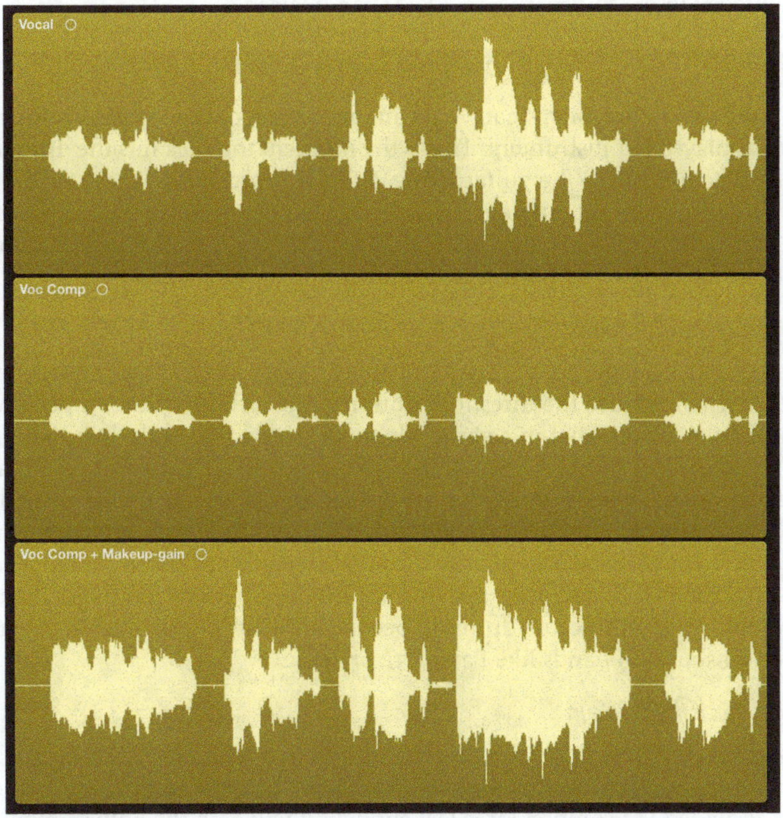

FIGURE 15.2
Compressor on a vocal, mild setting. The upper waveform represents the signal without compression, the waveform in the middle indicates the compressed version, while the lower waveform represents compression after using makeup gain. Note that soft words have become louder.

Now, let's fine-tune attack and release. Dial in a large gain reduction, that is, a **high** ratio (10:1–20:1) and a **low** threshold. Once the process is clearly audible, it will be easier to determine the right settings for attack and release. Play with release first: it should not be too long but just short enough for the GR meter to regularly return to zero. Then, attack can be adjusted. When set too long, the compressor will grab the words after the fact; when set too short, it may cause individual words to sound aggressive, unnatural or "spikey." Generally, the tempo of the notes dictates settings for release (and attack), and cause the GR meter to move rhythmically. Now that attack and release are set, ratio and threshold can be set to a milder setting. As a last step, adjust makeup gain until the signal has the same audible volume, regardless of the compressor being on or off (Unity Gain).

Due to its natural and forgiving behavior, the famous Teletronix LA2A is a popular choice for this application. In **Logic,** it can be found under Compressor ->"Vintage Opto." In **Pro Tools,** "BF2A" can be used.

2. Compression for Sound

It's in the signal's transient that we get an impression of how aggressively a musician has played her instrument. This is the moment the stick hits the drum head, or the plectrum picks the guitar string. With compression, we can either **reinforce** this aspect or **attenuate** it. In order to hear changes, we'll need more gain reduction and a higher ratio. Small changes to the attack and release time will make all the difference in this application.

a. Increasing Attack

Let's say a kick–snare track needs extra attack. First dial in a high ratio, like 10:1 or 20:1. Then, reduce the threshold until the GR meter indicates a peak reduction of 10 to 20 dB. Dial in a short release time and an attack time that's short but not zero. This causes transients to slip through, while **the compressor attenuates the sustain portion of the notes** (see Figure 15.3). With sustain being softer, the attack is now (relatively) louder. In fact, the signal has become more dynamic! This seems contradictory if you think of compression's main purpose (. . .). As the peaks are left intact, this application requires hardly any makeup gain. Settings like this cause drums to become punchy and aggressive, not unlike the drum sound of bands like Paramore or Muse.

b. Increasing Sustain

Let's again compress the kick-snare example with a high ratio, like 10:1 or 20:1, and quite a bit of gain reduction (10–20 dB). Now, set the compressor's attack to **zero**. Provided the compressor is fast enough, it will grab the note immediately. By dialing in a (very) fast release, the compressor has already stopped working at the moment the note enters its **sustain phase**. As the attack is attenuated, **large amounts of makeup gain** are needed to bring the signal back on level. This application causes the sustain of the note to become loud (see Figure 15.4). Sustain is where the **tone** of the note is.

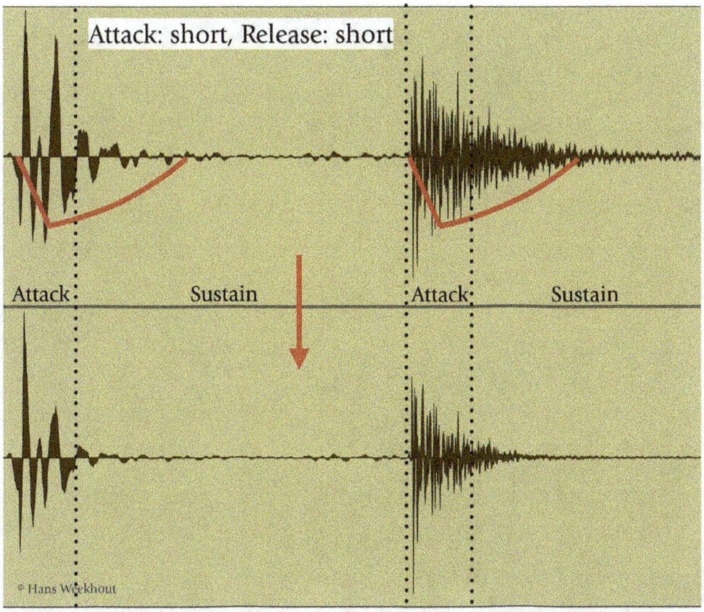

FIGURE 15.3
Compression for increasing attack on kick and snare. The note's attack is allowed to pass, while its sustain portion gets attenuated.

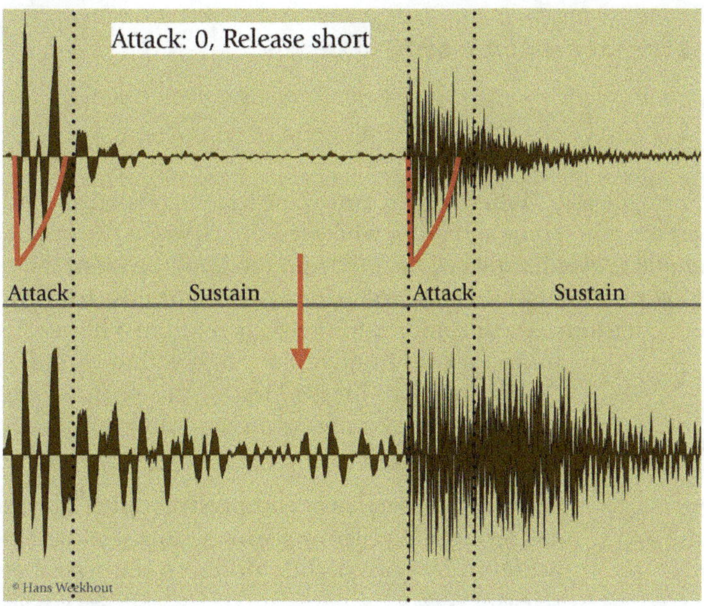

FIGURE 15.4
Compression for increasing sustain on kick and snare after using makeup gain. The (fast) compressor grabs the transient instantly. Due to the short release, the compressor has already stopped working at the moment the note enters the sustain phase.

FIGURE 15.5
John Bonham–style compression with Logic Vintage FET compressor.

Now you might ask, "When would you want less aggression in pop music?" Well, there are quite some situations where less attack works better. For example, when a guitar player has picked his plectrum too hard, or when the picking or slapping of a bass guitar was too aggressive, you might miss **tone**, or **body** in the signal. On drums, these compression settings result in a hotter signal, with more tone, not unlike the drum sound of Led Zeppelin's "When the Levee Breaks." In case the recording was done in a proper room, compression can "blow up" the acoustics in a beautiful way.

Only **slightly** different settings of attack and release result in a **completely** different sound.

As you can see, two compressors with almost identical settings lead to opposite results. Either goal can be achieved by only slightly different settings of attack and release.

For aggressive compression on drums you'll need a fast compressor. In **Logic** you can choose "Vintage FET" in the "Compressor" plugin (see Figure 15.5), in **Pro Tools** there's "Purple Audio MC77" and "BF76."

▶ HOW MUCH?

The harder you drive a compressor, that is, with higher ratios and lower thresholds, artifacts like pumping and distortion will increase. Although this can add great character when executed on the right instrument, it's all too easy to go overboard. This will result in sound that's either artificial, lifeless, spikey or undersized—or all at the same time. Apart from the occasional compressor as a gimmick, professionals usually aim for inaudible compression. This requires choosing the right compressor type and dialing in the right settings. Distributing the load over two different compressors could also help for the process to take place relatively inaudible.

▶ PROS AND CONS OF COMPRESSING INDIVIDUAL INSTRUMENTS

There are great advantages to using compression on instruments:

- If the listener can hear every note of a recording, it will be easier to stay in contact with the musician and his performance.
- A dull and lifeless sound can be given punch and aggression.
- With reduced dynamics, the instrument can sit more stable in the mix.
- Overloads are prevented.

At the same time, when softer soft signals have become louder due to compression, the following side effects can occur:

- Bad acoustics become audible (e.g., with drums or vocals)
- Louder crosstalk, louder ghost notes and louder cymbals (e.g., with drums)
- Louder finger squeaking when sliding over guitar strings
- Louder hum and noise (for example with electric guitar)
- Louder breathing, *s*'s and *t*'s of a vocal (in Figure 15.2 "Mild compression," breathing is clearly visible)

Compressing the Mix

Compression on the mix can provide "glue" for individual instruments, creating a compact and tight sound with more energy and urgency. However, a compressor will react aggressively to the full frequency spectrum of a mix (see Figure 15.6). As this can easily result in distortion and/or pumping artifacts,

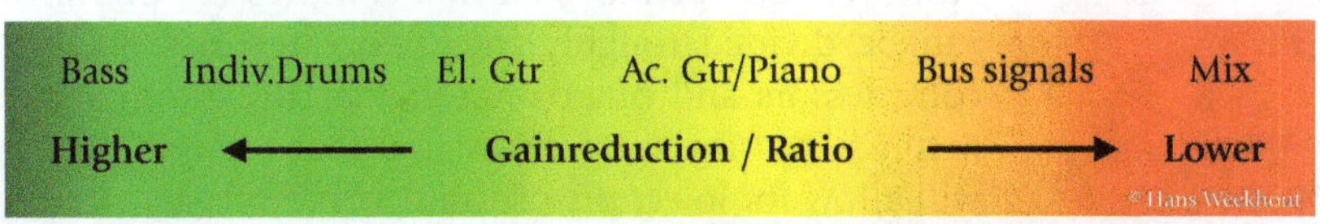

FIGURE 15.6
The fuller the spectrum, the more aggressively a compressor responds. To prevent negative side effects, sources that appear right are generally compressed with lower ratios and less gain reduction.

low ratios must be used, like 3:1, 2:1 or even 1.5:1. Generally, 2 to 3 dB of gain reduction is the maximum before artifacts become apparent.

▶ ATTACK AND RELEASE TIME

Similar to compression on an instrument, an attack time too short will cause valuable transients in the mix to be lost. With attack settings of 25 to 100 ms, punch is preserved as transients can pass. When the release time is too long, the compressor will also reduce the volume of the following note. When release is too short, distortion enters the signal and the compressor will start pumping. Common release times for a full mix are 200 to 800 ms. Here are some additional tips:

- Before finding the right attack time, it may be necessary to first find a proper release time.
- When attack and release time are too short, distortion increases and the compressor starts pumping.
- Keep an eye on the GR meter: in case it doesn't return to zero regularly, this could indicate either too much compression or release time being too long.

▶ PROS AND CONS OF MIX COMPRESSION

Compression on the mix has many advantages:

- It provides mix glue for the individual instruments.
- It will make the mix sound tight and compact.
- It provides excitement, urgency and brutality.
- A higher volume yields more attention.

Although these are no small feats, there are strong **disadvantages** to compression on the mix too:

- Listening fatigue: overcompressed music is tiring to listen to.
- Without any soft notes, loud notes will fail to impress.
- With less and less dynamics, the mix may appear undersized, or even flat like a pancake.

Overcompressed music is **tiring** to listen to.

Not every compressor is a mix compressor. Compressors can be divided roughly into two categories: instrument compressors and bus compressors. The latter category is better suited for the dense signal that the mix bus is and will allow for lower distortion figures.

In **Logic**, "Studio VCA" and "Vintage VCA" are suitable for use on the mix bus. **Pro Tools** has "Impact."

WHAT CAN YOU DO TO MAKE YOUR MUSIC SOUND GOOD ON THE RADIO?

Before being played on the radio, the mix passes many stages. First stop is the mastering studio. Here, the mastering engineer is likely to apply compression (and limiting). At the radio station, aggressive compression is used to impose the station's sound and identity on the mix. Finally, the mix is broadcasted through an FM transmitter, which is protected with—you guessed it—another compressor/limiter. It is by no exception that the average snare sound has seen five, six or more different compressors before it is heard on the radio! In case of an overcompressed mix, (heavy) broadcast compression will squeeze out the last signs of life in the mix and make it sound fuzzy. In order to prevent this, it is better to not overcompress the signal during mixing and mastering and preserve a certain amount of punch and dynamics.

▶ COMPRESSION AND EQ

In case both EQ and compression are needed, the question arises in which order you should apply these effects. A good rule of thumb is: "cut before, boost after." Using EQ **before** compression allows removing frequencies that you won't use, which reduces the compressor's load. After compression, a second EQ can be used to **boost** frequencies. By doing so, the amount of compression will stay the same, regardless of the EQ settings.

Note

Compressor presets often boost volume. As a result, the effected version may seem to sound better than the bypassed version. Only when working **unity gain**, a fair comparison can be made.

Advanced Compression Techniques

▶ PARALLEL COMPRESSION

Parallel compression works by first duplicating the original track and then compressing it. Then, the compression effect can be dialed in to taste, conveniently with a fader. For this technique to work on multiple signals, such as the different mics of a drum kit, the individual channels must be routed to an **aux** (Chapter 16, "Organizing a Project"). After creating this regular aux, the parallel compressor can be inserted in a second aux, that has the same input as the original aux. In case the compressor has a **dry/wet** knob, parallel compression can be applied without duplicating a channel or aux.

> Applying compression in parallel lets you add the effect **without any** of its negative side effects.

Parallel compression allows the original signal to flow into the mix unaffected, which is a great advantage (see Figure 15.7).

FIGURE 15.7
Parallel compression: although the peaks of the compressed signal have been squashed, they are still intact in the original signal. Once both signals are mixed, *the soft notes have become louder.* Parallel compression is therefore called "upward compression," while regular compression is called "downward compression."

Transients are intact while the effected signal can be added to taste. Just because it is added, the amount of compression in the parallel aux can be exaggerated. Virtually every instrument can be parallel-compressed and even the mix itself! But the technique is probably most popular on drums.

▶ KNEE

Compressors can be soft knee, hard knee or variable knee. Hard-knee compression causes a peak in the signal to be grabbed right at the moment it crosses the threshold. With soft-knee compression, the compressor starts working subtly before the signal crosses the threshold, and harder after exceeding it (see Figure 15.8). In other words, the louder the signal, the higher the ratio. Modern compressors sometimes offer a variable-knee setting.

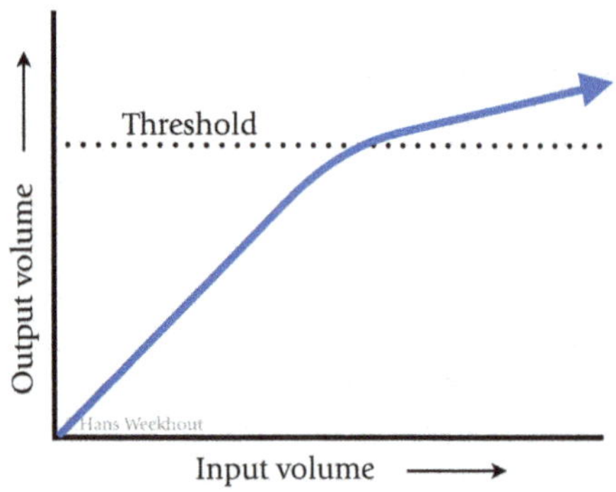

FIGURE 15.8
Soft-knee compression.

▶ SIDECHAIN COMPRESSION

Many compressors offer a sidechain, key or trigger input. This can be used to **trigger** the compression process. When you want to improve the audibility of the kick drum, for example, sidechain compression can be applied to the bass. First, insert a compressor on the bass track. Then, after sending the kick signal to the sidechain input of the bass compressor, the kick drum will trigger the bass compressor, thereby reducing the volume of the bass upon every kick hit.

The well-known rhythmical pumping effect in electronic music is achieved with sidechain compression too. For this to work, all instruments and effect channels must be routed to an aux, except for the kick. After compressing the aux, the kick's signal is sent to the sidechain input of the compressor. This will cause the aux level to be reduced (*ducked*) whenever the kick plays. As a positive side effect, the kick is allowed maximum space.

▶ WHAT IS AN EXPANDER?

While a compressor is used to **reduce** the dynamic range of a signal, an expander (actually "uncompressor") **increases** this. Any signal **below** a certain threshold

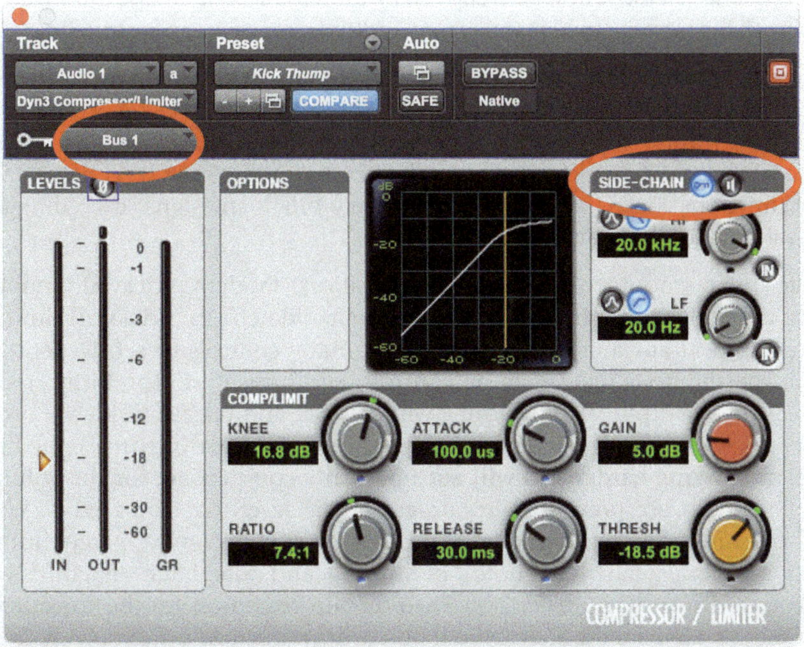

FIGURE 15.9
Sidechain compression in Pro Tools with Dyn3 Compressor/Limiter: the "Sidechain" knob (upper right) is switched on, "Bus 1" is selected as the key-input signal. In Logic, you can do without a bus; just select the desired instrument or audio track from the sidechain's drop-down menu.

is attenuated with a ratio between 0 and 1 (like 0,5:1, or 0,7:1). Closer to zero, expansion increases. What's the use of an expander?

1. **To clean signals.** An expander attenuates low-level signals gradually. Here are a few practical examples:
   - To attenuate unwanted noise between the individual phrases of a vocal
   - To reduce kick and hi-hat bleed on the snare microphone
   - To reduce cymbal bleed on the kick-out microphone
   - To shorten the decay of the bass drum
2. As an "**uncompressor**." An expander can be used to bring back life into signals that have suffered from excessive compression. When soft notes are attenuated, dynamics will increase.

Logic has "Expander," while **Pro Tools** has "Pro Expander." And a noise gate that can work as an expander too: "Dyn3 Expander/Gate."

Limiting

When a compressor is set to a ratio of 20:1 or beyond, the process is called *limiting* instead of compression. Not only does a limiter work with high ratios; it also has fast attack and release times. Limiting can be used to radically shape the envelope of a signal, for instance, when you want to add aggression to an instrument.

Just as often, limiting is used as a technical tool to efficiently reduce the peak level of signals, for instance, with the mix. Software limiters use ratios of up to 1000:1 and work with extremely fast attack and release times. They have a *lookahead* function too, which allows the limiter to "foresee" the level of a signal. These functions cause the limiter to effectively **block** any signal exceeding the threshold, hence the name *brickwall limiter*.

Due to their smooth and advanced operation, modern brickwall limiters reduce dynamics very efficiently with minimal side effects. Brickwall limiters have a practical advantage too: there aren't many controls to set. Depending on the specific unit, there may be controls for "**threshold**" (or "drive") and "**out-ceiling**" (output volume) only, while attack and release times are taken care of automatically. By lowering the threshold (or increasing drive), the limiter starts working harder and **will automatically compensate for the volume loss**. Overloads are prevented, regardless of the level that you feed it. With small amounts of gain reduction, the process may not become apparent immediately. Now exaggerate the amount of limiting, and note that even a brickwall limiter can distort. Generally, 2 to 5 dB of gain reduction can be considered safe. Once you bypass the effect, you'll notice how much volume you've gained. That's why brickwall limiters are commonly used on mixes. In fact, they have been largely responsible for pop music getting louder and louder over the years (the so-called Loudness War).

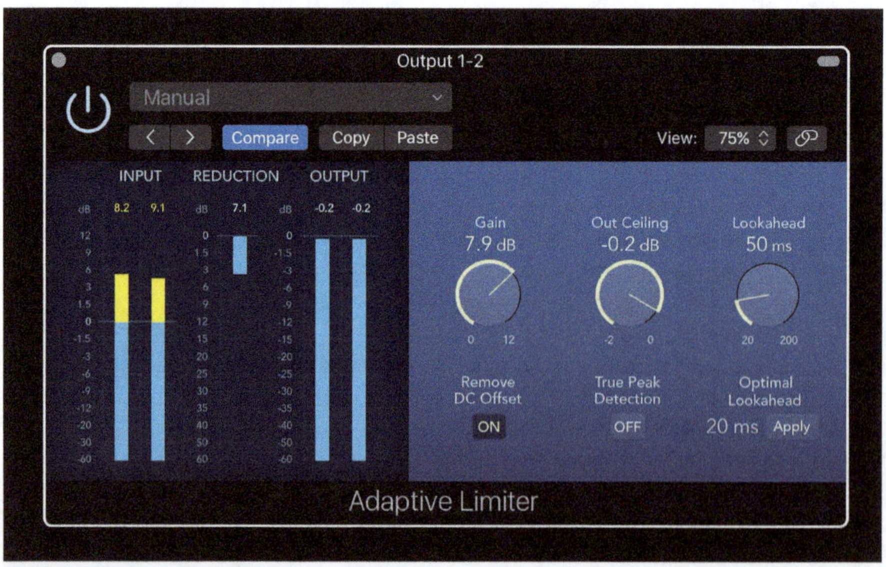

FIGURE 15.10
Logic: Adaptive Limiter.

The advanced and smooth process of brickwall limiting may give you the impression that volume comes for free. But it is important to realize that a brickwall limiter will **always** reduce dynamic range and therefore introduce **distortion**.

Note: no other processors may follow a brickwall limiter; otherwise, digital clipping will occur (see Chapter 33, "Just One Louder"). That's why **a brickwall limiter should always be last in the chain!**

Brickwall limiting in **Logic** can be done with "Adaptive Limiter" (see Figure 15.10); **Pro Tools** has "Maxim."

CHAPTER 16

Organizing a Project

Although The Beatles were quite happy with just four tracks, modern pop projects easily contain a hundred tracks or more. Managing and overviewing such a project requires proper organization. When the project is accessible, better decisions are made. During the mix, you don't want to ask questions like "I'm hearing a twang guitar, but where is it?" or "Are we in the first verse or the second?" or "Did the guitarist play his melody in the first chorus too?" Another aspect is time. If every single technical operation takes time to complete, momentum is lost, and it will be harder to hold on to a vision for the project. A project should almost invite you to tweak it. This chapter contains all the tips and tricks for working happier, faster and more efficiently.

▶ FIND TEMPO

In case the song was recorded to a click track, it's not difficult to determine its tempo and then visually drag all audio on the beat. Once the song is on the **grid**, it's easy to execute operations like copying parts, exchanging song sections or adding quantized MIDI parts. Working on the grid has another advantage: rhythmical effects like delays are automatically synced to the "**host tempo**."

Note: always set tempo **before** cutting and moving audio regions. Otherwise, audio will be out of sync.

After tempo has been established, here are the steps for organizing the project:

Sort tracks. Vertically sort all tracks that belong to an instrument group. Kick, snare, hi-hat and similar tracks belong in the drum department, vocals should be in the vocal department et cetera. Try to stick to the same ordering for every project.

Short, descriptive track names. Although a track name like "23_ElecGtr03_SM57_44,1kHz" is very informative about the recording, it's of no use in the mix. Besides, long labels clutter up the screen or get abbreviated automatically by the program. Short track names that describe the character of the sound or its **function in the arrangement** allow for quick reading. The shorter, the better, like BD, SD, Fuz1, SoloGt or SynthBass.

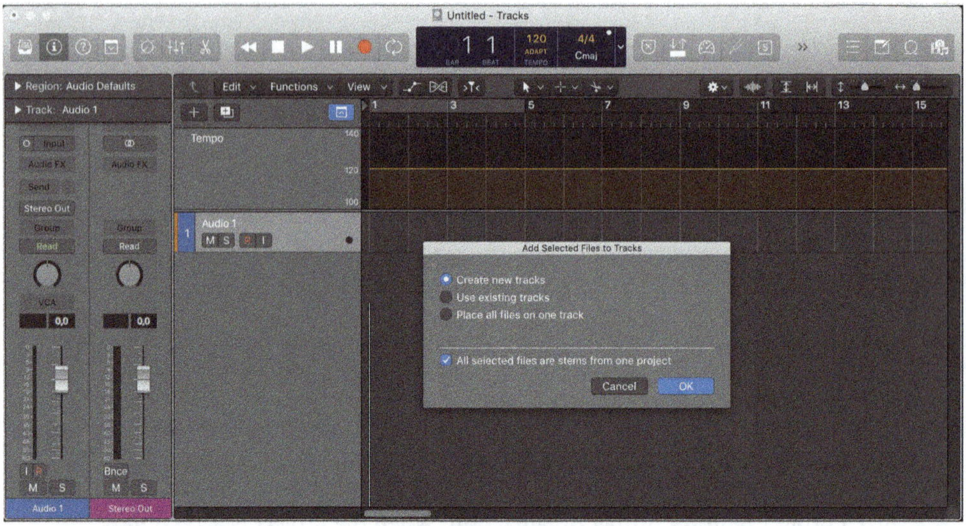

FIGURE 16.1
Importing audio files in Logic. Even if the song wasn't recorded to a click track, Logic allows finding tempo with "Smart Tempo." In the "Control Bar," change "Keep" into "Adapt" and drag all the audio files into the project. After selecting "All selected files are stems from one project" in the "New Tracks" dialogue, Logic will generate a tempo track automatically.

A proper project organization allows **fast** and **intuitive** decisions.

Color tracks and regions. Van Gogh was successful with colors; you can be too! My drums are always yellow, bass is brown, guitars are green, keyboards are blue and vocals are red. Once channel strips and regions/clips have equal colors, instruments can be found instinctively.

Add markers. In order to navigate quickly through the project, it's helpful to add colored markers for song sections such as verse, chorus and bridge. Again, short names like Vrs-1, Cho-1 and M8 (middle eight) allow for easy reading, even when the project is zoomed out. In **Logic**, markers can be colored by using the color panel. In **Pro Tools**, markers are colored by ticking "Always display marker colors" in Pro Tools->Preferences->Display.

Strip Silence. Often, a mix project consists of many audio files with the same length. After dragging them into the project, the long rectangles lack any visual information. Some tracks might contain only short fragments, like the fuzz guitar in the chorus or the organ in the bridge. For a proper overview of the arrangement, it is helpful to remove silence from the audio files. This can be done by hand, but it may take some time. Fortunately, "Strip Silence" (see Figure 16.2) can do this automatically.

Note: the emptier arrangement, the less appropriate Strip Silence is. It could result in unnatural and sudden cuts in a vocalist's breathing, the ending of a note or the room tone of a recording. Strip Silence is nondestructive, so it's always possible to fine-tune results or revert to the original audio file.

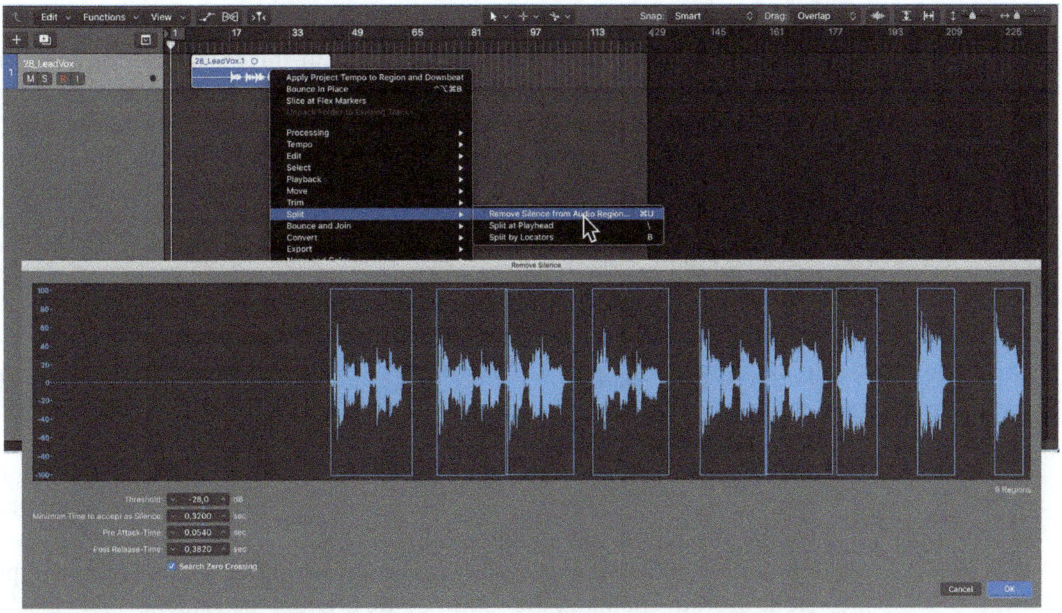

FIGURE 16.2
In Logic, "Remove Silence from Audio Region. . . " can be found in the "Split" department, after Ctrl+clicking a region. In Pro Tools, Strip Silence (Command + U) can be found in the Edit menu.

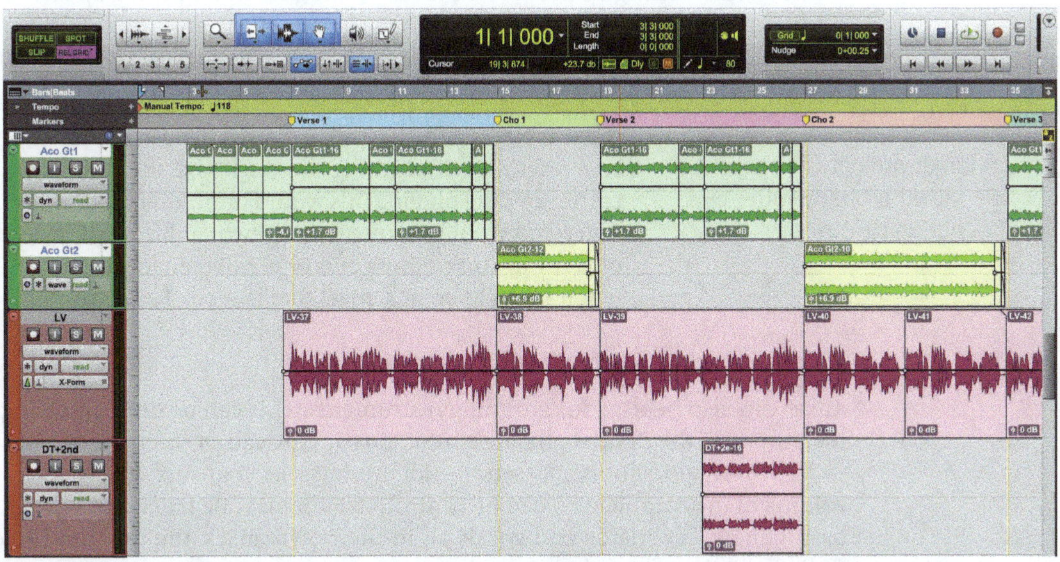

FIGURE 16.3
Project organization: the project is locked to tempo; it has markers, colors and descriptive track names. Empty spaces in the audio files are erased.

▶ AUXES FOR EFFECTS

Let's say, 16 backing vocal tracks need reverb in the mix. After finding a suitable plugin setting for one track, you copy the reverb plugin to the other fifteen tracks. Undeniably, this will result in reverb on the backing vocals, but it comes at the "cost" of 16 plugins. The exact same result can be had by using just one reverb plugin in an *aux* channel. An aux channel is a channel that can be used to feed the mix with additional signals, like effects or signals from external hardware. In this application, aux channels receive their signals from "**Buses**" (**Pro Tools**; see Figure 16.4) or "**Effect-sends**" (**Logic**; see Figure 16.5).

Auxes for effects offer great advantages:

- **Reduced CPU load.** In the preceding example, the computer must calculate only one reverb instead of 16.
- **Accessibility** and **speed** of operation. In the previous example, what if the reverb time turns out just a little too long later on in the mix? The original setup would require you to alter the reverb time of 16 plugins. With an aux for the effect, only one plugin can be adjusted.
- An aux effect can be processed with **additional effects**, like EQ or compression, without changing the sound of the original instrument.

After inserting an effect in the aux, make sure its Dry/Wet balance is **100% Wet**. Otherwise, dry signal will leak into the mix.

Note 1: DAWs sometimes automatically create *mono* auxes, while a stereo-aux is what's needed. In Pro Tools, this can happen if you overlook the mono/stereo setting in the "New Tracks" dialogue. In Logic, it happens after activating a send on a mono track. When the aux is mono, reverb will sound mono in the mix (!). Therefore, always verify that your auxes have the correct format.

A larger project cannot be mixed properly without using auxes.

Note 2: Never add effects like phaser, flanger, chorus or delay and reverb by duplicating the track and then inserting the corresponding plugin. This would unnecessarily tax the computer and would require editing two audio lanes instead of one. Always insert these effects as either an insert on the original track or as an aux effect that is set to 100% wet.

▶ AUXES FOR INSTRUMENT GROUPS

Auxes can also be used for grouping instruments. This can be done by changing the output of the desired channels into "Bus 1," instead of "Stereo-out" (Logic) or "Out 1–2" (Pro Tools). By sending all drum tracks to an aux, for instance, the volume of the complete kit can be controlled with just one fader. After assigning bass, guitars, keyboards and vocals all to their own auxes, the mix can be controlled with just five faders.

As aux channels have insert slots too, such a group can be processed with effects. You could, for instance, brighten up the complete kit by inserting an EQ in the

FIGURE 16.4
Effect aux in Pro Tools.

- Select: [New Track]->[Stereo] and [Aux Input].
- Insert a reverb in the Aux.
- Click and hold on the [Sends] field in the source channel, choose [Bus 1].
- Select [Bus 1] as an input on the [Aux 1] channel.

The bus serves as the connection between Aux 1 and all channels that contain "Bus 1" as an effect send. The "Send fader" (right) controls the level of the signal that feeds the Aux (reverb). A bus can be renamed by right-click-holding its text label.

FIGURE 16.5
Effect aux in Logic.

- In the [Mixer], choose: Options->Create New Auxiliary Channel Strip.
- Insert a reverb plugin in the Aux.
- Click-hold the [Sends] field in the source channel; choose [Bus 1].
- Select [Bus 1] as an input on the [Aux 1] channel.

The bus serves as the connection between Aux 1 and all channels that contain "Bus 1" as an effect send. The circular send knobs control the level of the signal that feeds the Aux (reverb).

aux. Last, auxes have effect sends too. Now you can first send all backing vocals to the BV aux and then add reverb via the effect send that feeds the reverb aux.

As you can see, working with auxes not only eases the load on your computer; it also allows for an easy and accessible workflow. They also let you process grouped instruments easily. Larger projects couldn't be mixed properly without using auxes.

Since **Logic** has "Track Stacks," auxes for instrument groups can easily be created by selecting the desired tracks in the workspace, and then choose Tracks->Create Track Stack. . . ->Summing Stack.

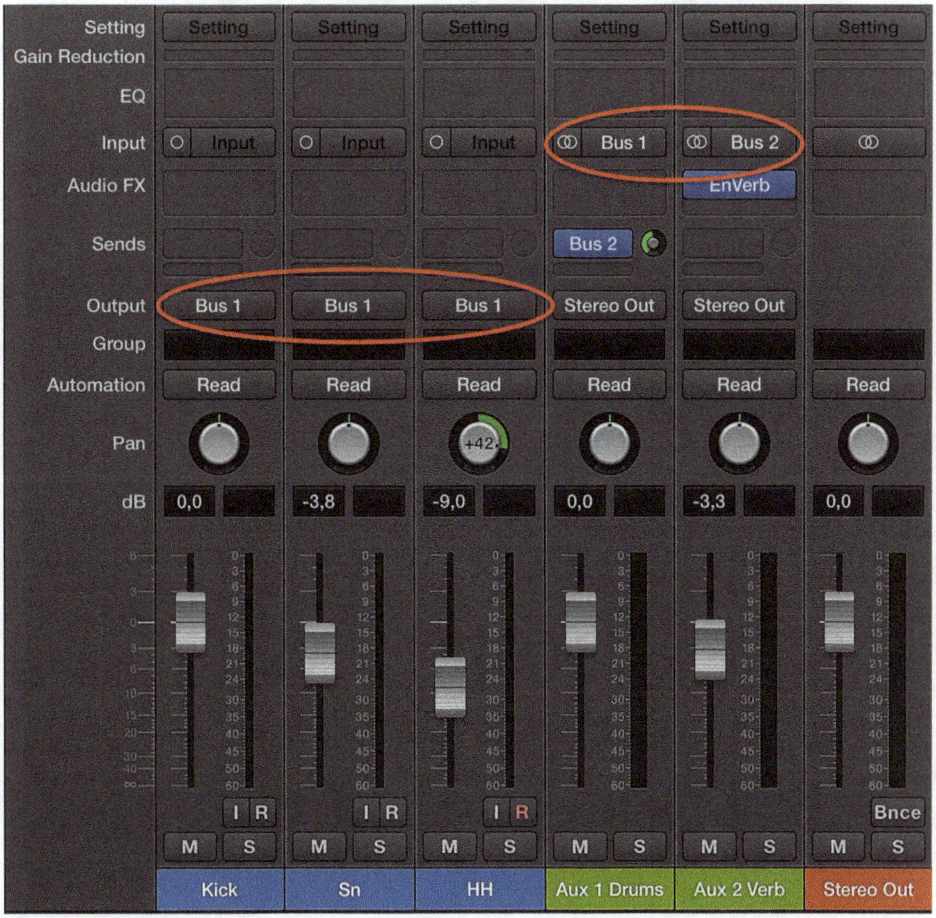

FIGURE 16.6
Mixer with both an effect aux and a group aux. All drum tracks are routed to "Bus 1," which feeds "Aux 1." Now, the aux fader determines the total volume of the drums in the mix. Additionally, reverb is added to the drum signal with "Bus 2," which feeds "Aux 2" (reverb).

Again, closely watch the mono/stereo format of the channels. In case the drum aux were mono, the hi-hat would end up in the center instead of right.

FIGURE 16.7

Auxes for separate headphone mixes in Pro Tools. Two channels at the left are ready to record and not only send their signal to the mix in order to monitor the signal, but also to buses that feed auxes. The auxes do not feed the mix but are sent to the independent outputs of a multi-output interface. This is where you'll hook up headphone amps. Other than sends for effects, headphone sends are usually "pre-fader." This allows changing the mix in the control room freely without altering the mixes in the cans. In order for the click to appear in the cans and not in the control room, for instance, simply pull down the click channel's big fader, and dial in the desired send level on the small fader. "Pre-fader" is enabled by switching on the blue "P" in the send. Effect sends can be changed into small faders in View->"Expanded sends."

> **TIP**
>
> **AUXES FOR VOCAL RECORDING**
>
> Auxes can come in handy for vocal recording sessions too. Applying effects like EQ, compression, reverb and delay to an aux, after assigning all recording tracks to the aux, saves you from copying effects to individual recording tracks. Changes made to an aux effect will be effective for all recorded tracks. You can consider one aux for the verse vocals, one for the chorus vocals, one for the ad-libs and one for the backing vocals.

Hide Tracks

Tracks that won't be used in the mix, can be hidden from view (see Figure 16.8). Old vocal takes, for instance, might be too valuable for a hard delete, but hiding them from view regains screen space. Even active tracks, for example, the backing vocals that are grouped into an aux, can be hidden from view. Hidden tracks will continue playing, while any leveling or processing can be done on the aux channel.

FIGURE 16.8
Hide tracks in Logic: Choosing Tracks->Show hidden tracks causes each track to display an "H" button. Every track that has its "H" activated will be hidden from view once the master "H" button (on top of the track list) is clicked. To hide tracks in Pro Tools, press Control and click the track name; then choose "Hide track."

▶ BUILD A TEMPLATE

Most pop songs contain a kick. And a snare, and a bass and a vocal. The same genre often uses similar instruments. But building a mixer with all tracks, effects and auxes can easily take hours. By instead working from a template, only the audio files need to be dragged on their respective tracks. As a great deal of the laborious mouse clicking has been done, you can start mixing right away. This leaves more time for making creative decisions. Professionals work from templates too, as it is the only way to have a high output.

What could be in the template? Well, except for audio, just about anything:

- Custom **screen sets** or windows positioned conveniently in the screen
- Predefined **colors** and **names** for tracks
- **Markers**. Almost every pop song has an intro, verse, chorus and middle eight. Once the markers are already there, they only need to be dragged to the right position.
- **Panning instruments** in the stereo image
- **Auxes** for instrument groups
- **Auxes** for effects: as a starting point, a long reverb (plate/hall) is handy, as are a short reverb (like a room), a short delay (maybe slapback echo) and a longer delay (a ping-pong delay, for instance).
- Insert your **favorite plugins** on the tracks. Then, assign them **settings** that could work as a starting point.
- A **click** (with an appropriate sound)
- A **spectrum analyzer** on the mix bus

In case you made a good mix already, that project can be used as a starting point for future mixes, after removing audio and automation.

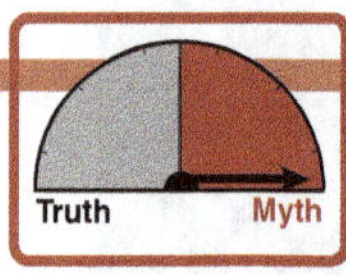

"WITH TEMPLATES, MUSIC ENDS UP SOUNDING THE SAME."

In the context of this book, it is tempting to overrate the art of mixing. But usually, the musician, his instrument, the acoustics and the mic are more important than any effect in the mix. For this reason, working from a template can never lead to mixes sounding the same. Besides, the template serves as a starting point only; you'll adjust settings as you go along. And, you'll update the template regularly with new techniques or plugins.

▶ KEY COMMANDS

Hovering back and forth with the mouse to reach for a menu takes time. Changing the cursor tool takes even more time, as it requires one hoover action to change the cursor (into scissors, for instance) and then performing the cut, while another hoover action is needed to change the scissors back into the cursor. There are only so many clicks you can make in one day. If simple actions take too much time, it will interrupt the flow. Therefore, it is better to learn key commands of often used functions, such as zooming, muting and the transport (see Table 16.1). There's also another advantage: it will reduce the risk for RSI (Repetitive Strain Injury).

Table 16.1 Often-Used Key Commands for Pro Tools and Logic

Often-Used Key Commands			
	Logic Pro	**Pro Tools Mac**	**Pro Tools PC**
Transport			
(Fast) Rewind	(Shift +) comma	1	1
(Fast) Forward	(Shift +) period	2	2
Record	R or *	3	3
Cycle	C	4	4
Metronome	K	7	7
Go to Start	Return	Return	Return
Play from Selection	Shift + Space bar	[	[
Windows			
Mixer	X	Cmd + =	Ctrl + =
Piano Roll	P		
Library	Y		
Zoom			
Zoom horizontally	Cmd + Arrow	R,T	R,T
Zoom vertically	Cmd + Arrow	Ctrl + Opt + Arrow Up/ Down	Ctrl + Opt + Arrow Up/ Down
Zoom selection	Z	E	E
Operation			
Mute Track	M	Shift + M	Shift + M
Solo Track	S	Shift + S	Shift + S
New Track	Alt + Cmd + N	Shift + Cmd + N	Shift + Ctrl + N
Duplicate Track		Shift + Opt + D	Shift + Alt + D
Mute Clip/Region	Shift + M	Cmd + M	Ctrl + M
Loop	L	Opt + Cmd + L	Shift + Ctrl + L
Repeat Clip/Region (s)	Cmd + R	Opt + R	Ctrl + R
Split at Playhead	Cmd + T	B (Separate)	B (Separate)
Join Clips/Regions	Cmd + J	Cmd + H	Ctrl + H
Bounce	Cmd + B	Opt + Cmd + B	Ctrl + Opt + B
Cursor Tools			
Fade/Crossfade	Shift + Ctrl + drag border of a region's title bar	With smart tool enabled (F6 + F7 + F8), click + drag over upper clip area. Or select area and press "f."	With smart tool enabled (F6 + F7 + F8), click + drag over upper clip area. Or select area and press "f."
Tool menu	T	F5–F10	F5–F10
Both Logic and Pro Tools	Alt + Click to return a fader, pan knob or other knob to its "normal" value		

Logic's Marquee Tool

Logic's mouse is not only a regular pointer, but it can also be assigned a "command + click" tool. After click-holding the command key, the cursor temporarily changes to the tool chosen in the command + click menu. Of all tools, the "**marquee**" tool needs some explanation. It can be used for a multitude of operations:

- Double-clicking a region will result in a cut.
- Dragging an area within a region (or multiple regions) and pressing [backspace] will erase the selection, while pressing Ctrl+M will separate and mute the section.
- Dragging an area within a region (or multiple regions) and then clicking results in three independent regions.
- Dragging an area within a region (or multiple regions) and then Option + dragging allows copying the selection, after which the cut is automatically healed.
- When recording starts, the marquee range automatically becomes an auto-punch area.
- Marquee selection can be adjusted by shift + drag.
- Marquee selection can be adjusted to previous/next transient by clicking arrow left/right.
- Marquee selection can be processed with several "Functions" by control + clicking the selection.

As you can see, the marquee tool is very powerful, and assigning it as the Command + click tool enables functions that would otherwise cost precious time.

FIGURE 16.9
Various companies offer custom labeled keyboards for both Pro Tools and Logic. There are also keyboard overlays.

Source: Photo courtesy of logickeyboard.com.

LOGIC TIP

Play/Pause

In Logic, the default behavior for the space bar is play/pause; the playhead will continue where it left off. Clicking and holding the "play" button in the [Control bar] allows selecting "Play from last locate position" instead. Now, Logic will start playing from the same position every time you press the space bar. This will prevent you from frequent relocating when working on a specific section or when rerecording the same section over and over again.

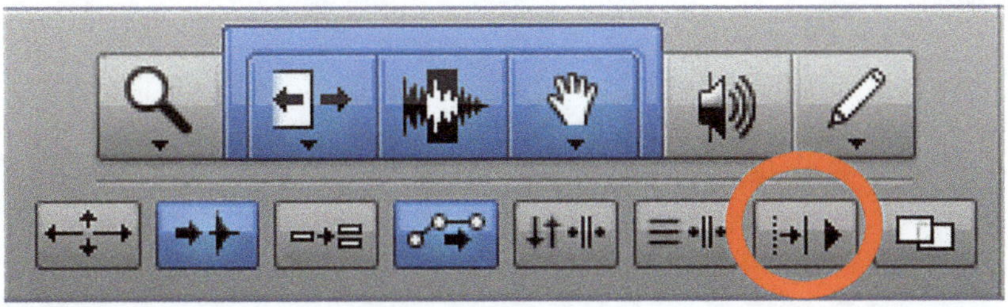

FIGURE 16.10
With [Insertion Follows Playback] off, Pro Tools will start playing from the same position every time you press the space bar.

CHAPTER 17

Setting Goals for the Mix

Some people might say, "The mix is a creative product, so its quality is just a matter of taste." Of course, taste is an important factor; either a record strikes you or it doesn't. But even then, it's possible to come up with objective criteria for a good mix. What are these criteria? What do you aim for and when is a mix a good mix? Read on to find out *why* we do things instead of *how* we do things.

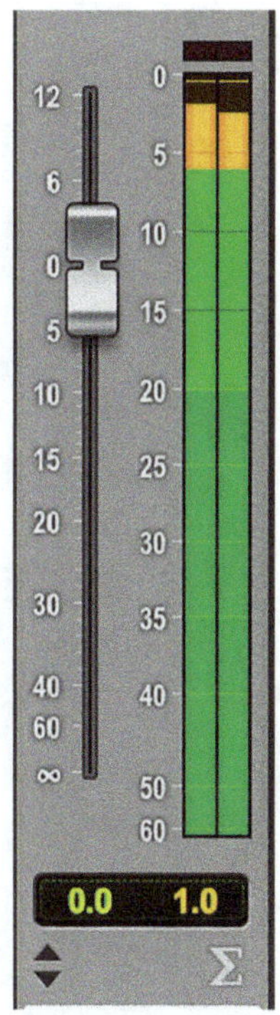

The Big Five

To communicate a song, the listener should be able to stay in contact with each instrument, as if he could touch it. This requires **separation and contrast** (see Figure 17.1). How can you achieve that?

1. THE RIGHT BALANCE

It may sound obvious, but finding the right volume for each instrument can be difficult. Without a proper balance, the arrangement can never flourish. One instrument has the lead and needs to be loud, while others are embellishments that reach the listener only at a subconscious level. The listener may become aware of them only after playing the song several times. Loud instruments are, almost without exception, bass drum, snare, bass and lead vocal. I call them the **"holy quaternity."** When in doubt about the level of one of these instruments in the mix, chances are that the louder option is the best!

2. THE RIGHT PANNING

There are no rules for positioning instruments in the stereo image, it is much about taste. That being said, there are pros and cons to certain choices. With stereo in its infancy, instruments like drums, bass and lead vocal got often panned off-center. Some current producers still favor extreme panning for these important instruments, as it makes the mix more edgy and specific.

In the mix, kick drum, snare, bass and lead vocal make up for the **holy quaternity**.

That being said, extreme panning may cause the song to sound less coherent, or even feel like the mix falls apart. Especially when the listener has positioned his speakers extremely left-right, or when listening on headphones. At a concert or in a club, the audience on one side may not be able to hear instruments coming from the other side. This is the reason why most pop music has the holy quaternity centered.

Every instrument of the holy quaternity has its own reasons to appear in the center:

- The lead vocal, as it communicates the song
- Bass drum and snare, as being the backbone of the rhythm
- The bass, as its low frequencies require a ton of energy from amps and speakers. On one speaker, only half of the power would be available. Apart from this, our ears get hardly any directional information from the low frequencies; this obviates the need for panning.

How can you position instruments like guitars, synthesizers, cymbals, toms, percussion or backing vocals? In most mixes, those instruments are evenly distributed across the stereo image. A hi-hat could be positioned halfway right; the shaker, halfway left. A muted pick guitar at the left can be balanced with a synth at the right. Stereo sources like stereo-miked piano, a Leslie box or a drum

kit's overhead mics may be hard panned, depending on phase coherency. In general, it's a good idea to start with extreme positions and see how it feels in the mix. In case the mix skews, or if the song ceases to sound coherent, you can relocate instruments toward the center. Remember, every panned instrument frees up space for the holy quaternity.

The stereo width of the mix can also help develop a song. You could, for example, use narrow panning for an intimate song section (maybe the verse) and more extreme panning for the most compelling section (maybe the chorus). The increased width will cause the chorus to open up and add to its dramatic impact.

FIGURE 17.1
One of the most important concepts of this chapter is contrast. Similar to a photograph, a mix can never appeal when it lacks contrast.

3. EQ FOR BALANCE AND SEPARATION

EQ is a mighty tool to achieve separation. How can you use it in the mix?

- **Take down frequencies** that aren't necessary, that hinder, or cause nasty resonances.
- Wherever possible, make an instrument's tone flourish by **boosting frequencies**. For well-recorded instruments, only a small boost at the right frequency may be required. In other cases, when looking for sound, you can try extreme boosts. The maximum is reached when the sound becomes either artificial, painful, boxy or boomy or when certain frequencies get in the way of other instruments. It may help to assign priorities to instruments: in case the vocal and the snare are the most important instruments in your mix, you'll fuel them with as much energy as possible. Other instruments should adapt to this and never get in the way.

EQing for sound is just as important as EQing for balance. What do you listen for when EQing for balance?

- In the **bottom end**, our ears are less sensitive, while speakers often have trouble reproducing bass. That's why achieving a solid low end is an art in itself. Fortunately, we're dealing with just two instruments here, namely, kick and bass. In many genres, other instruments are hi-passed in order to prevent any low-end rumble from blurring kick and bass.
- In theory, finding the right balance in the **mid-frequencies** should be easy, as our ears are most sensitive in that area and most speakers reproduce them well. I say "in theory," as most instruments in the pop arsenal derive their specific character from the mids. This means that the mids are crowded. **Adding** mids to one instrument

for extra aggression or to make it stick out will cause other instruments to be masked. Reversely, **cutting** the mids of one instrument in order to make space for others could harm its character.

- In the **highs**, our ears are less sensitive than they are in the mids. This is why it's easy to overlook this frequency band. To achieve balance in the highs, it may help if you ask yourself questions like "How are the highs of the cymbals compared to the band?" "How are the highs of the cymbals in relation to the lead vocal?" or "How are the highs in the lead vocal in relation to the backing vocals?" If you have trouble answering those questions, temporarily add a little treble on the mix with a hi-shelf EQ. Do not overdo (and don't listen too long), as it might cause you to lose your reference point. Now, any instrument that benefits from the correction can be made brighter, while instruments that sound painful, hyped or artificial can be made less brittle.

EQing Dynamic Mics

Dynamic microphones have a tendency to emphasize the mids while downplaying highs and lows. Because they are preferred choice for many pop instruments, the mix can easily turn out middy, thin and dull. If that's not what you want, then your general EQ strategy should head toward cutting mids where necessary and adding highs (and optionally lows) where possible.

4. VARIED EFFECTS

Be wary of excessive **mids**, cherish the **lows** and don't forget the **highs**.

Panning causes the mix to be 2D instead of mono. Reverb and delay can make that same mix a **3D** experience, not unlike depth of field of high-quality cameras. A varied use of effects will help you achieve depth. By adding room reverb to the drums, you'll position the kit **behind** the dry vocal. The vocal will appear even drier now, while contrast increases. One instrument has a slapback delay, the other has a long hall, while a third instrument has a spring reverb. Modulation effects like flanger, chorus and phaser can also be useful for separating instruments. The typical cycle of those effects may cause an instrument to appear in its own secluded space.

In professional mixes, it's often a question of **many effects doing a little** rather than one effect doing a lot.

Apart from the obvious effects, like the big reverb on the piano or the slapback echo on the guitar, the average pop mix has often many effects going on, although their individual impacts are (very) subtle. Only when bypassed, you may hear a difference. All the small improvements add up to a substantially better mix.

5. THE RIGHT DYNAMICS

In case you want a delicate song to sound intimate, it's important to not overdo compression. If you want to prevent distortion and other compression artifacts, you must use gentle compressor settings. That being said, the compact sound that's so characteristic for pop music can only be achieved with sufficient compression. Too little compression can cause a band to sound less coherent. Undercompressing a rock song could result in a lack of urgency and aggression.

Professional mixers achieve large amounts of gain reduction by distributing the load over several compressors. For example, after using two different compressors on an individual snare track, the total drum signal in the aux will be treated with a third compressor. This aux may be parallel-compressed, while the mix bus has both compression and limiting going on. Cascading several gentle working compressors will result in fewer side effects than doing the heavy lifting with just one unit.

Apart from this, the song's dynamic development should be addressed. We call this **macro dynamics.** If you want the chorus to kick in, the verse should not be too loud. Note that this sounds more obvious than it is: after compressing individual instruments, natural dynamics are destroyed. To reinstate dynamics, fader automation can be used.

With **micro dynamics**, you guide the listener's ear to a certain instrument at a specific moment. Again, fader automation can be used to temporarily ride the volume of instrumental licks of guitar, piano or organ, for instance in between

FIGURE 17.2
Neumann cutting lathe for vinyl records.

the vocal lines. Important accents of instruments like crash cymbals or bass drum can be ridden too. Andrew Scheps (Red Hot Chili Peppers, Blood Red Shoes) and Chris Lord Alge (Biffy Clyro, Snow Patrol) are known practitioners of this technique. For instance, by riding the first kick of the chorus. Even with extreme compression on single instruments, riding the faders will prevent the mix from becoming lifeless.

Production Styles

> I don't care much about music. What I like is sounds.
> —Dizzy Gillespie

Every engineer has his own approach to achieve previously mentioned goals. Through the years, roughly two different production styles have developed:

1. **Slicing instruments.** By means of EQ, "unnecessary" energy is removed from an instrument. The resulting space can then be assigned to other instruments. After all instruments are focused at frequencies that represent their character best, the mix can be built by stacking "slices." Hi-pass

filters are essential for this to work. As you know, our ear will automatically reconstruct any missing low frequencies. So there's really no penalty to hi-passing. Even after boosting the lows of the kick and the bass, rumble can be cleaned with hi-pass filters. It will improve the low end's focus and definition.

2. **"More is more."** There are also producers with a more organic approach. They'll only clean up signals or cut frequencies when it's clearly bothering. In case an instrument starts to disappear in the mix, they'll turn up the fader, or boost certain frequencies with EQ. This results in a less polished or even edgy sound. It also allows for a little more mystery in the music. Producers that work according to this style are Daniel Lanois, T-Bone Burnett, Black Keys/Mark Neill, Bon Iver/Justin Vernon, Jack White and Steve Albini. The emptier the arrangement, the easier it is to work along the more-is-more production style. It's by no coincidence that the work of mentioned producers often makes use of open arrangements, with a limited number of instruments and fewer notes.

The **emptier** the arrangement, the **more obvious** it is to work according to the "more is more" production style.

NOTE

For the rest of this book, I'll demonstrate how to achieve contrast and separation in dense arrangements, as this is technically the most challenging. In case you're working on an emptier arrangement, certain techniques can be left out to taste. Thereby, you'll automatically start working toward the more-is-more production style.

CHAPTER 18

Workflow of the Mix

Now that we know the general direction of the mix, it's time to make our hands dirty. Step by step, we'll go through the different stages of the mix. With which instrument or section should you start? How can you support the musical flow of the song? Along the way, we'll find out how overloads can be avoided through proper gain staging. That timing and tuning corrections can be more important than regular sound manipulation tools. When all the work is done, you'll take a rest and check the mix. How can you get the most from that sacred moment of review?

▶ WHERE BRAINS COLLIDE

Music production is about translating emotions into technical actions. But when building an aux for the drums, it is hard to judge the emotional impact of the vocal. In a way, technical actions distract us from the creative and emotional aspects and it appears difficult for our brain to switch. Unfortunately, both jobs are equally important. During the mix, it is important to continuously be aware of this conflict and group tasks as much as possible. After performing the technical operations, you'll try to stay in the groove and oversee the mix from a musical perspective as long as possible.

IS MIXING IN THE STUDIO THE SAME AS MIXING LIVE?

No. Although gear and techniques are similar, mixing is different. Why is that? Let's use a metaphor to find the answer. For a theatrical scene to communicate, sounds and gestures of the actors must be loud and clear. Otherwise, the audience far away in the theater will not be aware of what's going on. With film, on the other hand, a single wink of an eye could reveal the plot. The same is true for music; live and studio compare as rough strokes versus detail. During a concert, the details in a performance may easily get buried, while in a studio production, 1 dB on a fader or EQ could make all the difference.

Overloads in the mix bus should always be **prevented**.

▶ HOW TO FIND THE MINIMUM AND MAXIMUM LEVELS OF AN INSTRUMENT

For instruments that must appear **soft** in the mix, it can be hard to find the minimum level. This may help: pull back the fader completely; then push it up slowly until the instrument is just audible. Verify the instrument's contribution by muting the channel. The **maximum** level of an instrument (like guitar) can be found by pushing up the fader until the instrument is loud—or even beyond loud. Now, by slowly pulling back the level, other instruments will gradually reappear. The right level is found once other instruments have become sufficiently audible. Often, this level is related to the amount of punch you owe the snare (and bass drum).

▶ PREVENTING OVERLOADS IN THE MIX BUS

As we've seen in Chapter 10, overloads during recording must be prevented at all times. Otherwise the signal will be damaged. When mixing, distortion can occur in the **mix bus** (Logic: "Stereo-Out," Pro Tools "Master Fader"). For instance, after pushing the faders too much,. clip indicators will show red (see Figure 18.1) and the transients that are contained within the mix signal will be decapitated (see Figure 18.2). This information is lost forever. To prevent this, best practice is to start the mix at a safe level, while leaving sufficient headroom in the mix bus for pushing faders a few times. "Start at a safe level" can seem easier than it is: you may have noticed that the individual channel levels can easily get out of hand, for instance when using compression, limiting, distortion or extreme EQ-settings. Also, the output level of software instruments can be (extremely) high. That's when you should lower volume at the source and stick to that level throughout the mixer. We call this *gain staging*. With proper gain staging, plugins should neither increase, nor attenuate volume (Unity Gain). In case you do notice overloads, best practice is to grab all channel faders and pull them down sufficiently, while leaving the master fader at 0 dB. Doing so will not alter the mix's balance.

FIGURE 18.1
Overloads in the mix bus of Pro Tools.

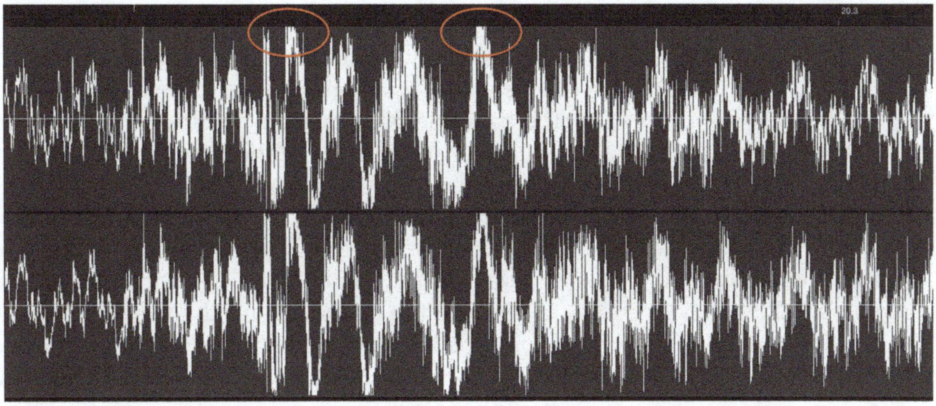

FIGURE 18.2
A Loudness War mix with digital distortion. If you look closely, it's apparent that peaks are not round but, rather, decapitated.

Stage 1: Setting Up the Mix

The first stage of the mix is the most technical part. With a quick balance on the faders you'll get an idea of where the mix might head to. Then you can start organizing tracks, build a mixer and correct timing and tuning if necessary.

▶ TIMING AND TUNING

In chart-based productions, many instruments are quantized and tuned. If you were to do this in other styles, it would suck out the life. On the other hand, if the bass doesn't lock with the drums, the rhythm section will never sound solid. If the trombonist plays out of tune, the horn section can never sound tight. To achieve mix glue, there are limits to the deviations in timing and tuning. Adjustments can improve the sound of the mix more than any effect could. In the production, the challenge is to separate the notes that need adjusting from notes that should be left alone. Just because of the profound impact on the mix in general, corrections should be made **before** starting the mix.

> The challenge is to distinguish the notes that need **adjusting** from notes that should be **left** alone.

▶ WITH WHICH INSTRUMENT SHOULD YOU START?

This is up to personal taste, do whatever feels good. That being said, most mixers start with the **drums**. Of course, the drums are the fundament of the song, and they can be hard to get a good sound for. In extreme cases, drums might account for 30% to 40% of the total mix time. Once you've nailed this instrument, you can rest assured that one of the hardest jobs is done. You could also opt for the **vocal** as a starting point, as it is the most important ingredient to communicate the song. Especially with ballads, the mix can be built around the vocal; other instruments may never get in the way.

BASS LAST

From the moment the bass enters the mix, all people in the room start shaking their heads. Although this is certainly a great effect, it might at the same time divert attention from aspects in the mix that need improving. Postponing the bass may urge you to work harder on other instruments. Then, the mix can only get better after this "feel-good instrument" makes its appearance. In the second half of The Beatles oeuvre, Paul McCartney didn't play bass before the rest of the arrangement was finished. Björk is also known to be a bass-last aficionada.

▶ WITH WHICH SECTION SHOULD YOU START?

There are no rules, but starting with the loudest, most dense section, has some great advantages. In pop music, this is often the last section of the song. At the moment many instruments play at the same time; it is technically challenging to achieve separation and tonal balance. Apart from this, compressors and limiters must be prevented from overcompressing loud signals. Lastly, the total mix level must never exceed 0 dB. Once the big section is done, a difficult job is out of the way.

Another reason for starting with the loudest song section has to do with **macro dynamics**. After establishing the upper limit of the song's loudness,

SOLO

When working in solo on a bass drum, you'll add EQ, compression and other effects in order to make it bigger and fatter than any kick imaginable. But chances are, that after un-soloing, the kick doesn't fit the music at all. Due to **masking**, it might lack contrast with other instruments. That's why professionals try to avoid the solo button; they rather judge an instrument in **context** of the mix. That being said, the solo button is indispensable for pointing your ear toward disturbing frequencies or the exact amount of effects.

the volume of other song sections can be made to follow logically and musically. This requires listening to the song in total, which is time-consuming. Eight plays of a 4-minute song takes more time than eight plays of a 30-second chorus. It also requires a certain empathy and sensitivity. The more contrasting the song sections, the harder it is to make the song feel like a unity.

▶ THE HOLY QUATERNITY AS A STARTING POINT

Every musician wants to hear his instrument loud. The tendency to keep pushing up faders can be endless, which will sooner or later result in overloads in the mix bus. Instead of mindlessly pushing faders, it may be better to establish a proper balance of the mix's main instruments: kick drum, bass, snare and lead vocal. During the mix, you stick to this balance by taking down instruments that tend to get in the way of the holy quaternity.

Stage 2: The Mix Progresses

Typically, the second stage of the mix is about smaller improvements, rather than rough strokes. By listening in context, it becomes apparent that instruments **relate**, rather than the mix being the random sum of multiple instruments. Balance improves. You'll find that changing one aspect on one instrument almost always requires adjustment of another instrument. For instance, by adding aggression to the guitars, you hear that the snare has suffered. To solve this, you add compression to the snare, but now the hi-hat (bleed) sticks out and so on. That's why in

Changing **one** instrument in the mix causes **other** instruments to be affected too.

this stage you'll find yourself going back and forth through the mix, hopping from one channel to the other. Here you see that a proper project organization pays off, as it allows for quick and effective adjustments.

The first stage of the mix has engaged you more and more, which causes an acceleration of inspiration and ideas. Excitement makes you go eagerly through the project doing many adjustments. In this higher state of consciousness, your mouse hand may hardly be able to execute the ideas that come to mind. In the meantime, the world outside has disappeared and you've entered a state of "flow." It is important to hold on to this state as long as possible and to not get distracted by other things that ask for attention, like telephone calls, mail or messages.

FLOW

In his seminal work *Flow: The Psychology of Optimal Experience*, Mihaly Csikszentmihalyi outlines his theory that people are happiest when they are in a state of "flow." This is characterized as a complete concentration and absorption with the activity at hand. In this state, people are so involved that nothing else seems to matter. There's a feeling of great fulfillment, engagement and skill, while temporal concerns (time, food, ego-self, etc.) are ignored. Musicians will certainly recognize the state of flow.

Stage 3: Finishing the Mix

The further the mix progresses, the more time has passed since you solo-checked certain instruments. But in the meantime, the landscape around such an instrument has changed. Chances are that quality can be added by revisiting EQ, compression and effects. Many small improvements will add up to a mix that's substantially better.

In the last stage of the mix, tempo decreases, and it's only in the details that progress is made. Improvements can only be made after a rest and require putting the mix away. With fresh ears, you can do some final adjustments. If the deadline allows, put the mix away for a few days (or even weeks). The longer the cure period, the better your ears will revert to a sound that's "normal" and, not unimportantly, the less you'll want to hold on to used principles and techniques. Kill your darlings.

▶ THAT SACRED MOMENT OF REVIEW

Upon review, make sure the mix has your undivided attention. This moment of consciousness only has a short duration, with the first play being the most important. Keep pencil and paper ready, and write down anything that stands out negatively. After that first play, commit changes **without listening**; try guessing the new fader positions, EQ settings and so on. This is your best chance to preserve objectivity while reviewing those changes in the second play.

Don't play the mix from your DAW, but rather open the **bounce file** in QuickTime, iTunes or other application. Not only will this guarantee uninterrupted playback; it also prevents distractions from moving faders, meters or screen dialogs. Even better, it can make you perceive the music as if it were a **finished product** rather than a **work in progress**. Using a different playback application may help you force into the role of consumer rather than being a professional.

Never underestimate the listener. People may not be able to describe things **technically**; they can certainly **feel** them.

CHAPTER 19

Mixing | Drums

Some people say that when the vocals and drums sound good, it's impossible for the mix to fail. While that may be a little overstated, it's safe to say that drums are vital to the mix. In practice, chances for the drums to sound good are limited by the recording. There are many variables that can make or break the sound: the quality of the kit, tuning of the heads, the choice of microphones, their placement and, last but not least, the acoustics. In case any of these elements falls short, drum sound will suffer. In the meantime, the standard for drum sound has increased considerably over the last years. So even with well-recorded drums, all tricks and techniques are needed for crafting a good drum sound. In this chapter, we go through that process step by step using the stock plugins of your DAW.

STEP 1: CHECK PHASE

The more microphones used, the more phase issues to deal with during the mix. By combining mic signals, certain frequencies die out while others get amplified. The resulting sound will then be **colored**. As EQ can at best *partly* solve phase problems, finding the right phase setting for each mic is our first mission. We'll use the overhead mics as a reference, and try to phase align individual mics accordingly.

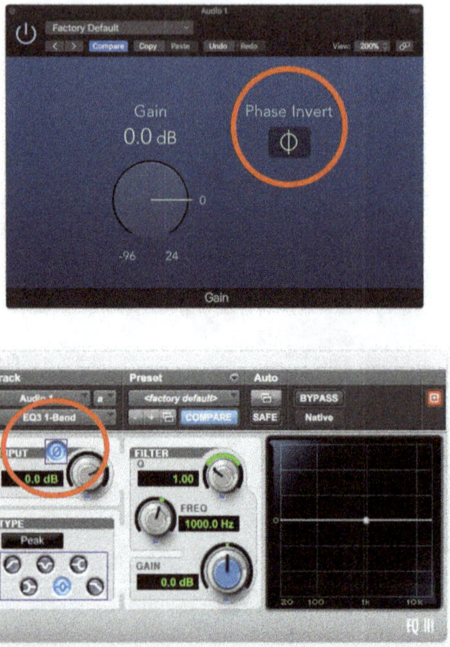

FIGURE 19.1
Switching polarity with (a) Logic's Gain plugin and (b) Pro Tools' EQ3–1 plugin.

Start by bringing up the overheads and **kick** equally loud. Then switch the polarity of the kick signal(s) (see Figure 19.1); *the state that yields the best low end is the right setting*. Now, the phase of **other mics** can be compared to the overheads. A special case is snare (and toms) recorded with both a top mic and a bottom mic. One of those signals will invariably be out of phase and requires reversing polarity. Sometimes when switching, you can hear the sound change, but it is difficult to choose which setting is right. Well, in our search for **sound**, there is no setting that's "right." In some cases, the "wrong" setting could cause a sound to fit the total picture perfectly.

Mixing drums is a **compromise**. After **improving** one instrument, you might **lose** quality on another.

STEP 2: ROUGH MIX

In order to get acquainted with the individual channels of the project, we'll make a proper working balance. Then, changes on an individual instrument can immediately be checked in context. The rough mix helps steer toward a specific target and will lead to better mixing decisions when working on individual channels. It's not a drum solo (or snare solo) you're mixing but, rather, an instrument that should fit a bigger picture.

- Make a proper balance, **pan** instruments and apply temporary **EQ** to correct obvious shortcomings on individual tracks. Don't use (too much) delay and reverb in this stage, as it will blur the image.
- **Route** all the individual channels of the drums into an **aux.** This allows for easier leveling, EQing or compressing the total kit. For instance, if the drums are too dull, a little air can be added with EQ in the aux.
- **Pan the drums.** Some mixers are drummers themselves, and they'll pan the hi-hat, overheads and toms according to the player's position. Others pan drum instruments as if they are part of the audience. There's no wrong or right here; do whatever feels good. As toms produce quite some low-frequency energy, positioning them closer to the center has the advantage of the opposite speaker helping with the bottom end. Kick and snare are part of the holy quaternity; they usually reside in the center. In case the hi-hat attracts too much attention on one side, it can be centered too.
- Insert a **spectrum analyzer** on the mix bus. This allows viewing the spectrum of either the mix in total or

FIGURE 19.2
Drum setup in Pro Tools.

individual tracks in solo. Along the way, it will help you recognize frequencies and spectra.

STEP 3: PROCESS INDIVIDUAL TRACKS

Even the best drum recording can be enhanced by processing individual tracks. With less-than-ideal recordings, it's not uncommon to make drastic changes. Before getting to work with EQ and compression, there are a few things to note:

1. In case serious hi-passing is needed, always be sure to not cut away valuable energy. In case you see activity on a spectrum analyzer at 40 Hz that is useful, the hi-pass filter's cutoff frequency should be set no higher than 20 to 25 Hz (depending on the slope). Otherwise, the low end will suffer.
2. When using compression on drums, "vibey" compressors are most popular. In **Logic**, "Studio Fet" (or "Classic VCA") can be used, in **Pro Tools**, "Purple Audio MC77" or "BF76" is a good choice.

▶ Bass Drum

EQ: the average bass drum has three hot spots: the lows (40–100 Hz), the (boxy) low mids (125–500 Hz) and the click (2 kHz and upward).

The bottom end of the kick must provide for the foundation of the song. This area can be adjusted with either shelf EQ or bell EQ. Although a shelf adds wide-band lows in a musical way, it might also boost (unwanted) sub lows. A bell-EQ on the other hand, adds low frequencies in a focused fashion. So which frequency should you work on? Although the kick's low end in most current styles focuses in the 40- to 80-Hz area, the *recording* may show a 100- to 120-Hz focus. To prevent the kick from sounding "undersized," the 120-Hz area can be attenuated with a bell-EQ, while the 40- to 80-Hz band can be boosted.

After the bottom end has the right **shape**, the fader can be brought up to the right **level**. This "right" level can be found by listening to **reference tracks** and viewing them on the analyzer. As the kick most often accounts for half the amount of lows in the mix (bass being the other half), you'll now have established half of the song's bottom end. Later on, when working on the bass, it might be necessary to review the kick EQ.

Now, we'll determine the amount of **highs,** which allows **hearing** the pulse of the kick in a busy mix. In case the click needs boosting, a bell EQ seems obvious, as it will cause less cymbal bleed than a shelf. Finding the exact frequency should be done in context of the mix. After boosting the mid/highs, sweeping through the spectrum allows finding the desired click frequency. In case the click has trouble showing up in the mids, try boosting the highs (up to 8–12 kHz). This relocates the click from the mids to the highs, thereby elevating the click beyond the spectrum of other instruments. This has become an art in rock and metal.

Always verify changes in **context** of the song. It's not a drum solo you're mixing but, rather, an instrument that should **fit** the bigger picture.

The **lower mids** (125–500 Hz), account for the warmth, or boxiness, of the kick. For a retro/vintage sound, sufficient energy in this band is needed. Vintage kicks provide for good audibility on small speakers, due to sufficient energy in the lower mids. On the other hand, vintage kicks may appear undersized and fail to impress on a bigger system.

That's why the lower mids often are cut, sometimes aggressively. After cutting, the sound will open up. At the moment the kick loses warmth, you've gone too far.

The kick signal can be cleaned by using **hi-pass-** and **lo-pass filters**. A lo-pass filter set at 4–14kHz will remove spill of cymbals, hi-hat- and snare. Once the kick's click is affected, you've gone too far. A hi-pass filter can be used to remove unnecessary low-end rumble. Along the way, it will cause a lo-shelf boost to focus. Be absolutely sure to not cut off essential frequencies. Settings between 10 to 50 Hz are usually safe. Shallow curves (12–24 dB/oct), have fewer side effects due to a better phase response.

All in all, most kicks can be thought of as two separate drawers, one for the lows, one for the highs. For the low end, both the fader and EQ determine how the kick is felt, rather than heard. This is probably a bit more scientific than it is artistic. The exact amount depends on the genre. After settling the lows, the kick's presence can be adjusted by EQing the (higher) mids. How much of the click can be *heard* in the arrangement is more of an artistic choice.

Compression

Sometimes, the drummer's bass drum pedal playing is inconsistent. As the bass drum accounts for about half of the bottom end in the mix, accidental weak bass drum hits may feel like the mix collapses. Compression to the rescue! With mild settings (3:1–6:1), and 3 to 8 dB of gain reduction on

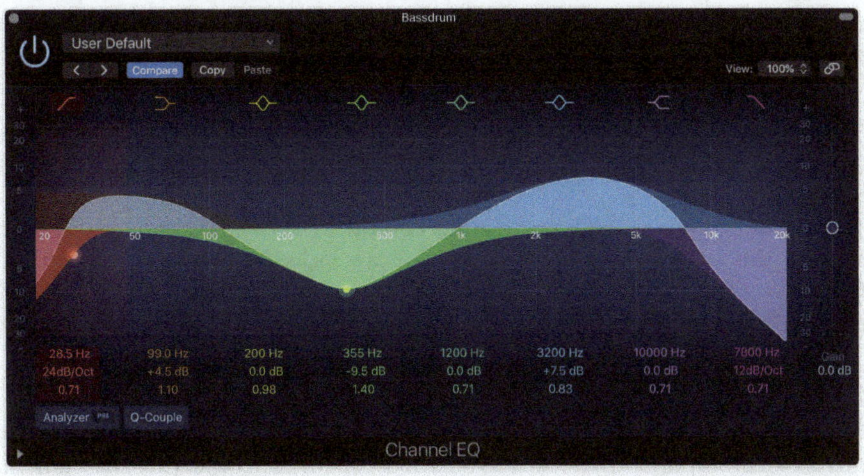

FIGURE 19.3
Sound depends on the instrument and mic used. Even then, most engineers will recognize this EQ curve as being a typical bass drum EQ.

A **weak** bass drum hit may feel like the mix **collapses**.

the meter, this can be solved. Always be careful with fast attack settings on low-frequency instruments. As sound waves are long, the compressor might grab an individual sine wave, resulting in distortion. Attack settings beyond 40 ms are "safe." For release, be sure that the GR meter is back at zero once the next note hits the compressor.

Effect

Long reverb is not common on bass drum, as it will cause mud in the bottom end of the mix. A short room or ambience could work better as it can provide **sustain**, **depth**, **size** and **width**. In case the drums were recorded in a proper room, then the signal of the room mics might provide you with the best reverb.

▶ Snare

EQ: The snare has a few hot spots. The low end (125–250 Hz) provides dimension and warmth, the mids (around 1.5 kHz) cater for aggression and the highs (4 kHz and upward) account for the snare's sizzle. With shelf EQ, the lows and highs can be corrected. A hi-pass filter at 80 to 150 Hz (12–24 dB/oct), can be used to remove rumble and bass drum spill. When the snare is too dull, the bottom mic's volume can be increased. Be careful, though; too much volume from the bottom mic can make the snare sound "military" or undersized. Any boxiness can be suppressed by cutting the lower mids with a narrow-band bell EQ (300–800 Hz).

In most genres, the snare **rules**. Other instruments should **accommodate** the snare instead of the snare accommodating other instruments.

As a snare drum produces energy across a broad spectrum, its frequencies can often be focused at will. By adding mids with a bell EQ and sweeping through the frequencies, it will not be hard to find a frequency that makes the snare stand out of the rest of the mix.

Compression

To even out unwanted dynamics, mild compression can be applied, similar to what we did on the bass drum. In case the snare lacks either punch or sustain, an (additional) aggressive compressor (10:1 and beyond) can be used to shape the instrument's **envelope**. Chapter 15 shows you exactly how to do this.

Effect

Reverb is a common effect for the snare, it will add **size** and **width** to the instrument. Not unimportantly, it will also detach the snare from other (dry) instruments. For slower songs, longer plate and hall reverbs (1.5–6 seconds) are suitable. The reverb time should be short enough to prevent blurring the next note. Short room or chamber reverbs (less than a second), can add size without blurring the mix (see Figure 19.4). Sending the snare signal to the kick's reverb will bring both instruments together in the same space, which might sound more organic. Using two reverbs, like a short reverb for size, and a longer reverb for depth may work well too.

Because a good snare sound can be crucial for the mix, the exact settings of EQ, compression and effect will need regular reviewing as the mix progresses.

TOP 3

DRUM REVERB

1. Room/chamber (short to very short)
2. Gated reverb
3. Plate/hall

▶ Toms

Toms are used to reinforce certain accents in a song or announce a new section. At such a moment, it's not uncommon for the guitars, bass or organ to play busy

FIGURE 19.4
Short room setting in D-Verb (Pro Tools).

slides, slurs or fills. Or maybe the vocal has an upbeat. That's the reason why toms often experience fierce competition. To make matters worse, their boomy character prevents them from cutting through the mix. Although simply pushing the faders on the toms will solve this problem, it might cost you a few dB of your precious mix level. That option doesn't sound attractive for an instrument that's only playing now and then. So the question is, How can we help the toms cut through the mix without raising their level?

EQ: First, verify if the toms contain sufficient low end in order to sound sturdy. Depending on their size, the area between 60 to 200 Hz can be boosted with bell EQ or shelf EQ. Be careful here, as every extra dB will eat up space of other instruments. With hi-pass (30–150 Hz) and lo-pass filters (7–14 kHz), unnecessary low end and spill can be removed. Boosting 2 kHz and upward adds attack and increases the chance for the toms to cut through the mix. By carving out the lower mids (200–600 Hz) with a narrow band, the sound will open up. At the moment the tom loses warmth, you've gone too far.

Compression may help the toms to sound more powerful and pronounced. Settings similar to those of kick or snare can be used.

Reverb/Delay

Sending the toms to the snare reverb will improve the suggestion of one kit in a room. A slapback delay (40–120 ms) could help the toms to become audible. Reverb and delay are very effective on toms: even a touch of the effect may help their audibility and detach them from other dry instruments. For bigger toms (and slower tempos), long reverbs can work well too.

CLEANING TOMS

The shells and heads of toms resonate constantly, even without being hit. Although this complex cluster of tones can be considered a natural byproduct, it could blur the rest of the drums. Verify this by muting the tom tracks in your own project. To get rid of the noise, tracks can be cleaned between the tom breaks (see Figure 19.5). Depending on the number of fills, cleaning might take some time, but it will give you the most controlled drum sound.

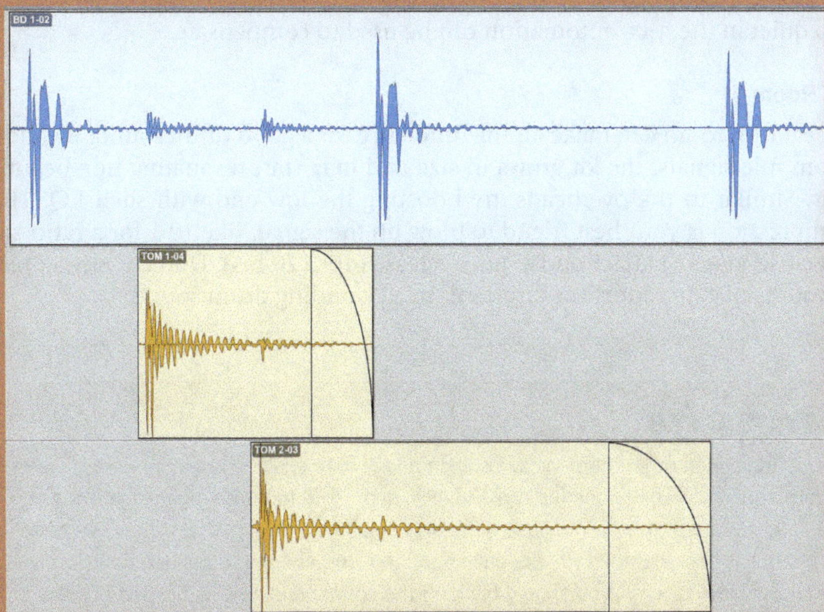

FIGURE 19.5
Pro Tools: tom tracks, cleaned. The upper track is the bass drum, the others are toms. Fade-outs prevent digital clicks. Lining up the ending of the tom with the bass drum will prevent kick bleed to interfere with the direct sound of the kick. The missing tail of the tom is usually compensated for by the tom's leakage in other mics, such as the overheads.

▶ Overheads and Hi-hat

EQ: Deciding on EQ for overheads and hi-hat depends on many factors. With a good recording, only small corrections might be necessary. In case condenser mics were used, a small dip with a bell EQ at 10 kHz might compensate for the nonlinear response of those mics. **Boosting** the lows of well-recorded overheads may make for a nice, wide and sturdy kit sound.

With less-than-excellent recordings, cutting off bleed (of kick, snare and toms) might be your best option. This can be done with a 12 to 24dB/oct hi-pass filter, up to 1 kHz. After hi-passing, pull back the overhead fader, and then slowly bring it back in until the cymbals are sufficiently loud in the mix. Hopefully, the signal is quiet enough now to not further deteriorate the total kit sound.

Compression

Uneven playing of the cymbals can be counteracted with compression, for example by using 2–6dB of gain-reduction at a ratio of 2:1 or 3:1 and medium to slow attack and release times. Along the way, compression will get you some aggression in the signal too. Be careful, the overhead signal is full of energy, causing the compressor to grab immediately. In case the crash cymbals end up too quiet in the mix, automation can be used to compensate.

▶ Room

A beautiful room can make all the difference for a good drum sound. By adding room mic signals, the kit grows in size and may start resonating in a beautiful way. Similar to the overheads, try boosting the low end with shelf EQ. Then, compression is your best friend to blow up the signal, like 10:1 for a ratio, zero (or close-to-zero) attack and a quick release (40–120 ms). Don't be shy; copious amounts of gain reduction can result in a pounding drum sound.

CYMBALFEST

Due to spill on other drum mics, cymbals and hi-hats easily end up too loud in the mix, especially when boosting highs in kick, snare and toms. Compression makes this even worse. In extreme cases, the only solution is muting the original overhead and hi-hat tracks. If the cymbals still sound too loud and unnatural due to EQ on kick, snare and toms, you should back off the corresponding EQs. Too-loud cymbals should at all cost be prevented.

STEP 4: FINAL ADJUSTMENTS

▶ Surgical Corrections

The acoustics of the room or the nonlinear response of mics may cause resonances in the drum mics. **Room** or **overheads** signals are particularly susceptible to this. Some snares produce resonances that are disturbing or clash with the rest of the instruments. In these cases, a (very) narrow-band bell EQ (*notch*) can be applied to cut the offending frequency (see Figure 19.6). Use search and destroy to scan the frequency spectrum; in case one specific frequency reacts more aggressively than others, you might have found a resonance. Such a peak can be recognized by a drone or ringing/whistling quality. Dipping the right resonances results in a signal that's less straining for the ear. As higher Q-settings result in added phase distortion, linear-phase EQ can be considered for this application too.

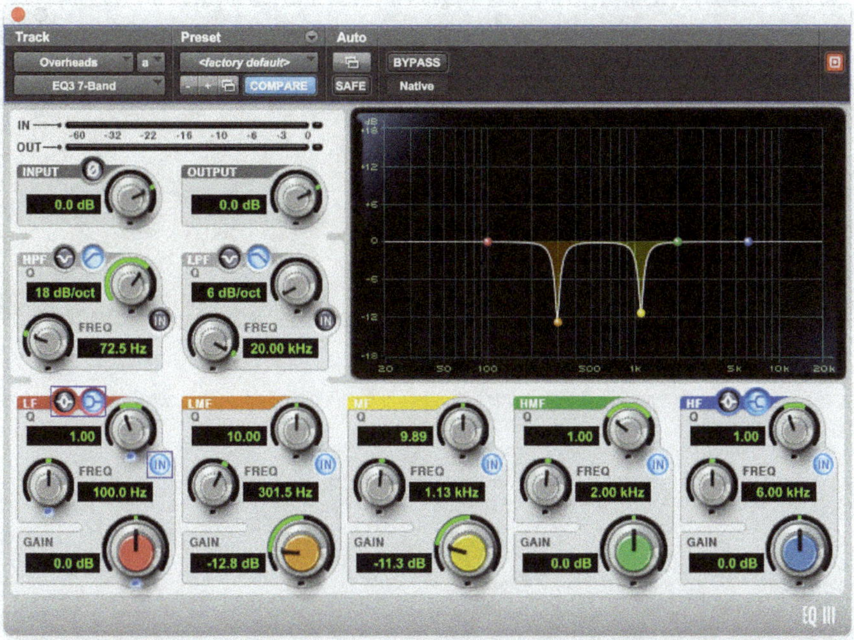

FIGURE 19.6
Ugly resonances or overtones in snare, overheads or room signals can be corrected with surgical EQ.

▶ Dynamics: A Final Check

Listen to **other music** in the same genre at a low listening level. Chances are that the levels of kick and snare are more consistent than you expected. They might even play at exactly the same level, no matter if it's the verse or the chorus. Now, check your own drums in context of the mix. In case the drums are too dynamic, there are a few things you can do:

1. Increase or add compression.
2. Match the levels of individual notes by separating softer notes in the audio waveform and then boosting them with "Region Gain" (**Logic**) or "Clip Gain" (**Pro Tools**) (Figure 19.7). This has the added advantage of individual notes being "prepared" for compression. With all notes at the same level, a compressor can treat them equally.
3. Soft notes that are boosted may still lack punch and impact. Use cut-and-paste editing to replace soft strokes with loud strokes.
4. Certain drum hits may still lack impact on important moments, for example, the first bass drum of the chorus. To solve this, lift the offending note with fader automation.

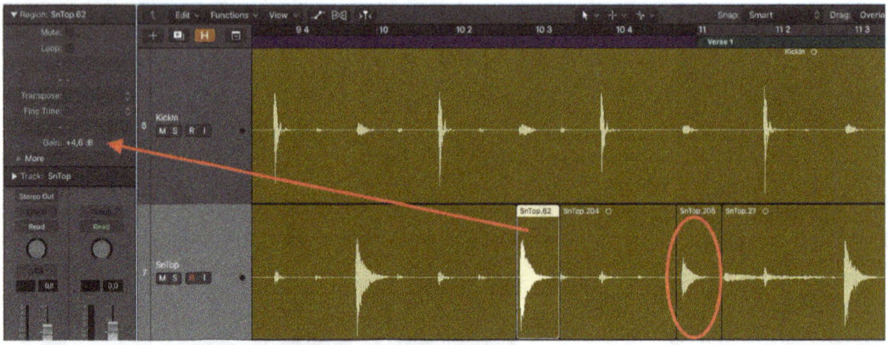

FIGURE 19.7

Snare performance, too dynamic. One soft snare stroke (left) has Region Gain applied, while the stroke at the right awaits gaining.

CHAPTER 20

Mixing | Bass

Working in the low end of the spectrum is often difficult. Not only do speakers have trouble reproducing the bottom end accurately; our ears are less sensitive in the low frequencies too. This is why in demo mixes, bass (and kick) is often "forgotten." But only with sufficient energy in the lows can a mix sound powerful and trigger people to start dancing. Although too much bottom end will "eat up" the mix, sufficient bass is needed for warmth. So the low end must be balanced with the mids and highs. This chapter shows you how to achieve that.

In the 1960s and 1970s, pop music contained less bottom end than current music does. Why is that? Well, the old microphones, EQs, compressors and tape machines cannot be the reason, as these are the devices that are still in use today. One area that has seen a big improvement over the years is speakers. Both studios and consumers own speakers that can reproduce more low end with better detailing. This has enabled producers not only to add more bass energy but also to sculpt the low part of the spectrum very precisely. Especially the area between 40 to 100 Hz has benefit. That may seem like a small band, but we're talking one-and-a-half octaves here!

▶ MIC OR AMP?

Often, bass guitar is recorded on two tracks: one for the amp signal and one for the DI. Since the microphone signal has to travel the distance from speaker to microphone, it will arrive a fraction later than the DI signal. As a result, phase differences may cause certain frequencies to die out when mixing the signals. This problem can be solved by visually aligning the microphone track with the DI track (see Figure 20.1).

In which proportion should you mix DI and amp? Well, the amp produces the most colored and distorted character, while the DI reproduces the bass signal as is. It has a straight frequency response, including sub lows and top highs. Depending on taste and style, a mix of these signals can be made.

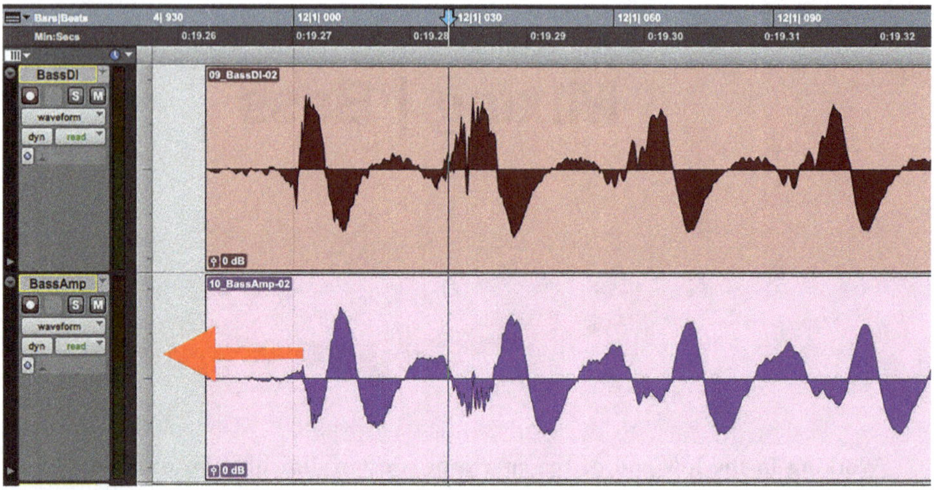

FIGURE 20.1
Visually aligning mic and DI signals can solve phase issues in the bass.

Note: Due to faulty speaker/mic combinations, the amp signal may lack sub low energy (40–80 Hz). In such a case, sufficient level of the DI should be used; otherwise, the mix misses out on those precious sub lows!

In guitar-driven music, the bass often plays in unison with the guitars. Only the root notes (40–120 Hz) are relevant for the bass then. Provided that the bottom end has been recorded properly, this can be a good reason to opt for the amp signal only, as the DI may reveal too much fret noise and rattling.

▶ EQING BASS

A four-string bass has three hot spots in the frequency spectrum. Most important is the **bass area** (40–100 Hz), which can easily be adjusted with shelf-EQ. So how much lows do you owe the bass? Well, speakers in combination with acoustics may have trouble reproducing this area properly, so we need other means for determining the right amount of bass. What can you do?

1. Use proper studio **headphones**.
2. Use **reference tracks** (and compare them at the same volume).
3. Use a **spectrum analyzer** in order to **see** how much bass is needed and at which frequencies.
4. Temporarily engage a **lo-pass** filter over the mix, for example, 24 dB/oct at 100 Hz. As your ears will no longer be distracted by higher frequencies, it will be easier to focus on the low end. This can also help establish a good balance between bass and kick.

TIP

Bass parts played in the high register of the instrument can easily prevent the bass from "carrying" the song. Temporarily adding (large amounts of) bottom end with a lo-shelf EQ might solve this.

Sufficient energy in the **low-mid frequency range** (125–250 Hz) is important for a 1960s' or vintage sound. It helps discerning note pitch and improves audibility of the bass on small speakers. However, with too much energy in the low mids, the bass may appear smaller in size than it really is. For a more modern and open sounding bass guitar, the low mids can be dipped with a bell EQ. Doing so directs our ear's attention to the notes' fundamental frequencies instead of the overtones.

The **higher mid frequencies** (250–1000 Hz) are responsible for definition. Depending on taste, these can be dipped or boosted (in context of the mix).

For the **high frequencies,** every style requires its own approach. The average pop/rock/indie bass stops being effective at 2 to 3 kHz; higher frequencies are usually less relevant. In certain styles, however, the high frequencies are important. If you think of bass players like Fieldy (KoRn), Tony Levin (King Crimson, Peter Gabriel) or Marcus Miller (jazz/funk), their sound is largely dependent on sufficient energy in the top end and an absence of lower mids. A DI signal could help achieve this specific character.

Compression

By nature, a bass guitar is a dynamic instrument. Depending on the skills of the musician, the volume of the notes may vary considerably. But the bass is responsible for at least half of the bass energy (the other half being the kick). That's why soft notes can cause the mix to collapse at certain moments. To prevent this, mild compression can be applied, for example, 3 to 8 dB with a ratio of 3:1 to 6:1 (see Figure 20.2).

FIGURE 20.2
In Pro Tools, "BF76" is a popular choice for bass.

FIGURE 20.3
"Vintage VCA" in Logic is a model of the famous DBX160 compressor that is often used for bass (and kick). Other popular models are "Vintage Fet" and "Vintage Opto."

How about the compressor's attack and release? For a round tone with less attack and more sustain, a short attack and release time can be chosen. Longer attack times allow for a punchier sound, while the rhythm of the notes is a good indicator for setting release time. The GR meter should regularly and musically return to zero.

When looking for a clean bass sound, you generally stay away from fast attack and release settings on the compressor. That's because the compressor could grab individual (long) sound waves, with distortion as a result. In case dirt is what you want, then you should definitely try (a close-to) zero settings for attack and release.

▶ SATURATION/DISTORTION

Even after app.lying large amounts of EQ in the mids, the bass might still refuse to become audible. That's due to the instrument's weak harmonic structure. Other instruments, like distorted guitars, have much more energy in the mids, causing the bass to be **masked**. In case you want the bass to survive the competition, saturation and distortion are the most effective solutions.

In **Logic** you can use "PhatFX," **Pro Tools** has "SansAmp PSA-1," "Air Distortion" or "Reel Tape Saturation."

▶ REVERB/DELAY

Due to the low nature of the instrument, articulation of bass notes is often difficult. Using reverb could make this worse or cause an unclear and muddy low end in the mix. So reverb and delay are not standard effects for bass. That being said, effects like spring reverb or slapback delay could add instant mojo to a bass part, especially when it's articulated or happens to live in an empty arrangement.

▶ BALANCING KICK AND BASS

As far as low end is concerned, kick and bass provide for roughly the same amount of bottom end in most pop music. This requires careful balancing the signals. Often, the kick can be thought of adding attack to the bass notes.

CHAPTER 21

Mixing | Guitar

Finding a good sound for guitar can be hard as it isn't always clear what direction to head for. There are a zillion different guitar sounds that may individually be good, but how they will fit the track at hand is something else. So what's the thought process for building a good sound? How can you make guitars cut through the mix without being harsh or painful?

Electric Guitar

▶ GETTING THE PHASE RIGHT

In case of a multi-mic recording, the first step for getting a good guitar sound is to minimize phase issues. A guitar's timbre often changes dramatically when adjusting the mix of the mics or when flipping the phase of one of the mic signals. Hopefully, flipping the phase brings out the best of two mics. Otherwise, you could try visually aligning the waveforms, similar to what we did with bass. Panning mics is another good solution, although the guitar sound will suffer when the mix is played on a mono system. While finding the right balance for the microphones, it may turn out that every mic combination has its own disadvantage. In extreme cases, this could leave you with no other option than to opt for just one microphone. There's nothing wrong with that; many classic guitar sounds have been recorded with a single microphone too!

▶ EQ

After settling phase, it's time to grab EQ. It's by no exception that extreme settings are required before the tone of the guitar starts to bloom. But even well-recorded guitars can often benefit from a hair of EQ.

As far as **highs and lows** are concerned, shelf EQ can be used for general tone shaping. As always, don't forget to **hi-pass** (6–12 dB/oct at 80–200 Hz), as this will make room for kick, snare and bass.

A guitar may sound **aggressive**, but it should never be **painful**.

Nasty Peaks

Certain amp–mic combinations produce narrow, nasty peaks in the 1- to 5-kHz area. To prevent these peaks from hurting, **search and destroy** can be used to clean the signal. The technique is similar to EQing the overheads (Chapter 19, "Mixing Drums").

Mighty Mids

Guitars are the most mid-heavy instruments in the pop arsenal. Their character is determined by the specific distribution of overtones and energy in the 800- to 4000-Hz area. As excitement, presence and aggression are all attractive qualities for guitar, you'll want the mids to have as much energy as possible. But there are other important instruments that need space in the mids too, like vocals, snare, piano, organ and synths. So the question is, How much energy do you owe the guitar? Let's try to find an answer by opposing two contrasting guitar tones:

1. "**Rootsy**" timbres of artists like Jack White or Keith Richards. The spectrum of this type of sound is limited to roughly 250 Hz to 4 kHz, with energy peaking between 800 and 2000Hz. These frequencies provide for that typical rock 'n roll bite and aggression. Due to the lack of low end, these guitars sound relatively "small."
2. **Fuzz** sounds of bands like Smashing Pumpkins, Deftones, Muse or koRn. Although the total spectrum of that sound could easily span a whopping 80 Hz to 10 kHz, the signal contains relatively **little energy in the mids.** By boosting the 800- to 1500-Hz area with a bell EQ, you may find that the spectrum shifts toward the rootsy tone. By dipping the mids instead, the sound becomes "big," as lows and highs gain importance. We call that **scooping**. Scooped guitars account for less fighting with vocals and snare in the mix. Not unimportantly, they can be played (very) loud without hurting the ear. Unfortunately, scooping cannot be endless, as the guitar will lose **aggression**.

Although we're largely talking **crunchy and distorted** guitars here, it's relatively valid for other guitar timbres too. It's impossible to predict if the mids need boosting or cutting, as it all depends on the recording and the sound wanted. A well-recorded guitar may require taming an ugly resonance only, while other guitars seem to come alive only after adding large amounts of EQ. The trick is to find the right frequency that needs boosting. This can be found by sweeping a bell EQ through the frequency spectrum until you find the spot that works best with the rest of the instruments. Stop boosting once other instruments suffer, or the guitar starts to sound harsh or artificial.

Scooped guitars leave **room** for vocals and snare in the mix.

Remember, our ear is most sensitive in the mids, so small corrections have great effects on sound. Not only does the guitar timbre change, but so does the balance of the mix. Dipping a guitar by 2 dB at 1 kHz could suddenly make space for the vocals.

Lower Mids

A guitar sound can be "cleaned" by cutting the 200- to 600-Hz area. Too much cleaning will make it sound "neat" or uncharacteristic. Boosting the lower mids, on the other hand, adds warmth, though too much of this good can make the guitar sound muddy or boxy.

WAKE-UP CALL

The longer you work on guitars, the easier it is to lose perspective. Resetting your ear by quickly playing a few classic guitar records (or outtakes of good guitar sounds) may work well as an inspiration then.

▶ COMPRESSION

The waveform of **distorted** guitars usually looks like massive blocks. This means that there are hardly any dynamic differences. That's why many mixers choose to not compress distorted guitars. Or maybe just a little, like 2 to 5 dB at a ratio of 2:1 to 4:1.

Clean guitars may be more dynamic, so compression could work well for evening out dynamic differences. With more aggressive settings, compression can be used to either strengthen the attack or attenuate it (as noted in Chapter 15, "Compression").

Which compressors are suitable? **Pro Tools** and **Logic** have clean compressors on board, respectively, "Dyn3 Compressor Limiter" and "Compressor" (preset "Platinum"). Although these will indeed reduce dynamics, they won't add any **color** to the sound. With guitar, the nonlinear behavior of vintage compressors like the Universal Audio 1176 and Teletronix LA2A often works well. **Pro Tools** has "BF2A," "Purple Audio MC77" and "BF76" for that, **Logic** has "Vintage FET" and "Vintage Opto" (see Figure 21.1) in "Compressor."

FIGURE 21.1
Model of the Teletronix LA2A in Logic: "Vintage Opto."

▶ PANNING

With the holy quaternity in the middle, guitars almost "beg" for panning. Moving them to the sides widens the mix and allows for sufficient space for kick, snare, vocal and bass. Note that extreme panning of a single heavy guitar could cause the mix to skew. Generally, another heavyweight on the opposite side is needed, such as piano, organ or other guitar.

▶ EFFECTS

From phasers to ring modulators and from bitcrushers to pitch shifters, many effects could suit guitar. And let's not forget the **Leslie** option: both distortion of the built-in tube amplifier and the rotating speakers cater for a complex and intriguing sound. **Logic** has "Rotor Cabinet" and "Scanner Vibrato," while **Pro Tools** has "Roto Speaker" and "Voce Spin."

But why reinvent the wheel? Usual suspects like **spring reverb** and **slapback echo** invariably work well on guitar. For a more sophisticated, big guitar sound, **ping-pong delay** could work as a sonic sweetener, especially on solos. Just a little bit of medium to long **plate or hall reverb** may be useful on power chords and single-string fuzz. It will widen the stereo image, add sustain, and separate the guitars from their background.

▶ DYNAMIC MIXING

To preserve the attention of the listener, specific guitar licks can be lifted with volume automation. Obvious spots are the pauses between the vocal lines. In the heavier genres, this technique may work well on rhythm guitars too. By riding the faders, you'll bring back life into the mix.

Acoustic Guitar

Most of the preceding (electric) guitar techniques are applicable to acoustic guitar too. However, a few things are specific:

- The hollow body of a nylon or steel string guitar produces large amounts of **bottom end**. To free up space in the mix, the signal can be cleaned by using a 6 to 12 dB/oct hi-pass filter at 80 to 300 Hz. The fuller the arrangement, the more you are likely to cut.
- Steel string ringing is often a useful ingredient in the mix. It will brighten up the **highs**, and add a nice sheen to the mix. Emphasizing the top end with shelf EQ causes the focus to shift toward the high register. Be careful, however; the spectrum of acoustic guitar is full of overtones, containing loads of energy. Especially when using a condenser mic, the 10-kHz area may already be represented sufficiently. Adding (too much) top end could make the guitar sound artificial and cause unbalance with cymbals, vocals or other instruments.

- Frequency-wise, an acoustic guitar can be positioned at will, due to the sheer amount of energy throughout the spectrum. Sweeping through the spectrum with a bell EQ will find you the best frequencies to focus on, again in context of the mix.
- Be careful with **compression** on acoustic guitar. Due to the full spectrum, a compressor will grab instantly. Use a low ratio and a slower attack, and keep a close eye on the GR meter to prevent the compressor from working too hard. "Just kissing the meter" is a good rule here. The previously mentioned compressors probably yield best results.

CHAPTER 22

Mixing | Keyboards

Fitting synths into a band setting can be difficult. Their direct character and artificial spectrum prevent blending with the rest of the (organic) band instruments. In the mix, our mission is to unite these worlds. There's also another issue to be solved: keyboards compete with guitars for a spot in the mix! This chapter looks at solutions, and we'll also go through techniques for mixing electromechanical keyboards and acoustic piano.

▶ ELECTROMECHANICAL INSTRUMENTS

Fender Rhodes, Wurlitzer piano, Mellotron, Hohner Pianet, Hohner Clavinet, Yamaha CP70/CP80 and the Hammond organ all belong to the family of electromechanical instruments. Other than synthesizers, sound is generated organically, which will generally fit the band picture easily. Especially when recorded thru an amp. As far as EQ, compression and effects are concerned, electromechanical keyboards can be treated similar to guitar, as both their spectrum and function in the mix are similar. In case of chords that support the song harmonically, it can be nice to widen a (mono) instrument with effects like phaser, flanger or chorus. In case the cyclic variation of these modulation effects is not what's needed, a harmonizer (see Chapter 30, "Advanced Mixing Techniques | Vocals") may produce a more stable stereo widening. For stereo widening, it's generally not a good idea to duplicate the track and shift it backward by a few milliseconds, as poor mono-compatibility will be the result.

▶ ACOUSTIC PIANO

Similar to acoustic guitar, acoustic piano has a full frequency spectrum and produces powerful transients. This has consequences for the mix:

- The more space in the mix, the bigger and more natural a piano can sound. Otherwise, masking effects of other instruments will cripple the piano and reduce its size.
- Be careful with compression on piano: a compressor might react aggressively, and destroy the sound. With ratios between 2:1 and 4:1, a slower attack and a gain reduction of 2 to 6 dB, you're mostly safe.

EQ-wise, the sweet spots of a piano can be treated to taste (see the figures in Chapter 12, "Effects | Equalizers"). In case you are looking for "sound," then there are many options, from extreme filtering to re-amping. Distortion boxes generally don't work too well with acoustic piano, due to its complex, full and irregular spectrum. A harmonizer or chorus can be used to turn the instrument into a boogie-woogie/honky-tonk piano.

Fitting Synths Into the Band Sound

The character of **synthetic** sounds can be very direct and up front, containing more overtones and frequencies than is the case with organic sounds. In the mix, they will push other instruments back. Synth sounds can make the average metal band appear like a jazz combo in a hotel lobby. What can you do to solve this problem?

1. **Limit the frequency spectrum** with (drastic) hi-pass and lo-pass filtering. The actual frequency range needed from a synth in the mix is often limited. Shrinking the spectrum makes space for other instruments, while the synth's function in the arrangement can be preserved.
2. **Re-amping** forces a synth's overtone structure to become similar to other amped sources. The air between speaker and microphone allows the synth to breathe. The amp and cabinet combination work as hi-pass- and lo-pass filters, thereby focusing the frequency spectrum in the mids. Last, re-amping will account for more personality and edge.

Synth Sounds As A Contrast With the Band Sound

The pure, synthetic and raw character of a synth may also be used as a **contrast** with the band. From this standpoint, there's the risk of the synth being "eaten" by other instruments. Like (distorted) guitars or piano. The powerful mids and highs of these instruments can mask the synth's high-frequency content, causing it to lose both power and character. What can you do? Well, with the maxim "less is more" in mind, leaving out competing instruments will be the most effective solution. In case band members are less enthusiastic about the idea (. . .), brute force is your only option left. Turn up the volume of the synth in the mix, add more top end with EQ or cook the sound with distortion. Panning the synth away from an important organic instrument (like distorted guitar) can prove a good solution too.

▶ DISTORTION

With extra overtones in the signal, it's easier for a synth to compete. There are many options for distortion. In the analogue domain, a mixer or pre-amp could be driven into overload, vintage compressors can be set to their fastest attack and release times, and there's always the option of inserting analog stompboxes in the signal chain. In the digital world, there is a great variety of distortion boxes too: tape emulators, clippers, waveshapers, bitcrushers and so on. Both **Pro Tools** and **Logic** have multiple plugins for distortion, experimentation is key here!

Distortion notes:

- Distortion may **never** hurt the ear. When it does, the aggressive mids (1–2 kHz) can be dipped with a bell EQ.
- Distortion evens out dynamics drastically; it works like compression at 11. This reduces the need for compression.
- As distortion tends to widen the frequency spectrum, it may be necessary to tame a distorted instrument with lo-pass filters. This will cause the spectrum to refocus.

TOP 3

SYNTH DISTORTION

1. Ohm Force Ohmicide
2. Decapitator (Soundtoys), Thermionic Culture Vulture
3. Tritik Krush, Tritik Fuzzplus (free), Audiothing Reels

Distortion works like compression at 11; it will obviate the need for any dynamics processing.

Free: amps, pedals and specialized distortion plugins in Pro Tools and Logic.

FIGURE 22.1
Pro Tools distortion: SansAmp PSA-1.

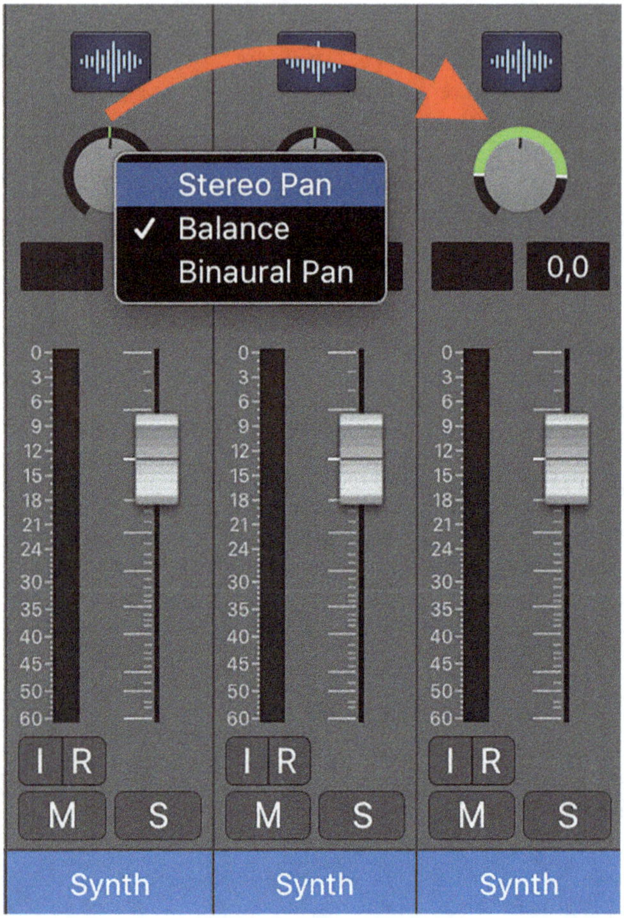

FIGURE 22.2
Panning in Logic. Normally, the pan knob of a stereo channel balances the left and right signal: the more you turn left, the lower the volume of the right channel and vice versa. By Control + clicking the pan knob, it can be changed into a "Stereo Pan" button. Now it's possible to pan the left and right channels independently. This can be useful for narrowing the image of a stereo synth, drum overheads signal, acoustic piano, backing vocal track or a ping-pong delay. In Pro Tools, a stereo track always has two independent pan knobs.

Tips and Tricks

- Before fiddling with EQ, always try to get the most from the synth itself.
- At the same time, synth sounds may seem "finished," but once you start playing with **EQ**, there's often room for improvement. Just a touch of EQ could cause a synth to cut through the mix.
- The preset sounds of **workstation keyboards** often include (large amounts) of reverb, chorus or other effects. Not only can these effects make the mix sound cheap; the synth sound could also take up too much space. Even worse, multiple dubs of such a keyboard could cause a "pseudo-stereo" image (also known as "big mono"). Although the mix sounds wide, it is impossible to pinpoint instruments. To prevent this from happening, bypass the effects and use just one channel (instead of two). This channel can be panned to either side of the stereo image and given its own specific effect. Never down-mix a stereo synth signal to mono, as phase canceling may occur.
- Synthesizers and software instruments can produce large amounts of **sub-low** or **top-high** frequencies, thereby taxing compressors, amplifiers and speakers unnecessarily. Use hi-pass and lo-pass filters whenever possible.

Oberheim

CHAPTER 23

Mixing | Vocals

The vocal is the ultimate instrument for driving the song home to the listener. That's why you'll usually want it loud in the mix. But not too loud; otherwise, the band will cease to make an impact. It's important for the vocal to sit stable in the mix, it must appeal and it should sound forward. The vocal must be part of a picture that the listener can believe in. Those are requirements that can take the most from our technical skills. Finally, we'll also mix doubles, triples, harmonies and choir. Fasten your seatbelts; mixing vocals is working inch by inch!

In this chapter, we'll discuss the following topics:

1. EQ for vocals
2. Dynamics for vocals
3. Punch/aggression for vocals
4. De-essing vocals
5. Delay for vocals
6. Reverb for vocals
7. Volume of the vocal

With EQ, we'll remove ugly frequencies and find the right color. With compression, we'll even out dynamic differences and add punch, if necessary. In case the *s*'s of the vocal have become too loud due to EQ and compression, we'll tame them with a de-esser. Last, the signal can be sent to a delay and/or reverb. This particular order can also serve as a good starting point for the corresponding plugins on the vocal track. That being said, every vocal is different and will ask for a custom approach. So never forget to experiment!

STEP 1: EQ

Hi-passing

The low frequencies of a vocal can be important for warmth and size, for instance, in the verse. But in the chorus, space is needed for drums, guitars and other instruments. Hi-passing the vocal allows for that space. It is relatively painless, as our brain makes up for any missing information.

How much can you remove before the vocal sounds thin? This can only be determined **in context of the mix**. Slowly increase the cutoff frequency of a hi-pass filter (12–24 dB/oct) until the vocal starts to sound unnatural or thin. That's the point to back off a little. Then, allow yourself some time to get used to the hi-passed sound. Depending on vocalist and register, 100 to 400 Hz as a cutoff frequency could work. When in doubt, always choose for a less radical setting, as a slightly boomy vocal will be less harmful than an artificial sounding vocal. Even though the low frequencies in the verse vocal may be just what's needed, hi-passing can still be a good thing, as it cleans any unwanted sub-low energy caused by tapping feet, contact noise or passing trains. Stealing away just a little bit of the audible low frequencies will cause more focus for the bottom end.

EQ

The best EQ curve is, of course, dependent on the vocalist, the way the vocal was recorded and the sound needed in the mix. Let's go through the thought process for making adjustments so that you can come up with a good EQ setting yourself:

- **Top highs.** Adding top end with a shelf or bell EQ (3 kHz and upward) adds breath and "identity" and causes the vocal to come close. Plus, it allows the vocal to escape from the crowded mids. The maximum is reached when the vocal starts to sound artificial, hyped or painful. Low vocals usually allow for more highfrequency boost, while high vocals can easily become harsh sounding. As a rich top end is often associated with a sound that's "chique," sophisticated or hi-fi, boosting highs may be more appropriate in genres like Top 40 or R&B and less appropriate in rock and indie. In the latter styles you might even want to back off the top end with a lo-pass filter (6–24 dB at 4–13 kHz), especially when the vocal is recorded with a condenser mic.
- **Mid frequencies.** In case there are ugly resonances in the recording due to bad acoustics or the microphone, these can be tamed by using search and destroy. By dipping the 800- to 1200-Hz area at a medium bandwidth, the vocal can be made to sound bigger, cleaner, less stressed and less painful. Such a mid-scoop will provide space for other instruments too. Once articulation and/or aggression suffers, cutting is too severe.

TIP

REMOVING PLOSIVES

In case the vocal contains plosives ("pops"), the cutoff frequency of a hi-pass filter (12–24 dB/oct, 100–300 Hz), can be automated to remove the bottom end of just that one *p* or *b*. How successful this is depends on the arrangement; in a dense arrangement, you might get away with it, while in an empty arrangement, even the smallest intervention may become audible. In such a case, your last option is copy/pasting a *p* or *b* from another spot.

- **Lower mids.** The area between 350 and 600 Hz is important for warmth in the vocal but may at the same time sound "boxy." Cutting this area will be beneficial for most other instruments and will open up the vocal sound. The art here is to find the right balance between warmth, on one hand, and a clean, open sound, on the other.
- **Lows.** In order to make the vocal sound bigger, the bottom end can be added using shelf EQ.

Note: in case you have trouble taming the harshness of a loud and high vocal, adding the low end (200–500 Hz) may help to distract the ear from the painful mids.

THE TELEPHONE EFFECT FOR VOCALS

The "telephone effect" is an often used effect for vocals. It adds instant vibe and identity and can be easily constructed with EQ.

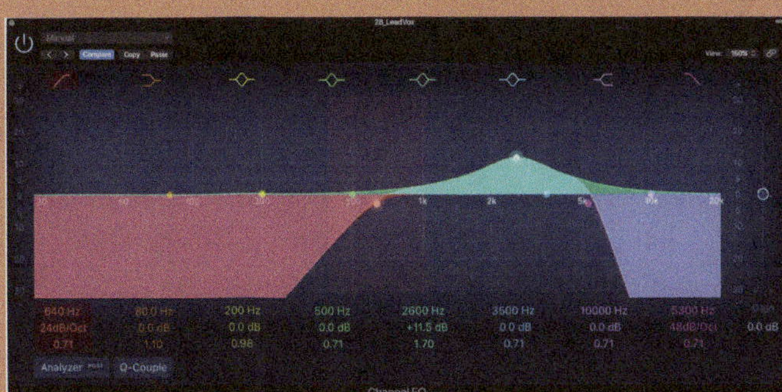

FIGURE 23.1

For a more extreme effect, a guitar amp, pedal or distortion plugin can be added, either pre- or post-EQ. Due to the narrow spectrum and confined dynamics, the signal allows for easy mixing. In case the effect causes the vocal to sound painful, an extra EQ with a mid dip (at approximately 1 kHz) may help.

STEP 2: DYNAMICS

As discussed in Chapter 15, "Effects | Compression and Limiting," we usually want to reduce the dynamics of vocals. With current records, vocal dynamics are often (very) constrained. Let's see how we can achieve this. In case you want to preserve natural dynamics, certain techniques can be left out at will.

a. Region Gain/Clip Gain

In case certain sections, sentences, words or even syllables need leveling; they can be separated and given their own volume with "Clip Gain" in **Pro Tools** or "Region Gain" (see Figure 23.2) in the Inspector of **Logic**. Leveling this way has great advantages:

1. **The vocal is "prepared" for compression**. By boosting soft clips with Clip Gain, the softer signals will trigger the compressor too. This causes more smoothing, as every note can benefit from the compressor's sauce.
2. **Fewer automation curves** need to be drawn, which also makes for a display that's easier to read.
3. **Individual clips** can easily be moved or copied.

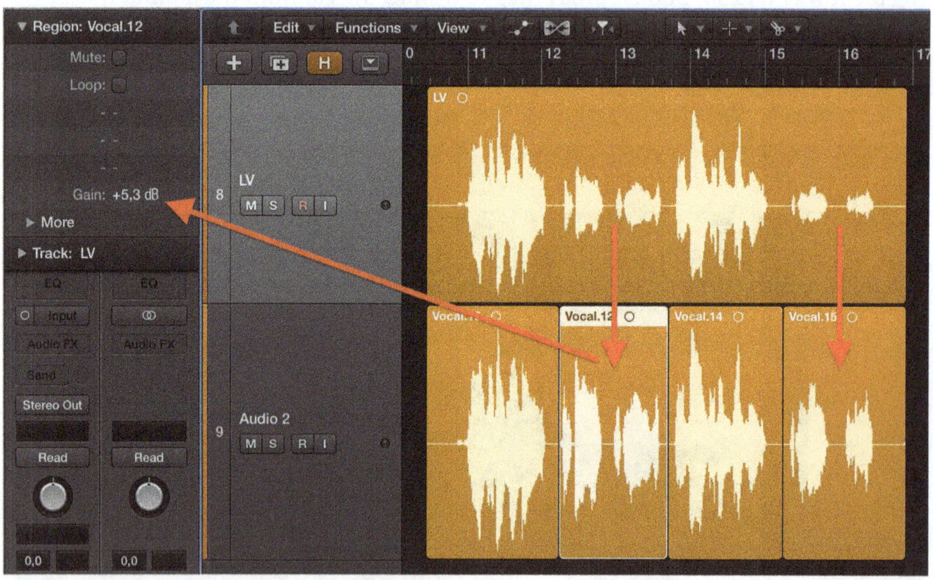

FIGURE 23.2
Logic: vocal leveling by means of "Region Gain."

b. Compression

A compressor can even out dynamic differences more precisely than you could ever achieve with Region Gain. Plus, it can add a rich and sticky quality to the vocal in case you use a vintage compressor. How much compression should you use? Well, the **lower** the vocal, the greater the dynamic differences, so the more compression is needed. Other than high vocals, low vocals usually **benefit** from the compressor's harmonic distortion. So 6 dB (or more) of gain reduction with a ratio of 4:1 to 8:1 could work well. However, **high** vocals are less dynamic, and can easily sound harsh with compression working overtime: a few dB of gain reduction with a ratio of 3:1 to 4:1 is often the maximum.

REDUCING DYNAMICS WITH FADERS

Engineers like Mark Neill (Black Keys), Mike Shipley (Def Leppard, Alison Krauss) and Bruce Swedien (Michael Jackson) are not big fans of compression. For leveling vocals, they'll rather use the fader of an analog mixer. Likewise, when working in the box, fader movements can be recorded by choosing "Touch" or "Latch" as an automation mode (see Figure 23.3). Although riding faders live may feel like an intuitive process, it's not very precise, especially when executed with a mouse. Live curves often need adjustment afterward, which can be cumbersome in case the screen is cluttered with automation points. Therefore, forgetting about the live thing altogether and drawing straight automation lines with the mouse might be the best option, as it will save you time and frustration, though it is, admittedly, less funky.

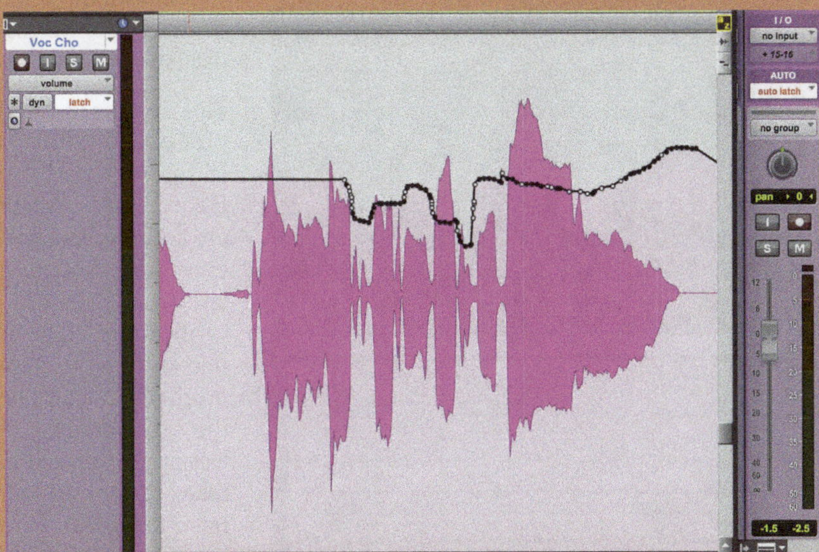

FIGURE 23.3
Fader automation: in "Latch" mode, the fader will remain at the same level after releasing. In "Touch" mode, the fader will jump back to the level where it came from before touching it.

STEP 3: PUNCH AND AGGRESSION

In case the vocal needs extra punch and aggression, a second compressor can be inserted. More aggressive settings are needed, like 8:1 to 10:1 (or higher) for a ratio and fairly fast settings for attack and release. The amount of gain reduction can be adjusted to taste. Always be sure to adjust makeup gain: the perceived volume should remain the same when switching the compressor off and on (**unity gain**).

Logic's "Vintage Fet" and **Pro Tools'** "Purple Audio MC77" and "BF76" are most likely to give good results.

▶ Breathing

Clicks, pops and saliva are common by-products of vocals. Whenever disturbing, these noises should be taken out. Breathing is something else, however. It's not only the natural by-product of human singing; it may also serve a rhythmical purpose. Even soft breaths can (unconsciously) secure the listener's connection with the vocalist. With more compression, however, breathing can become so loud that it sounds unnatural and disturbing. To solve this, loud breathing can be attenuated with volume automation or Region Gain/Clip Gain. The latter option saves you from drawing automation lines. Whenever necessary, use (short) fades.

Tip: to speed up the process of cutting, use "Strip Silence" (see Figure 23.4). As this is an automated process, every clip's start- and end point should be verified and adjusted if necessary. Note that the more organic and open the arrangement, the less appropriate it is to take out "empty" audio from the vocal. For the listener, the sudden disappearance of noise or room tone can be disturbing.

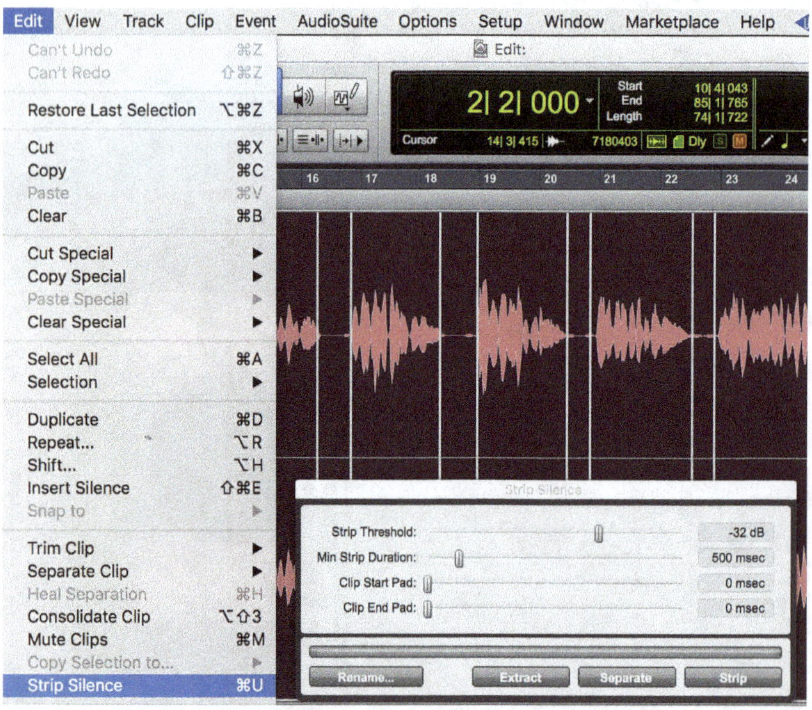

FIGURE 23.4
"Strip Silence" in Pro Tools will automatically cut one clip into multiple clips, based on the "Threshold." "Pad" offsets the start and end of the clips so that attack, breathing or a note's tail are included.

STEP 4: DE-ESSING

After using compression, it's inevitable that *s*'s and *t*'s ("*sibilance*") have become louder. Although these sounds are necessary for the vocal's articulation and aggression, they can be disturbing when too loud. Moreover, a mix with too much sibilance will distort on FM radio and vinyl. How can you tame sibilance?

- Use volume automation. Zoom in sufficiently and draw volume dips.
- Separate the *s*'s and *t*'s into individual clips and attenuate them using **Region Gain** or **Clip Gain**. Values of 6 to 15 dB can be expected here, as the compressor will counteract attenuation.
- Use a **de-esser**. A de-esser is a compressor with an EQ in the sidechain. The EQ passes high frequencies to the compressor's detection circuit so that the device will react only on *s*'s and *t*'s. Be careful with a de-esser, though; when set too aggressively, the entire vocal's clarity will be affected. Or worse, every *s* may turn into an *f*! **Logic** has "DeEsser," while **Pro Tools** has "Dyn3 De-Esser."

It's by no exception that a combination of techniques is needed.

STEP 5: DELAY

Despite the huge supply of effects nowadays, it's still good ol' delay and reverb that provide for the majority of vocal effects. But their application is often more refined nowadays. Let's look at some useful delay settings first:

Slapback echo (90–170 ms) is still a very popular vocal effect, although it is usually less explicit than it was in the rockabilly days. Slapback echo will sound more interesting and subtler when it has a lo-fi quality. Modern echo plugins either have a lo-fi character by nature or they offer knobs to "cook" the signal. **Logic** has "Tape Delay," in **Pro Tools** you can choose either "Tel Ray Variable Delay," "Moogerfooger Analog Delay" or "BBD Delay."

Longer delays (300 ms and higher) can be useful for the vocal too. With feedback, the cascading repeats add **depth** to the vocal. Most plugin delays will automatically conform to the project tempo, allowing rhythmic patterns of (punctuated or triplet) quarters, eighths and sixteenths.

Although a mono vocal with mono effects will cater for a strong center appearance, stereo effects can be attractive too. For that purpose, we'll build a **ping-pong delay** (see Figure 23.5). Let's say for a 100 BPM (beats per minute) song, you set the left channel of a stereo delay to 600 ms (one quarter note) and the right channel to 450 ms (punctuated eighth note). Dial in sufficient feedback for the delay to decay within one or two bars. Hi-pass and lo-pass the return signal of the delay so that the echoes take up less space in the mix. In case the delay pattern is too busy, try 300 ms (one eighth note) for the right channel. Is the delay still too busy? Then, use equal delay times for both channels, but subtract/add 10 ms to one side. This will cause stereo widening for every single repeat.

FIGURE 23.5
Pro Tools: MOD DELAY III ping-pong delay.

Although large amounts of ping-pong delay can make the vocal sound "big," it doesn't necessarily have to sound like a singer in Madison Square Garden. With judicious amounts, a ping-pong delay can subtly separate the lead from the rest of the instruments. As a last refinement, try riding the delay fader on certain phrases ("delay-throws"), and record the movements with automation. A ping-pong delay can also be useful on guitar solos or synth leads.

Logic offers "Stereo Delay"; **Pro Tools** has "Air Dynamic Delay" and "MOD Delay III."

STEP 6: REVERB

Reverb will make the vocal sound bigger, wider and detach it from the background. Similar to delay, "just" using reverb may easily timewarp the production to the 1980s or 1960s or push back the vocal in the mix. This conflicts with the idea that the vocalist should appear up front. Let's look at options for a refined approach:

- **Use less reverb.** Find the minimum level; then switch the reverb off and on. You may find that only little of the sonic sweetener is needed for eliminating the vocal's dry character.
- **Use pre-delay.** Pre-delay (20–150 ms) causes the direct signal to disconnect from the actual reverb. The vocal **appears** dry, while it's not. The reverb floats like a mystic layer in the background of the mix.
- **Use a less bright reverb.** Or EQ a brilliant reverb. Reverb will become less audible, and more intimate. As a positive side effect, it will suppress *s*'s and

t's. Be careful for dull reverb to get in the way of other instruments. Many mixers will cut the reverb's bottom end too. Cutting both lows and highs from reverb is known as the "Abbey Road reverb trick" (see Figure 23.6).

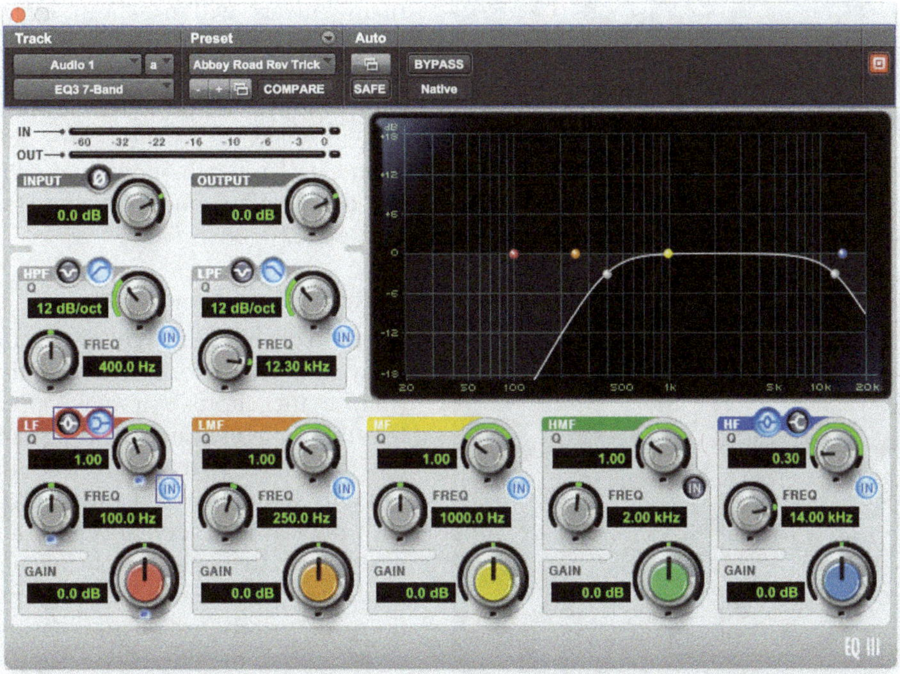

FIGURE 23.6
Abbey Road reverb trick: attenuating lows and highs causes reverb to focus in its most effective frequency range.

▶ Which Reverbs Are Suited?

Plates and halls must be the most popular vocal reverbs in pop music ever. A hall has a warmer, darker and more classic character, with more energy in the bottom end. As plates lack early reflections, they not only sound smoother but also brighter and more artificial. Their metallic sheen may, at first sight, seem unsuitable, but in fact, it often blends well in the mix. In case you're up to making a statement, spring reverb can instantly induce atmosphere and character to the production.

In **Logic**, "Space Designer" and "Chromaverb" are the best sounding reverbs, while in **Pro Tools** that is "Reverb One" (or "Space," see Figure 23.8).

FIGURE 23.7
Sending the delay signal to a reverb can make the delay sound smoother.

▶ Combining Delay and Reverb

Delay only may be inadequate for separating the vocal from other instruments, while **reverb only** could sound too simple or unexciting. That's why many mixers will use both effects. And there's another trick: in case the individual echoes of a delay sound too "choppy," the delay signal can be sent to a reverb. As the reverb adds a little tail to each individual echo, the delay will sound smoother (see Figure 23.7).

▶ Riding Reverb and Delay

After establishing the right amount of delay and reverb for a dense section (maybe the chorus), it may appear that the more delicate sounding verse has too much effect. Such a section can only sound personal and intimate with only limited levels of delay and reverb. To attenuate effects, the new levels can be recorded into automation. After each section has its own custom level, the effects will support (or even intensify) the musical flow of the song. **Riding** levels is probably more important than the exact right color of the effect!

STEP 7: VOLUME

Despite our efforts with Region Gain and compression, the vocal might still refuse to sit stable in the mix. Certain words may be unintelligible due to other instruments hindering. Let's look at some notorious spots:

- The **entry of the vocal** may get buried when the drummer plays a crash on the "one." Or, instruments and effects from the previous section carry over into the new section.

FIGURE 23.8
Pro Tools convolution reverb: "Space."

Especially at those moments, the vocal should be in command. Volume automation can be used to temporarily lift the vocal. Make sure that the level is back to normal directly after the entry: this will leave headroom to further build the song.

- **Low notes**, in general, can be unintelligible and may have trouble cutting through the mix. They often appear as an upbeat or at the end of sentences. Always be suspicious to low notes in the vocal, and lift them with the aid of automation, if necessary.
- For notes that propel **emotion**, maybe the crackle at the start of a word, or the pitch drift at the end of a word or sentence, temporarily adding volume may reinforce the delivery.

Note. Region Gain takes place *before* compression. In case one specific note lacks loudness in the mix but you notice that the compressor reacts normal, then you shouldn't touch Region Gain but use volume automation instead.

How loud should the vocal be? Well, at this point, you've probably heard the song a thousand times (or more). As you know the lyrics by heart, there's the risk of the vocal ending up too soft. But the vocal is a pivotal part of the mix! In many genres, vocals are loud or at least beyond intelligible. Once the band loses power, the vocal is too loud. Checking existing productions in the same style may help determine the right volume. When in doubt, choose for the loudest option; a vocal that's too loud by 2 dB is always better than a vocal that's soft by 1 dB!

▶ Split the Vocal

As you've seen, soft vocals require (totally) different settings for EQ, compression and effects than loud vocals. Of course, this can be catered for with automation. But the more parameters involved, the more curves to draw. As a good alternative, you can duplicate the vocal track and make custom settings for each song section. Now that the channels are liberated from automation, you can freely experiment with dedicated plugin settings and volume per section.

Mixing Doubles and Harmonies

How to treat doubles, triples and harmonies in the mix?

▶ Doubles

If you were to pan two tracks of a doubled lead vocal hard left and hard right, any differences in timing could become apparent and cause an uneasy perception. More common it is to center both tracks, with slightly more volume of the track containing the best performance. This balance can be adjusted to taste: the softer the double, the more personal the lead.

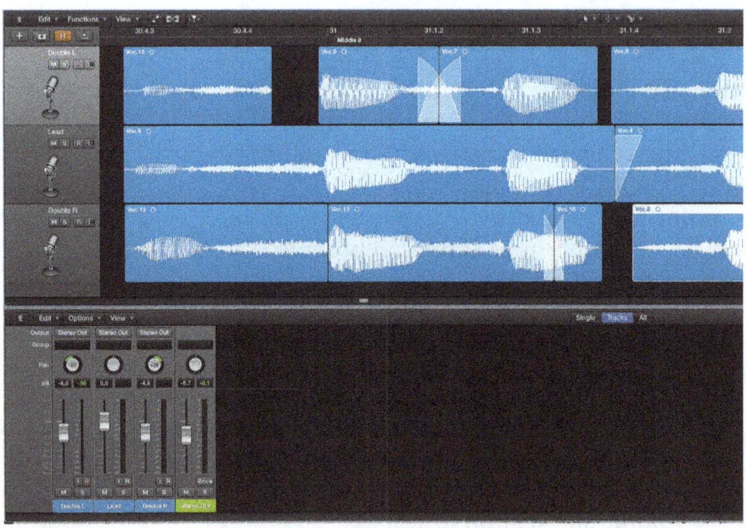

FIGURE 23.9
Lead vocal, tripled. The doubles are panned slightly off-center and lower in volume.

▶ Triples

Tripled lead vocals can either be mixed in mono or stereo. Closer to the middle, the vocal sounds more solid, while separation between the vocals (mono and centered) and other instruments (stereo) is at its best. For a stereo impression, the doubles can be positioned around the lead (see Figure 23.9). For more finesse, the doubles' hi-pass frequency can be set higher than the lead. This will lead to a tighter low end and more refined doubling. Always make sure that the loudest track contains the best performance.

Note! Duplicating the lead vocal track and applying a short delay is not the same as doubling! In fact, after mixing the copy with its original, phase cancelation occurs. This changes the **color** of the vocal. For a mechanical doubling to work well, the second track must deviate sufficiently, for example, by using AutoTune or distortion. Most other effects can be applied to the original vocal track as an insert.

▶ Second Harmony

Volume

Usually, harmony voices are mixed under the lead. This means, if the harmony vocal is too dynamic, certain harmony notes could become inaudible. Always make sure that the harmony's dynamics are sufficiently constrained, or lift any soft words with volume automation.

Panning

Although there are no rules, harmony vocals are commonly positioned at the same position as the lead. That's because a harmony vocal is usually intended for harmonic coloring only. Hearing it in isolation might sound incoherent.

Backing Vocals

Most of the preceding techniques also apply to backing vocals. Here are a few specific tips:

- Almost as an automatism, many people pan the individual tracks of a choir hard left, hard right. Of course, a wide stereo image is a good thing. But what about the "no-man's land" halfway the center and extreme ends? Combining hard-panned choir tracks with tracks positioned at 9 o'clock, 3 o'clock, 11 o'clock and 1 o'clock, might result in better separation. It may be beneficial for the lead vocal too, as it will get support from the choir.
- As our ears are less sensitive for directional information in the bass area, lower vocals allow for more extreme panning than do higher vocals.
- Secretly increasing the level of the higher harmonies in the last chorus could add to the excitement and dynamic development of the song.
- Frequency-wise, backing vocals should never be brighter than the lead vocal. Then, they can live **behind** the lead vocal, hence the name "backing vocals."
- Apart from regular long reverb on the choir, a **room or chamber** effect may work well. Such a short reverb widens the stereo image and makes the choir sound bigger, while the reverb can nicely cover up any irregularities of the individual voices. With large amounts of the effect, the choir gets a gospel-like quality. The difference between the lead's reverb and the choir reverb will cause separation and contrast to increase.

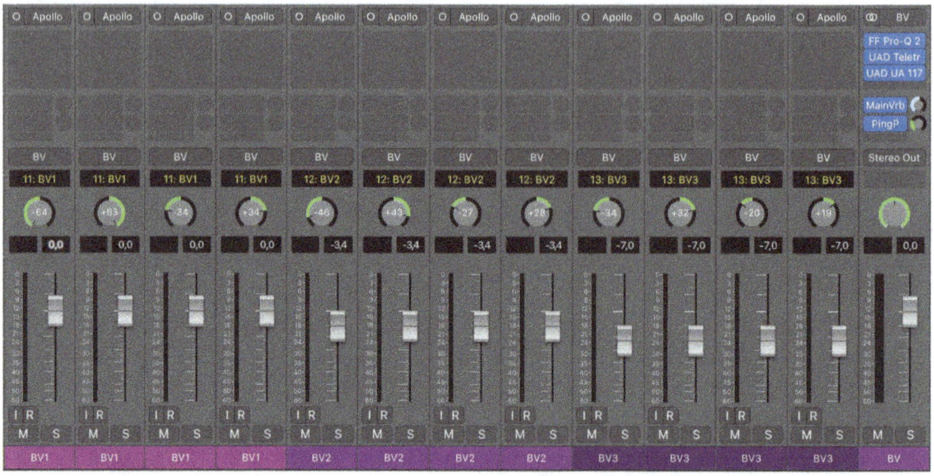

FIGURE 23.10
Typical mixer settings for a choir with multiple harmonies: lower harmonies require more panning and more volume usually.

▶ Multi-Harmony Choir

With a multi-harmony choir, it is important for each and every note of the harmony to be audible so that the chords can sound complete. In case of notes dropping out, these can be corrected with Clip Gain or volume automation.

Balancing multiple harmonies of a choir is done by first setting the lead at its proper level. Then, find the appropriate level of the lowest harmony. Last, fade in the remaining harmonies while ending with the highest. Every note of the previous harmony should remain audible; a new harmony may never obscure other harmonies. Lower harmonies have more trouble cutting through the mix. That's why they need more volume and possibly extra EQ in the mid-highs. To review the final balance, mute individual harmonies one after another, to find out if this causes the expected effect. If not, the balance should be adjusted (Figure 23.10).

CHAPTER 24

Getting More From the Mix, Common Mistakes

Now that the mix is done, it's time to review your work. Are there any objective tools and techniques that can help reveal shortcomings in the mix? How can you get closer to the sound of other music in the same genre? In order to prevent mistakes, this chapter lists 10 notorious mistakes and shows you how to solve them.

Four Techniques for Checking the Mix

1. CHECK WITH REFERENCE TRACKS

Our ears have a peculiar characteristic: they'll quickly get used to a certain spectrum and consider this as "normal." So at the start of the mix, you might push up the first faders and hear the dry sound of a bunch dynamic microphones. Before you know it, you'll use that as a reference. But this "reference" lacks fundamental amounts of bass, treble, reverb or a proper vocal level. If you don't escape from this restricted space quickly, it will prevent you from coming even close to a good sound. Taking breaks will certainly help, but even more important is to play **reference tracks**. They will reset our ears, and provide a clear and specific goal to aim for. Reference tracks can reveal shortcomings in the mix and minimize the chance of comparing unfavorably to other music in the same genre.

Reference tracks **prevent** your mix from comparing **unfavorably** to other music in the same genre.

Beginners are not the only ones who suffer from **tunnel vision**: professionals may have it too, although with a smaller error margin. Even then, most professionals use reference mixes, especially when working in an unfamiliar studio.

There's another reason for using reference tracks: the nonlinear response of a given speaker will become less relevant, as the reference track is filtered exactly the same as your own mix! This means a good mix can be made even on bad speakers. Of course, this cannot be the whole truth, as lower-quality speakers won't reveal details in the mix. But even then, the better you know your speakers from your favorite productions, the better the mix will turn out. Last, reference material will cause you to **learn** from other people's productions. They make you an informed person and shape your vision along the way.

By using reference material, the **nonlinear response** of your speakers is less of a problem, as the reference track gets the same **treatment** as your own mix.

Reference tracks shouldn't be carbon copied, but rather used to get an idea of the playing field. How much bottom end do you hear on average? How wide is the stereo image? How dynamic are mixes in general? How loud are lead vocals? How loud are snares? Of course, the files must be of good audio quality and should be compared **at the same volume** as the current mix.

▶ Out-of-phase Mono Technique

The "out-of-phase mono" technique can be helpful for analyzing reference tracks. It works by switching the polarity of either side of the stereo signal and then summing both channels into mono (see Figures 24.2). This will cause all mono and centered instruments of the original mix to disappear (see Figure 24.3). Most often, this is the holy quaternity (kick, snare, bass and lead vocal). What remains is all signals that are **unique** to one of the sides, such as drum overheads, stereo synthesizer sounds, backing vocals and effects, such as reverb. And, of course, mono sources that were panned off-center, like guitars, percussion and so on.

Now that the loud (!) instruments of the holy quaternity have been taken out, the softer signals in the mix suddenly become audible, like the little embellishments of an arrangement or the reverb on a vocal. This is great, as we often wonder about the little secrets of a good production. It's like a sneak peek in the producer's kitchen.

SAMPLE MAGIC AB

FIGURE 24.1
Sample Magic's "Magic AB" makes comparing with reference tracks a breeze. After inserting it in the mix bus, files can be dragged into multiple slots. Then, the volume can be adjusted and specific sections can be looped. Finally, the "A–B" knobs allow easy switching between the reference track and the current mix. Melda "MCompare," Mastering the Mix "Reference" and Plugin Alliance "ADPTR Metric AB" are comparable plugins.

EFFECTS ON THE MIX BUS

By using EQ, compression and limiting on the mix bus ("Stereo-Out" in **Logic**, "Master Fader" in **Pro Tools**), it's possible to add sound quality to a mix that cannot be achieved when working on individual instruments. This can propel the mix one step closer to professional productions. Of course, the mix signal is precious, so you have to work with subtle settings, like one dB here, and one dB there. Effectively, you're entering the world of mastering, which is normally the domain of a professional. In case you lack the budget for professional mastering, I encourage you to go ahead and try to get the most of the mix yourself.

Compression: with compression on the stereo bus, mix glue improves while the band sounds more compact (as noted in Chapter 15, "Effects | Compression and Limiting"). Because it is likely for soft signals (such as reverb and delay) to become louder, the mix can be adjusted accordingly.

EQ: in case other music in the same genre has more highs, lows or mids, the mix can be adjusted with EQ. Even small changes can dramatically improve appeal for the mix. Note that big changes (3 dB or more) indicate an underlying problem in the mix: always try to find solutions in the individual tracks first.

Limiting: when you want to arrive at the same volume as other music, you can use brickwall limiting (as discussed in Chapter 15). Similar to compression, soft signals in the mix will become louder; adjust their levels accordingly. Be careful not to lose the transients of kick and snare!

FIGURE 24.2
Out-of-phase mono technique in Pro Tools with the "Downmixer" plugin (in "Sound Field"). In Logic, the "Gain" plugin (in "Utilities") can be used. Just phase-invert one channel and push the "Mono" button.

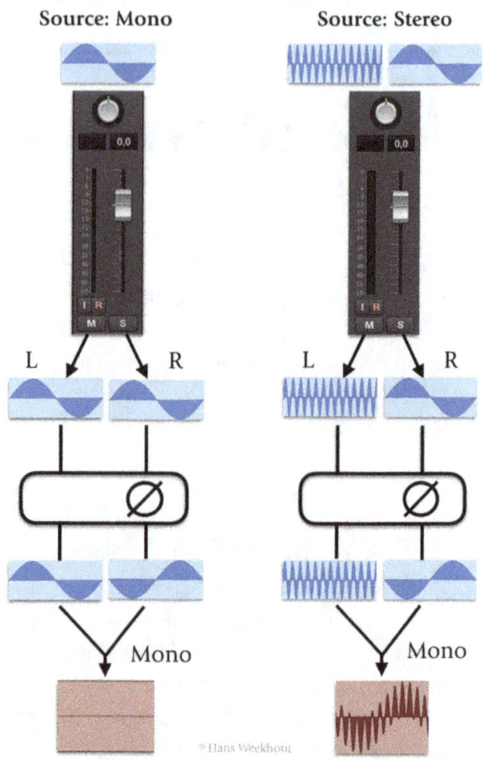

FIGURE 24.3
Out-of-phase mono technique: mono sources that were panned in the middle will cancel out, while any signal with a left–right difference will continue to sound. Note that it doesn't make a difference which channel you switch polarity on.

2. CHECK THE MIX ON VARIOUS SPEAKERS

You'll want your mix to sound good on all speakers. Professionals check their mix on big speakers, small speakers, headphones, laptop speakers, car stereos, sound docks, ghetto blasters or EarPods. As every speaker's particular frequency response emphasizes certain elements, this can point your ear to shortcomings in the mix.

Don't switch speakers when starting the mix though. In this phase, you're still trying to get to grips with the project, and no fixed reference point exists. Switching speakers could set you off then. Only in the later stages of the mix, **compatibility** with other speakers can be checked and the mix can be adjusted accordingly.

LISTENING ON DIFFERENT SPEAKERS IS AN EXPERIENCE

Cheap and small speakers are anything but linear. Their frequency response varies largely, while top highs and sub lows aren't reproduced at all. Fortunately, studio monitors reproduce the "small speaker frequencies" easily and very linearly. In a way, they represent a weighted average of all small speakers. This means that, once the mix is balanced in the mids, it should be compatible with any small speaker. That's the reason why many mastering studios work with just one pair of (big) speakers. That being said, listening to different speakers is an **experience**. They could reveal an underlying problem in the mix. But changing the mix just because you can't hear a certain instrument on one specific laptop speaker might not be a good idea.

Note that it is impossible for a mix to sound good on all speakers. For instance, a sine bass may go by unnoticed on a small speaker, while it could just be the element that makes the mix sound impressive on a big system.

▶ How Come Compressed Music Sounds Good on Small Speakers and Worse on Big Speakers?

When playing **uncompressed** music on a small speaker (e.g., in a living room), soft notes will be (almost) inaudible. The low listening volume (40–70 dB), causes soft signals to disappear in the ever-present background noise. With **compressed** music, however, soft notes become audible, causing the musical picture to be more or less complete. For the listener, this is an advantage. Note that it is impossible for the mix to be painful, as the living room volume is relatively low.

Now let's play **compressed** music on a big system (of a club or festival, for instance). With all frequencies equally loud and no space between the notes, this could easily result in a painful and tiring experience, as we will normally play loud on such a system. On the other hand, with **uncompressed** music on a large system, it's easier to stand the high playback volume due to the space in between frequencies and notes. Not only are our ears allowed to "breathe," but soft notes will be audible too.

Now you see why it's impossible for a mix to sound good on all speakers. Physical differences and corresponding listening levels simply require different characteristics of the mix.

THERE'S A PLUGIN FOR THAT

FIGURE 24.4

Audified MixChecker is a plugin that simulates the character of certain speakers or listening environments, like home speakers, car speakers, earbuds, laptops, tablets and phones. Never forget to bypass the plugin when bouncing the mix though!

3. CHECK ON A SPECTRUM ANALYZER

Although good ears cannot be fooled easily, there are limitations to what we can hear. Besides, speakers and acoustics color our perception, especially in the lows. A visual representation of the spectrum can then be of great help. Similar to recording, an analyzer (see Figure 24.5) will warn you for unusual amounts of sub lows or ultra highs during the mix.

Overcompression causes music to sound **better** on small speakers and **worse** on large systems.

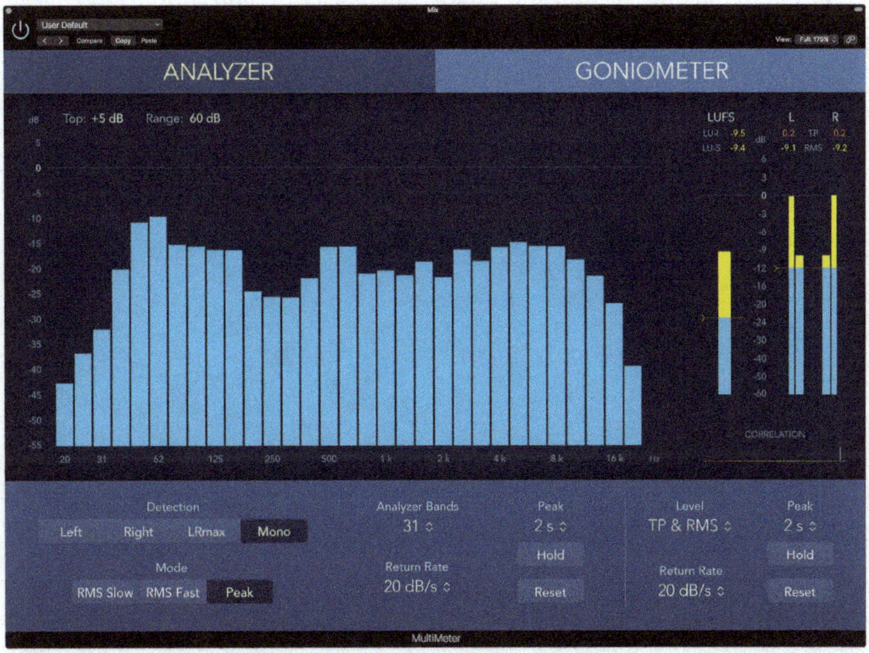

FIGURE 24.5
Logic has a good spectrum analyzer: "MultiMeter" (in "Metering"). In Pro Tools you can use third party plugins like "MAnalyzer Free" (Meldaproductions), or "Span Free" (Voxengo).

But there's more: if you analyze your favorite productions (within a certain genre), the analyzer might show a certain consistency. Generally, it is harder to draw big conclusions from the mids and the highs, but the lows may be easier to read. Maybe you can find a recurring pattern. Now compare this picture with your own spectrum. In case the bottom end focuses in a different area, you might want to change EQ or fader levels for kick and bass.

Disclaimer: ears are first; a spectrum analyzer can only help. Don't take decisions by looking at the picture only: two different mixes with a similar spectrum can sound (very) different!

4. CHECK MONO COMPATIBILITY

Although most music devices are stereo, consumers listen to mono sources too, for example, mono televisions, mono sound docks, bed radios, telephones or ceiling systems in shopping malls. In case the mix contains out of phase signals, mono playback will cause the mix to sound different or instruments to even disappear. Apart from this, a mix with phase issues cannot be cut onto vinyl. In case mono playback causes the mix to suffer audibly, phase problems should be addressed in the mix. Technically, phase coherence can (and should) be checked with a "phase correlation meter" (see Figure 24.6).

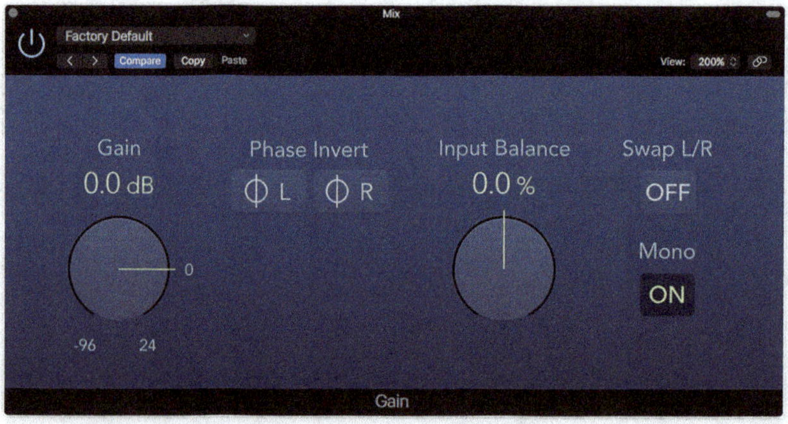

FIGURE 24.6
Checking mono compatibility of the mix. In Logic you can use the "Gain" plugin (in "Utilities"). In Pro Tools, you can use the "Downmixer" plugin (in "Sound Field").

Common Mix Mistakes

Even after reading the book so far, there's a fair chance of mistakes sneaking into the mix. This is nothing to be ashamed of; it just shows that you're in the process of learning. Music production is never the same and will always pose you with challenges you've never dealt with, causing you to make mistakes. So here we go; fasten your seatbelts and check your mix for the most notorious mistakes!

1. DUPLICATING IS NOT DOUBLING

Duplicating a track will only increase its volume. Even after panning the signals, an instrument will still appear mono. Only when two tracks are sufficiently dissimilar (timing and/or tuning), an instrument will sound **doubled**.

What if you drag the second track to the right in order to create a slapback delay? Although there's technically nothing wrong with that, it would be easier to insert a slapback delay plugin on the original track. This saves you an extra track, while any edits or changes to automation requires editing only one lane instead of two. Duplicating tracks is fine for parallel techniques (with distortion or compression).

2. MONO AUXES INSTEAD OF STEREO

Mono auxes may creep into your project when overlooking the "stereo-mono" menu in the "New Track" dialog in **Pro Tools**, while **Logic** automatically creates a mono aux when activating an effect send on a mono channel. But with mono auxes, your effects will be mono too! As are auxes that are used as instrument groups. So always make sure your auxes are stereo.

3. OVERLOADS IN THE MIX BUS

As noted in Chapter 18, "Workflow of the Mix," overloads in the mix bus (see Figure 24.7) cause transients to be decapitated. The mix signal will forever be compromised, as no magic process exists to restore the original peaks. In case of overloads, keep the master fader at 0 dB, and pull down all channel faders by the same amount until the clip indicators are off.

4. TIMING OR TUNING OFF

While recording, musicians can sometimes be reluctant to redoing parts, although timing and/or tuning was off. On a human level, this is understandable; when you agree on doing a retake, you acknowledge that the previous take was no good. Besides, there's no guarantee that new takes will lead to improvement. Last, rerecording takes extra time and energy from all people involved. It's by no exception that a note is called "cool," while in fact, it was **off**. But untimed or untuned notes might leave an unprofessional impression with the listener and may cause him to skip the track.

In the recording stage, it's important to pinpoint any "wrong" notes, and rerecord until they're good. Redoing parts is no shame; many big artists aren't one-takers either. Any remaining weak notes should be corrected in the mix. Only with limited deviations of timing and tuning can the mix sound tight and professional.

5. INSUFFICIENT BASS DRUM AND/OR BASS

These low instruments tend to be "forgotten." But sufficient bass energy is required for the mix to sound energetic and warm; it will prevent the mids/highs from being overly aggressive. When your speakers can't reproduce the lows properly, use a spectrum analyzer—or decent studio headphones, of course.

6. STEREO IMAGE NOT WIDE ENOUGH

Too narrow a stereo image is a pity, as instruments will clutter in the middle. Panning is actually the easiest way to achieve separation. Although positioning instruments like bass drum, snare, bass and lead vocal off-center is less common, all other instruments "beg" for panning. Always start with extreme positions. Only when the mix is skewing or when it starts falling apart, you'll revert to positions closer to the center. With every instrument panned, separation increases while space is created for the holy quaternity.

7. TOO MUCH MID

For recording pop instruments, dynamic mics are often the preferred choice. But these mics usually have a strong mid emphasis, with less highs and lows. If that response isn't counteracted with EQ, the mix can easily turn out **middy, thin and dull**. If you don't want that to happen, your general EQ strategy should head toward a "happy-face" or "smiley" EQ curve, that is, cutting mids, while adding lows and highs. Never mind extreme settings, if it sounds good, it is good.

8. HIGH FREQUENCIES UNBALANCED

This one is a little harder. As we've seen, cymbals can easily end up too loud and brittle in the mix. As such, they might compensate for other, dull instruments. The mix's **appearance** may be bright, but it is only the cymbals that supply top end. This mistake actually indicates two problems: not only are the cymbals too loud, other instruments lack top end too! Only with sufficient top end can an instrument open up and excite. To solve this problem, temporarily mute the overheads and verify the amount of top end on other instruments. In case they lack, add treble one by one. Imbalance of the high frequencies is usually an issue with cymbals, sometimes with vocals. Coincidence or not, these sources are often recorded with condenser mics!

9. OVERCOMPRESSION

With compression, you aim for an instrument to sound big, impressive and punchy. But with too much of this good, you'll cross a line where you actually achieve the opposite. This results in a wimpy, dead sound and instruments that shrink in size. In case you're not completely sure and still getting to grips with compression: **use presets**. Make sure the GR meter regularly returns to the zero position and moves within limits. If necessary, use a lower ratio and a higher threshold.

10. UNDERCOMPRESSION

This is a hard one. In case the dynamics of an instrument varies too much, it will lack stability in the mix. For the listener, it will be hard to stay in touch. Undercompression can cause the mix to sound weak or inconsistent.

How do you know if your mix is undercompressed? Check with reference tracks! Investigate the reference track's dynamic differences within single instruments: start with the bass drum, then proceed to the snare, then the bass and, last, the vocal. Now review the same instruments in your own mix. Maybe, dynamic differences are smaller than you expected. In many current genres, it's not uncommon for dynamic differences within an instrument to be small, or even close to zero. Why didn't you notice this before? Because of the advanced and sophisticated application of compression! Professional mixers have refined their techniques to reduce dynamics. They'll not only choose the best compression device per application; they'll also use parallel compression and cascade multiple units in order to minimize artifacts.

Preventing this mistake requires acknowledging the problem first (hearing!). Then you'll need to decrease dynamics, without any negative by-effects. Don't forget to use volume automation for the final touch!

Note that mistakes 5 through 10 can be prevented by using **reference tracks!**

CHAPTER 25
Bouncing the Mix

Performing the mix used to be one of the most exciting phases in the production process. It requires participation from all people available: the guitar player might turn up the delay on the guitar solo, while the engineer was riding the vocals. In case someone made a mistake, the mix had to be done all over again. Nowadays, bouncing the mix is nothing more than a formality; you'll just click "Bounce," while a few moments later you're presented with a perfect version of the final product. Even though the action may be simple, there's more involved here. Like, what format should you bounce to? Which mix level should you use? Last, we dive in the world of outtakes and stem mixes.

When bouncing the mix on the computer, the output of the channels is summed and the result is transferred into an audio file. Most DAWs let you choose between an "offline bounce" and a "real-time bounce." What are the differences? During a real-time bounce, the screen is "frozen," and you can't make alterations to the mix anymore. This option is handy for listening to the project a last time; otherwise, most people will prefer an offline bounce.

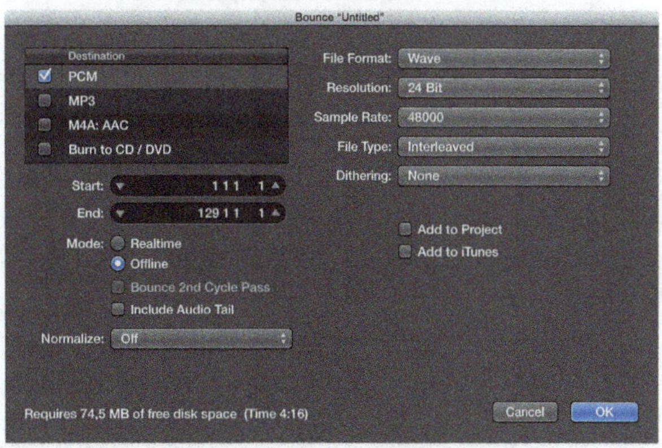

FIGURE 25.1
Bounce window in Logic Pro.

With an offline bounce, the computer maxes out CPU and proceeds through the project as quickly as possible. In a dense section, progress can be slow, while in an empty section the computer might advance at double speed, triple speed and so forth. Although it might sound counterintuitive, an offline bounce leaves less room for errors, at least theoretically. That's because the playhead will proceed to the next bar, only after calculating the previous bar correctly. Even when a virus scanner decides to start scanning the hard disk, this will, at worst, slow down the bounce process, but not allow for errors. Then again, before sending the mix off, you should always check it from start to finish.

▶ WHICH FILE FORMAT IS SUITABLE?

The bounce should at least consist of a lossless file, such as AIFF or WAV, even for demos. Of course, WAVs or AIFFs take more disk space than MP3s, but with current hard disk prices in mind, this can hardly be called an issue. You never know what a demo can be good for ("The Early Tapes"?). Most DAWs allow for mp3 and AAC as an additional bounce option: these are handy for sharing as a listening copy.

▶MAKING THE RIGHT SETTINGS

In general, you should first check the delivery requirements of the party you'll be sending the mix to. In case the mix stays in-house, always opt for the best settings. What are the "best" settings? As far as **bit depth** is concerned, the higher the better. That being said, most DAWs allow for high bit settings (like 32 bit), but not all professional parties will accept that format. As a **sample rate**, always choose the sample frequency of the project (or lower, if that's the requirement). Choosing higher sample rates makes no sense, as it is impossible to add quality in the bounce; it would only cause the file to become unnecessarily large.

As far as the **bounce range** is concerned, there are two possibilities. Either the bounce will serve as a **"pre-master"** that you'll send to a mastering studio. Then, allow for some margin at the start, in order for the first breath or transient to be complete. At the end, make sure to leave some extra space for the last note or reverb tail to end properly. This leaves options open for the mastering engineer to decide on the exact start and end point.

In case the of a **final master,** carefully check the start and the end of the project for any unwanted noises, breaths or ticks. Then apply a fade-out and determine a tight bounce range (Figure 25.2).

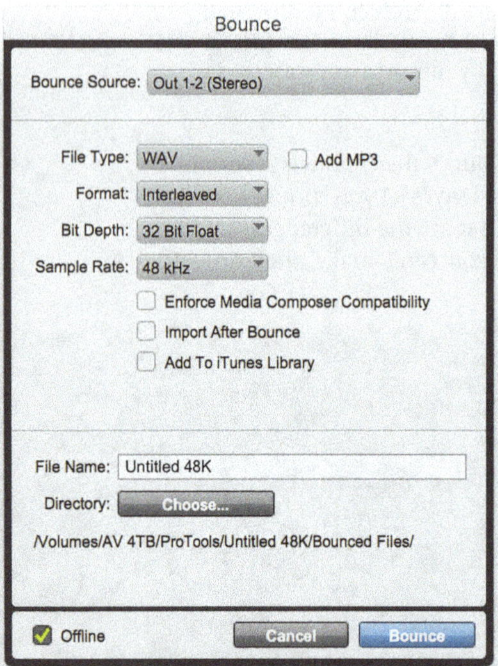

FIGURE 25.2
Bounce window in Pro Tools.

▶ HOW LOUD SHOULD YOU MIX?

It depends. As a **final master**, levels up to 0dB are allowed. As a **pre-master** that you'll send in for mastering, it is better to keep 3- to 6-dB headroom. Then, the mastering engineer can start his job right away, without any concern for overloads. What about the "**normalize**" option? When enabled, the DAW examines the bounce file after bouncing and calculates how much headroom is left. Then, the file's total level will be cranked by that same amount. Although there's nothing wrong with that in case of a listening copy, we've seen that high levels are not always an advantage (Chapter 18, "Workflow of the Mix"). Therefore, best practice is to decide on the correct level in the mix.

▶ WHAT ARE OUTTAKES?

After performing a regular bounce, it is useful to include alternative versions or *outtakes*. For example, an *instrumental* version, and an *a capella* (only vocals). Instrumental versions can be used for games, web or A/V purposes (broadcast, film), while a capella versions can come in handy for remixes or dance versions. Outtakes allow for easy mix revisions in the future too. For instance, if you want the vocal to be louder, you can drag both the instrumental and the a capella audio files into an empty project and start a new mix right away.

Apart from an instrumental and vocal only, it's good practice to bounce a *vocal-up* version (and possibly a *vocal-down* version). Record companies often ask for loud vocals, while during mastering it may appear that the vocal suffers from certain processes. It's handy if alternative versions are available then.

Master processes such as compression and limiting may **only** be applied to a **final product**.

When bouncing outtakes, processes in the mix bus such as EQ, or compression/limiting should be **bypassed**. Master processes are allowed, only for a final product ("production master"). As soon as outtakes are combined into a final product, master processes can be used again.

▶ STEMS

The idea of outtakes can be taken one step further by making "stems." *Stem* refers to grouped instruments that have a similar function in a classical score, like wood, copper, violins or percussion. Pop songs usually consist of five different stems: drums, bass, lead vocals, backing vocals and guitars/keyboards. In case any new mixes are needed, stem files can easily be dragged into a new DAW project. With five or more stems, drastic changes can be made to the sound or balance, without having to recall the complete project. When bouncing stems, always make sure the files have the same starting point!

> **ANALOG MIX**
>
> In order for an analog mix to be recalled properly, not only the knobs on the console but also the outboard gear and the patchbay should be returned to their original position. This makes the exact recall of an analog mix hard, if not impossible. Therefore, when mixing in an analog studio, always make sure to make individual stems, as it might be your only option for a proper recall.

All in all, stem mixes can be useful in the following situations:

- When a rented studio will be no longer available
- When creating backing tracks for live performances
- When making (dance) remixes
- For use in games
- For a "stem mastering" (see Chapter 31, "Mastering")
- As an insurance policy: your project will sooner or later refuse to open.

This last reason requires some explanation. Not only is DAW software updated regularly, but so are the plugins, computer operating system and computer hardware. New standards replace older standards. So eventually, you will not be able to open older projects. Stems will provide you with a safe haven.

FIGURE 25.3
The National Academy of Recording Arts & Sciences not only hands out the Grammy Awards but also advises on technical standards.

▶ AND WE'LL CALL HIM. . .

Proper and consistent naming avoids confusion. Even better, it saves time (and money) in the follow-up process, for example at the mastering studio. The National Academy of Recording Arts & Sciences, which organizes the annual Grammy Award show, has created a standard for this purpose (www.grammy.org/files/pages/deliveryrecommendations.pdf). According to this standard, a mix can be called:

LH_BodyAndSoul_Master_96k24_R01.wav

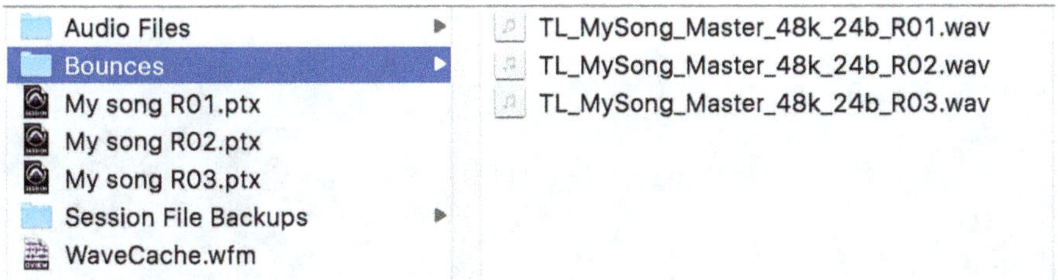

FIGURE 25.4
Identical names for both project and bounce allow for easier recall of earlier versions.

"LH" is the abbreviation of the artist name, "BodyAndSoul" is the song title, "96k" and "24" mean—you guessed it—96 kHz and 24 bit. "R" stands for revision and "Master" means that it is the final, complete mix. Other possibilities are VocUp, InstOnly, NoLdVoc, LdVocOnly, BgVOnly, Drums, Bass, Gtrs, Keys and so on. Spaces are not recognized within some systems (e.g. internet servers), therefore the words must be separated by underscores.

▶ HYBRID WORKFLOW

Both the traditional way of working and the "in-the-box" method are extremes. By combining elements, you get the best of both worlds. When it comes to mixing, many engineers prefer an analog console. Not only does the analog circuitry in the channels add to the sound; the mix bus adds to this too. That's the place where voltages of individual channels add up to a combined signal. We call that **summing**; every mixer brand is known for a specific coloration that happens in the mix bus. Apart from this, performing the mix on physical faders is an intuitive and hands-on experience. Settings can be dialed in physically, even by grabbing two knobs at the same time. Now try that with a mouse!

For this to work, an audio interface with multiple outputs is needed, as all DAW tracks must be sent to the individual mixer channels. Surgical editing and volume automation can be done in the DAW, while the mix itself can be "performed" just like they did in the old days. The stereo output of the mixer can be recorded to a separate track in the DAW, alongside the rest of the project.

You don't necessarily need a large format console for this workflow; there are dedicated **summing mixers** too (see Figure 25.4). Summing mixers are cut-down and cheaper versions of their large, legendary brothers. They may offer limited channels and lack EQ, but they will add that revered sound. Note that with a hybrid workflow, offline bouncing is not possible, as the mix is summed in the analog domain.

Summing mixers are manufactured by Dangerous, Chandler, Thermionic Culture, Neve, SSL and others.

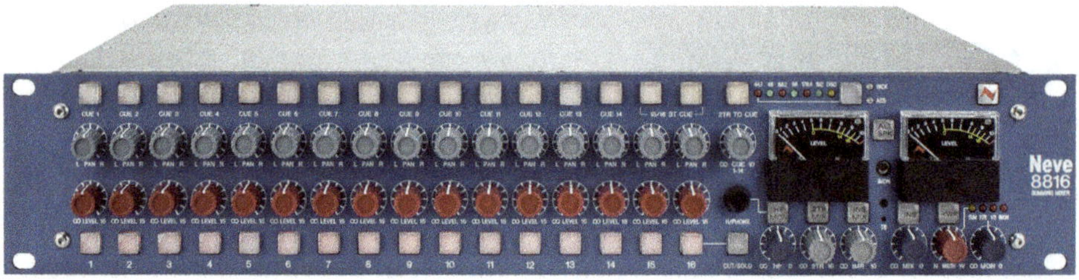

FIGURE 25.5
Neve 8816 summing mixer: settings can be stored in the DAW project through USB.

Source: Photo courtesy of ams-neve.com.

PART III

Advanced Mixing Techniques

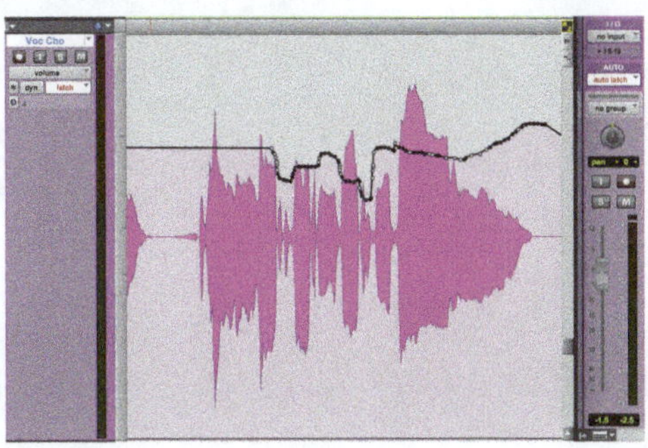

CHAPTER 26
Vintage EQ and Compression

Professionals often prefer the sound of classic EQs, pre-amps, tape machines, plate reverbs, tube compressors or tube microphones over modern equipment. If you consider the technological advances of the last few decades, this is nothing less than remarkable. What exactly is the reason that we favor those distorting, noisy old beasts? Which devices are the true classics? And, in case you lack the budget for the real thing, which software is good?

Provided that this is no hard science, let's first try to explain why vintage gear is so popular.

- The sound of classic equipment has become part of our collective consciousness. Whether it's a Stratocaster, a Linndrum, a Roland 808 drum computer or an EMT140 plate reverb, these devices will forever be associated with the great pop classics. The same goes for studio equipment: each time we hear the specific color of a classic device, it feels like coming home.

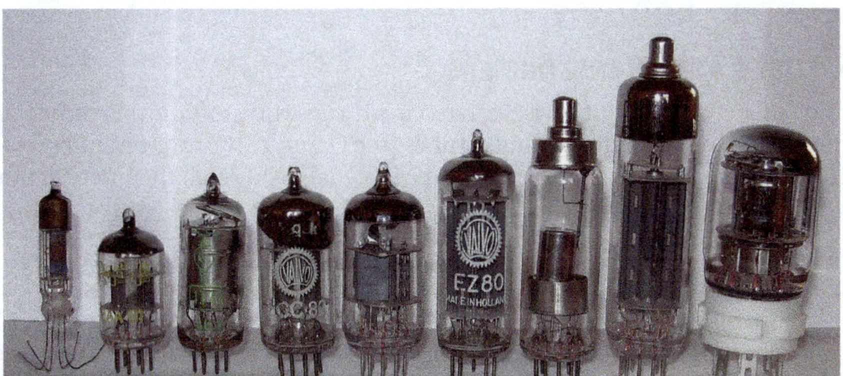

FIGURE 26.1
Various tubes.

Perfection in pop music is **boring**; it's the anomalies that add **personality**.

- As vintage gear has proved to be a recipe for good sound, it gives confidence to both artist and engineer. Especially when working on a tight schedule, it's great if you can grab a classic device knowing that it won't let you down.
- Perfection in pop music is boring, personality is what we're looking for. The idiosyncrasies that artists like Bob Dylan or Charlie Watts add to music is similar to what classic gear can add to sound. Distortion of vintage equipment can turn out crucial for making a sound interesting.

▶ WHERE DOES THE SOUND OF VINTAGE EQUIPMENT ORIGINATE?

The electronic components of modern devices are combined in IC's (Integrated Circuits, or "chips"). IC's are small, cheap, efficient with power and can easily be manufactured in large quantities. In the old days, the individual components of a circuit (such as resistors and capacitors) were soldered separately on a printed circuit board. This is called a **discrete circuit**. Discrete circuits can be fed with high voltages and are easy to service and maintain. Components can be chosen for quality, and the manufacturer can specify the exact behavior of the circuit. As chips are designed for general use, manufacturers have to settle for certain compromises.

Vintage equipment never has a straight frequency response. Often, high frequencies roll off, while low frequencies might get boosted. Therefore, yesteryear's devices sound "warm." And they also distort. The added harmonics cause a denser frequency spectrum, which can make an instrument sound more urgent or cut through the mix easier.

▶ TUBES AND TRANSFORMERS

Every single component in a discrete circuit causes distortion, but **tubes** and **transformers** account for the bulk. Tubes (see Figure 26.1) are glass bulbs by the size of a thumb and can be used to amplify signals. Although their performance is clean with soft signals, they'll add even and odd overtones when working at higher levels.

In discrete circuits, tubes often get accompanied by **transformers**. They consist of large windings of copper wire and typically exhibit a nonlinear response. Transformers distort, suffer from phase shifts and behave less linearly when you drive them harder. Last but not least, they flatten transients. It won't get more vibey than this! Famous transformer brands are UTC, Triad, Marinair, CineMag, St Ives, Carnhill, Jensen, Lundahl and Sowter.

Due to the higher working temperatures in tube equipment, the electronic components age. As a result, circuits behave less linearly and might even start to behave outside specification. This is the reason why an original piece of vintage equipment will sound different from a reissue.

Vintage EQ not to Be Missed

▶ PULTEC EQP-1A

This tube EQ entered the market in 1951, the same year that Les Paul used sound-on-sound to fit 24 different instruments on a 3-track recorder (see Figure 26.2).

Even after 65 years (!) the smooth and silky sounding Pultec is a popular choice to sculpt the sound of just about any instrument or mix. Because of its passive

FIGURE 26.2
Pultec EQP-1A: the lo-shelf section at the left has individual knobs for boosting or attenuating lows. The middle section is a hi-shelf, while the right section is a lo-pass filter. All sections offer selectable frequencies, "CPS" meaning "Cycles Per Second" (Hz).

FIGURE 26.3
The Pultec MEQ-5 is the three-band tube-amplified brother of the EQP-1A. It is often used on guitars and vocals.

FIGURE 26.4
The Neve 1073 class-A mic pre-amp/EQ has a low shelf with selectable frequency (35, 60, 110 and 220 Hz), a bell EQ for the mids (360, 700, 1.6K, 3.2K, 4.8K and 7.2K) and a hi-shelf fixed at 12 kHz. The outer rings select frequency while the inner knob is used to control gain (±18 dB).

design, EQ can be applied in generous amounts, without too many negative side effects. With the hi-shelf at 10 to 16 kHz, a signal can be given just that little touch of air. Engaging the lo-pass filter at the same time focuses that boost. By increasing the lows on kick or bass, earthshaking bass can be the result. Strangely enough, dialing in the attenuation knob does not cancel the boost but, rather, results in a low-mid dip (200 to 300Hz). Coincidence or not, boosting lows and cutting low-mids is what many mixers do on a bass drum and bass, almost as an automatism.

The Pultec EQP-1A is re-created in hardware by Warm Audio (EQP-WA), Tubetech (EQ1A), Acustica Audio (Cooltec EQP-1) and Ignite Amps (PTEQ-X) and others. In software, this EQ is modeled by Universal Audio, Waves (PuigTec EQP1A), iKMultimedia (T-Racks P-EQ 1A), Softube (PE1C), Overtone (PTC-2A) and the free Analog Obsession (MPREQ). **Logic** has "Tube EQ," while **Pro Tools** has "Pultec EQP-1A."

▶ NEVE 1073

Rupert Neve (1926) is a legendary name in audio. In 1964 he built his first, transistor-based mixer. In the early seventies, he built a high-quality mic pre-amp/EQ for UK-based Wessex Studio. This turned out to be a hit. The news about this excellent sounding "1073" (see Figure 26.4) spread quickly and established Neve's reputation.

While there's no need to be shy with the Pultec, the 1073 can change the sound by only the touch of a knob. The lo-shelf produces full-bodied lows, while the mid section can be very effective to add bite or sheen to sources like guitars, vocals or snare. The hi-shelf sounds airy and can cause instruments to open up, although the Pultec may sound a bit smoother and silkier in this area. With the 1073, you can change sound more drastically than with the Pultec. But really there is no wrong or right here; different applications call for different characters.

The Neve brand gets associated with almost all genres but especially with rock. Other than "fast" and linear modern equipment, the Neve sound can be described as rich, thick and warm.

In recent years, the 1073 has been cloned by various manufacturers, both in hardware and in software. In the latter category you can get this EQ from Universal Audio (1073), Waves (VEQ-3 and Scheps 73) and iKMultimedia (T-Racks 73) and Slate Digital (FG-N). **Logic** has "Console EQ."

▶ SOLID STATE LOGIC EQ

Solid State Logic (SSL) is an English brand that became a name after releasing its first automated mixing console in 1979. For the next three decades, this piece of gear by the size of a battleship was destined to become the standard for mixing and recording pop music.

The original SSL E-series console has both small and large numbers attached to it. A full-size model (96 channels) with all options could easily cost a million pounds. It "featured" the power consumption of a small village, and due to the heat generated in both control room and computer room, a good air-conditioning system was a must. By storing a "Total Recall" snapshot on the computer, the exact position of all knobs could be recalled at a later time. As this had to be done by hand, knob by knob and channel by channel, recalling a complete mix could easily take hours.

Mix automation allowed mute buttons and faders to be recorded dynamically into computer memory (128 kByte) and then stored on 8-inch floppies. Although the faders didn't physically move on the early models, they were represented as vertical lines on the built-in phosphor screen. This may sound somewhat archaic, as we're used to modern technology now. But in fact, SSL automation was fast, powerful and reliable.

Contrary to what you might think, controlling the mix on a large format console is easy and fast. As there are no hidden knobs, you get an overview of **all** controls while being able to instantly grab **any** knob. A mix can be "performed" with several hands at the same time, something that can't be done with a mouse. The tactile experience, ergonomics and immediacy are an overlooked aspect in the hardware versus software debate.

Many professionals, such as Phil Tan (Rihanna, Chris Brown), Chris Lord Alge (Rolling Stones, Alanis Morissette), Joe Barresi (Queens of the Stoneage,

FIGURE 26.5
Solid State Logic G-series console (64 channels) at Onkio House, Tokyo, Japan.

Wolfmother, Soundgarden), Mark "Spike" Stent (Björk, Massive Attack, Chvrches), Michael Brauer (John Mayer, James Bay, Kaiser Chiefs) and Joe Barresi (Kyuss, Tool, Queens of the Stone Age) still favor the sound and ergonomics of an SSL over other equipment.

FIGURE 26.6
SSL 611 4-band EQ module in hardware: two shelfs with selectable frequency that can be switched to peak/bell, plus two fully parametric bell-EQs for controlling the mids.

Source: Photo courtesy of solid-statelogic.com.

Compared to other equipment discussed in this chapter, SSL is least vintage sounding. It's relatively direct, clean and fast. Although the older E-series EQ is called slightly aggressive sounding by some people, this can be called an advantage in many pop applications. SSL consoles have undergone a number of revisions: E-series, G-series (1980s–1990s) and J- and K-series (end of the 1990s–2000s).

SSL EQ is simulated in software by Universal Audio (E-Channel), Waves (E-Channel, G-EQ), iKMultimedia (British Studio Series), URS (SSL EQ), Softube and others. SSL itself has also released plug-in versions of the hardware (see Figure 26.6).

Vintage Compressors not to Be Missed

Compressors can be divided into four groups: vari-mu, optical, FET and VCA. We'll find out that their working principle has great consequences for the way they sound. Once you know a bit about how they work, you can foresee how they will behave on a specific instrument. Let's take a closer look at some legendary compressors.

▶ UNIVERSAL AUDIO 1176

Designed by audio legend Bill Putnam in 1967, the 1176 must be the most popular compressor ever. Large numbers of 1176's have been produced, in 13 different revisions. The "Bluestripe" (see Figure 26.7a), which was the first revision, has slightly higher distortion figures than later models and is for many engineers a favorite for vocals. The "Blackface" (see Figure 26.7b) is a popular choice for drums, bass and other instruments. The 1176 was the first compressor to work with transistors instead of valves. The FET transistors used in its gain-controlling element can react extremely fast to incoming peaks. That's why the 1176 is a "go-to" unit for drums.

Notable properties:

- Attack and release knobs work in reverse: turning the knob clockwise **decreases** attack and release time.
- There's no threshold control but, rather, an input volume control. By increasing this, softer signals qualify for compression too, as they cross **the fixed threshold**.
- The "meter" buttons indicate whether the meter displays the input level, output level or gain reduction.
- Ratios: 4:1, 8:1, 12:1 and 20:1
- Soft knee

FIGURE 26.7
(a) Universal Audio 1176 "Bluestripe" in hardware; (b) Waves CLA-76 "Blackface" in software.

> **TIP**
>
> **BRITISH MODE (aka "NUKE" MODE)**
>
> By pressing all ratio buttons simultaneously, the compression curve of the 1176 changes drastically. This leads to over-the-top compression, which can be effective in a drums parallel setup for instance.

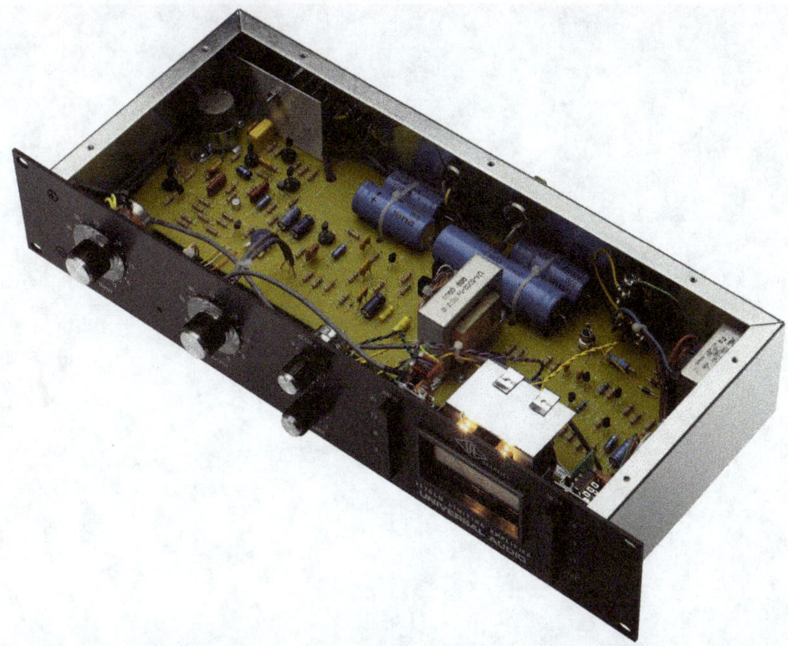

FIGURE 26.8
Guts of the Universal Audio 1176.

Source: Photo courtesy of uaudio.com.

The 1176 continues to be manufactured by Universal Audio and is cloned by numerous other hardware companies. Plugin versions are from Universal Audio themselves, Waves (CLA76), iKMultimedia (TR Black 76) and Native Instruments (VC76). **Logic** has "Vintage Fet" in "Compressor," and **Pro Tools** has "Purple Audio MC77" and "BF76." "Nuke" mode in BF76 can be activated by holding "shift" while clicking on the ratio buttons.

ALL BUTTONS OUT

Even if the 1176 is not compressing, the signal passes through nonlinear electronics. The added distortion (and noise) can be used for tone shaping. By increasing the input, distortion increases proportionally.

▶ TELETRONIX LA2A

The Teletronix LA2A (see Figure 26.9) is another Bill Putnam–designed compressor from the early 1960s. Due to its optical working principle, it has a soft knee and a smooth and musical character. The optical circuitry works as simply as it is effective: when sending a signal to the compressor, a lightbulb projects light

FIGURE 26.9
Teletronix LA2A.

on an optical sensor (the so-called T4 cell). Then, the sensor's resistance increases, causing attenuation of the incoming signal. Although it might seem restricting that the LA2A has no attack and release controls, the time it takes for the lightbulb to fade in and die out allow for a very musical behavior, albeit slightly slow. Release time is dependent on how hard the LA2A is driven: longer and louder signals result in a longer release, up to several seconds.

The absence of controls makes operating an LA2A easy. A dedicated switch lets you decide on whether to use compression or limiting, while the desired amount of compression can be dialed in with the "Peak Reduction" knob. Last, makeup gain is provided by the "Gain" control.

The LA2A is probably most popular for vocals, bass and guitar, and it even gets used in mastering. Albeit with decent amounts of gain reduction, as in "Compress" mode, the ratio is still around 3:1. For mastering purposes, that's considered high.

Software versions of the LA2A are from Waves (CLA2A), iKMutimedia (White 2A), Native Instruments (VC 2A), and Universal Audio (LA2A). **Pro Tools** has "BF2A," while in **Logic** you can choose "Vintage Opto" in "Compressor."

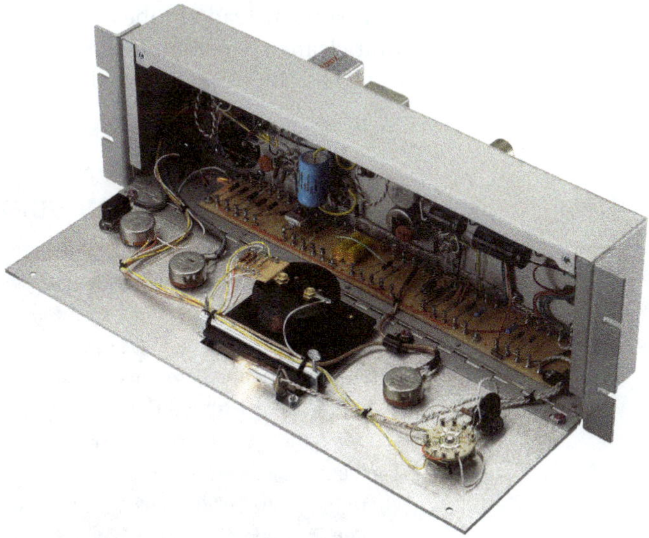

FIGURE 26.10
Inside the Teletronix LA2A.
Source: Photo courtesy of uaudio.com.

▶ UNIVERSAL AUDIO LA3A

In 1969, Universal Audio released LA2A's modern little brother, the LA3A (see Figure 26.11). Although it is a member of the optical family, its amplifier circuit is based on

FIGURE 26.11
Chris Lord Alge's LA3A in software: Waves CLA3A.

transistors instead of tubes. Therefore, the LA3A reacts slightly faster and sounds a bit cleaner. Most engineers favor it for use on piano, guitar and drums. Pro Tools has "BF3A" as a virtual model.

UNIVERSAL AUDIO IN SOFTWARE

Universal Audio has not only produced legendary hardware, but it has become one of the leading software brands too. Similar to Avid's Pro Tools, Universal Audio offers its own ecosystem of plugins for Mac and PC that must be powered by its "Apollo" audio interfaces (see Figure 26.12). Plugins on an Apollo interface offer near-zero latency monitoring, irrespective of the number of instances used. The amount of CPU power can be extended with either additional interfaces, or little black boxes, called "Satellites." The reputation of Universal Audio plugins is close to indisputable, so in practice, there are few professionals that don't work with Universal Audio plugins. Are there any downsides? Yes, for one, Universal Audio is not cheap. And, although interfaces can be expanded with extra Satellites, the plugins are slightly CPU-hungry. This limits the number of instances that can be used in a mix, or it requires you to invest in extra processing power. Note that Universal Audio plugins don't tax your computer's CPU, but there will always be the roundtrip latency of both computer and interface.

FIGURE 26.12
Universal Audio Apollo X8 interface.
Source: Photo courtesy of uaudio.com.

▶ THE MOTHER OF ALL COMPRESSORS: FAIRCHILD 670/660

An early version of the Fairchild compressor (see Figure 26.13) was conceived when Les Paul asked Rein Narma to build a compressor for his recording console. After that, the Fairchild Equipment Corporation acquired the rights and brought a final version to the market at the beginning of the 1950s. In those days, engineers wanted their recordings to sound as natural as possible. That's why the Fairchild was designed to add the least amount of coloration to the signal. Radio stations used it to protect their transmitters against dangerous peaks. Years later, The Beatles discovered that you could make "sound" with the Fairchild. For *A Hard Day's Night* in 1964, vocals were processed by the Fairchild. Because of its creamy sound, The Beatles started using it on drums and guitar too. Then, the word spread, and other producers and artists also started to make sound with the Fairchild.

FIGURE 26.13
Universal Audio Fairchild 670 in software.

Notable facts of the Fairchild:

- Contains 20 tubes and 11 transformers
- Weighs 30 kilos
- A well-kept secondhand unit may cost US$20,000 to US$30,000
- Member of the "vari-mu" family. Vari-mu compressors typically have a soft knee: their ratio increases as the input signal becomes louder. The Fairchild's ratio varies from 2:1 to 30:1.
- Attack and release times (see Table 26.1) are relatively slow.
- By increasing both "input volume" and "threshold," the amount of compression decreases, and the unit will produce distortion.

The Fairchild's frequency characteristic shows a little bump around 1 kHz, causing instruments to appear closer. In the studio, the Fairchild gets used on just about any instrument, and also in mastering. Rather than changing sounds drastically, it is probably loved best when used in a subtle way.

Nowadays, the Fairchild is re-created in hardware by Undertone Audio (UnFairchild), Anthony Demaria Labs (ADL670), Analogue Tube (AT101), Pom Audio

Table 26.1 Time-constant settings of the Fairchild 670/660

Time Constant	Attack Time	Release Time
Position 1	200 μs	300 ms
Position 2	200 μs	800 ms
Position 3	400 μs	2 sec
Position 4	800 μs	5 sec
Position 5	200 μs	Dependent on signal: 2 sec for transients, 10 sec on regular peaks
Position 6	400 μs	Dependent on signal: 300 ms for transients, 10 sec on regular peaks, 10 sec on continuous loud signal

and others. In software, the unit is available from Universal Audio, Waves (Puig-Child F670), Overtone DSP (FC70), and Steven Slate Digital (FG-Mu). **Pro Tools** has "Fairchild 670." The Fairchild 660 is the mono version of the 670.

▶ SSL BUS COMPRESSOR

SSL consoles have helped shape the sound of pop music. Since the 1980s, **every single** instrument could be compressed by the channel's built-in compressor. Another important factor was SSL's "**Bus Compressor**" (see Figure 26.14) on the mix bus. With mild settings (slow attack, low ratio and auto release), this soft-knee VCA compressor can do its magic to the mix by gluing individual instruments together. When driven hard, the mix will react to the specific behavior of the compressor, while the balance can be adjusted accordingly. Some mixers engage Bus Compressor right at the start of a mix. Although VCA compressors are faster than optical and vari-mu compressors, attack is slightly slower than FET compressors. After 40 years, the Bus Compressor has settled its reputation, and SSL has released it as a separate device too.

FIGURE 26.14
SSL Bus Compressor in software.

Other brands cloning the SSL are Alan Smart C2, TK Audio (BC1 MKII), Morbin Audio (Mix Bus Compressor), Dramastic Audio (Obsidian 500) and Serpent Audio (SB4001). In case you aren't afraid of using a soldering iron, electronic schematics of this compressor can be found on the internet. Due to its relatively simple circuitry it shouldn't be too complicated (or expensive) to assemble one yourself.

For SSL compression in software, there are many options: Waves (G Master Buss Compressor), Universal Audio (SSL G-series Bus Compressor), iKMultimedia (Bus Compressor) and, of course, the version of SSL themselves (Bus Compressor). In **Pro Tools**, the SSL character is mimicked by "Impact," while in **Logic** you can choose "Vintage VCA" in "Compressor."

Advanced Mixing Techniques | Introduction

Before jumping into the advanced mixing section, this chapter contains general notes for an advanced approach of the mix. How can you get that revered sound of vintage gear in your project, and what are ideal pairings? And, the more we'll work with vintage plugins, the more important it is to use proper gain staging. Professionals are meticulous on this subject. Why is safeguarding volume so important, and how exactly does it work?

VINTAGE PLUGINS

Vintage plugins will turn out crucial for our advanced mixing method. Although Avid and Apple stock-plugins can give you excellent results, other companies model vintage gear as a core business. Not only can these plugins provide better sound quality; there are also more devices to choose from. In the advanced approach in the rest of this book, vintage plugins play an important role. Apart from a unique sound, their often simple controls allow for intuitive operation. If required for sound, you'll easily set a knob to "11," while a modern plugin might make you hesitate, just because the displayed curve **looks** so drastic.

There are a limited number of companies that bring good-quality plugins to the market. Among them are Universal Audio, Waves, Softube, Soundtoys, Solid State Logic, Slate Digital, Eventide, Lexicon, Plugin Alliance, Audio Damage, Audiothing, Native Instruments, iKMultimedia and McDSP. Often, these brands offer their own take on the same classic device. For instance, the Universal Audio 1176 compressor is modeled by Waves under the moniker "CLA76." Slate Digital call it "76," Native Instruments has "VC76" and so on. To allow for easy reading, I will call devices by their original name and not mention every manufacturer's variation. A simple search on the internet will find you the relevant device from a certain manufacturer.

Ideal matches. In practice, it often appears that certain devices work remarkably well on specific instruments. Professionals know this, and they might have a certain device in mind even before they start working on an instrument. We'll call these "**go-to**" devices. In the following chapters, every instrument goes accompanied by a top 3 of favorite gear. Of course, this is no hard science; good results can be achieved with other plugins too. The top 3 should be taken as a starting point only—never forget to experiment!

> **TIP**
> Keeping the top 3s at hand while mixing allows for quick comparing of alternatives.

MODERN EQ

Although vintage EQ works often well for sound, it lacks precision. Modern EQ, on the other hand, lets you change the frequency spectrum precisely, without any character added. To get the best of both worlds, many professionals use modern, surgical EQ, **in addition** to vintage EQ. Apart from "Channel EQ" and "EQ3-7" from Logic and Pro Tools, respectively, FabFilter, McDSP and other brands make good EQs too. Which brand you choose is up to personal taste; there's more differences in operation than there is in sound quality.

Advanced Gain Staging

As you know from Chapter 18, proper gain staging is necessary to prevent overloads in the mix bus. What can you do in the event of overloads in the mix bus, and is the "no-overload requirement" also valid for channels and auxes?

Although audio files commonly have a resolution of 16 to 24 bit, it is in the mixer that the signal gets an upgrade to 64 bit. With channels and auxes working at a resolution this high, it is virtually impossible for the signal to distort, even in case of severe overloads. But the 64 bit upgrade can only be temporary, as the mix signal eventually has to take on the resolution of the outside world (16–24 bit). So in case of overloads, information get lost. The bounce can be seen as a "recording of the mix:" similar to a regular digital recording, you wouldn't want your signal to be decapitated.

▶ How Can You Prevent Overloads in the Mix Bus?

Well, there are a few different solutions, varying from convenient to best:

1. Pull down the **master fader** until overloads disappear. Although this will prevent the mix from distorting, it will decrease the digital resolution of the mix signal. Truth be told, the issue is a bit theoretical, as the dynamic range of a 24-bit bounce is 144 dB. Pulling down the master by 3 dB will leave you with a dynamic range of 141 dB, which is still respectable. But there are other disadvantages too. First, a low master fader level is impractical for fade-outs. Plus, any compressor or limiter inserted in the mix bus will already start working, even with the threshold at its maximum.

2. Reduce gain in the mix bus by inserting a **gain plugin** in the first insert slot. This will also properly prevent overloads, but it conflicts with the thought that we should use similar levels throughout the mixer.
3. **Rubber-band all channel faders and pull them down until the clip indicators in the mix bus disappear.** The faders maintain their relationship, so the mix will stay the same. Note that you should only include those channels assigned to the mix bus, not the channels assigned to auxes. Effect auxes should be excluded too. Any automation data for channel faders must be decreased by the same amount. Technically, this is the best method, although selecting the right faders for attenuation takes attention.

▶ What About High Levels in the Channels?

The 64-bit resolution of the mixer has the great advantage that no distortion in the audio can occur, even with extreme overloads. So you **could** theoretically use high levels in the channels, though it is **impractical** at the same time. Why is that?

1. If you want to prevent overloads in the mix bus, you'll end up with **low fader positions**. But low fader positions are inconvenient, as it will be hard to make fine adjustments. Faders are designed to have their finest resolution around the 0-dB mark.
2. Just like their analog counterparts, many **vintage plugins** are sensitive to the level you feed them. Higher levels will change the sound and cause distortion.

HOW CAN YOU PREVENT HIGH LEVELS IN THE CHANNELS?

1. **Start with the right level.** If necessary, lower the volume at the source. In case of a software instrument, decrease its output level. In case of an audio track, decrease the level by using "Region Gain" or "Clip Gain" (Chapter 23, "Mixing: Vocals"), or use a gain plugin.
2. **Work unity gain.** Similar to hardware units, plugins should neither increase nor decrease volume. When bypassing a plugin, the audible volume should stay the same.

Healthy signal levels will not only preserve audio quality but also allow exact and effective adjustments of the mix. The "Shangri-La" of gain staging is all faders at 0 dB while building the mix by adjusting volume in the plugins. Of course, this theoretical optimum is not very practical. But you may keep it in mind as a general objective.

Note

The knowledge and techniques shown in Chapters 19 **through** 23 **provide the foundation for the advanced approach in the next section. Therefore, reading and learning the earlier chapters is recommended for advanced readers too.**

CHAPTER 27

Advanced Mixing Techniques | Drums

Even if the drum sound didn't turn out top notch yet, there are many tricks and techniques that can be used to propel the sound to the next level. In this chapter, we'll largely hold on to the EQ and effects settings of Chapter 19 while replacing stock compression with third-party compression. Changes applied in this stage are often small, but with many of them, the drum sound will improve substantially. Every percentage must be cherished . . . and battled for!

These are the techniques that we'll go through this chapter:

1. Vintage plugins for drums
2. Advanced effects for drums
3. Working in parallel
4. Reducing bleed
5. Adding samples
6. Changing timing

1. Vintage Plugins for Drums

As far as **EQ** is concerned, we'll stick to the modern, surgical EQ applied in Chapter 19 and **add** vintage EQ. Two approaches are possible here: either you'll add vintage EQ for the final touch or you'll use vintage EQ to **replace boosted bands** of the modern EQ. According to the principle "cut before, boost after," it's most obvious to apply vintage EQ post-compression.

As far as **compression** is concerned, we'll generally replace stock compressors with plugins from one of the previously mentioned plugin manufacturers. That's because third-party plugins generally allow for more gain reduction, with fewer side effects.

▶ BASS DRUM

So which vintage EQs can work well for bass drum? As a go-to device, many engineers will opt for the Pultec EQP-1A. The lo-shelf of the Pultec can sound bigger and fatter than any other EQ. Boosting lows **and** attenuating them at the same time, results in the famous Pultec low-mid dip. Another popular brand for EQing drums is API. This brand's devices are perceived a little "faster" than other vintage gear, though they will definitely add color. Different types are available: semi-parametric (API 550A, 550B) and graphic (API 560). Last but not least, the Neve 1073 EQ is a firm favorite for bass drum. It can produce a full-bodied bottom end, while the mids and highs can emphasize the click very effectively.

TOP 3

CHARACTER EQ BASS DRUM

1. Pultec EQP-1A
2. Neve 1073
3. API 550, 560
4. Free: TDR VOS Slick EQ

▶ CONSTRUCTING A TWO-WAY BASS DRUM

Sometimes, the kick's mids need drastic cutting. But extreme EQ settings can cause phase issues that result in less punch. As an alternative, the bass drum can be split over two tracks. For this to work, duplicate the kick track, then **lo-pass** the signal in one channel (12–24 dB/oct, 100–200 Hz), and **hi-pass** the signal in the other channel (12–24 dB/oct, 500–1000 Hz). Now, the cutoff frequencies determine the total amount of mids in the combined signal. In the mix, two faders allow for convenient control of a two-component bass drum. Even better, the limited frequency range of an individual channel eases the load on a compressor. Distortion can now be targeted to one component only.

Splitting signals can also be useful for **bass guitar**, **bass synth** or **snare**. In case such an instrument was recorded with two microphones, it is even more obvious to apply this technique.

TOP 3

CHARACTER COMPRESSION BASS DRUM

1. DBX-160
2. Empirical Labs Distressor
3. Universal Audio 1176, Fairchild 660

▶ COMPRESSION

Although we've come to know the 1176 and the Fairchild already, the DBX-160 is new. It's a vintage VCA compressor that's known for its beautiful pairing with kick, snare and bass. The Empirical Labs Distressor (see Figure 27.1) is a modern classic that basically emulates an 1176 (although it offers some extra features). In "Nuke" mode, compression is so strong that it will almost work as a brickwall limiter.

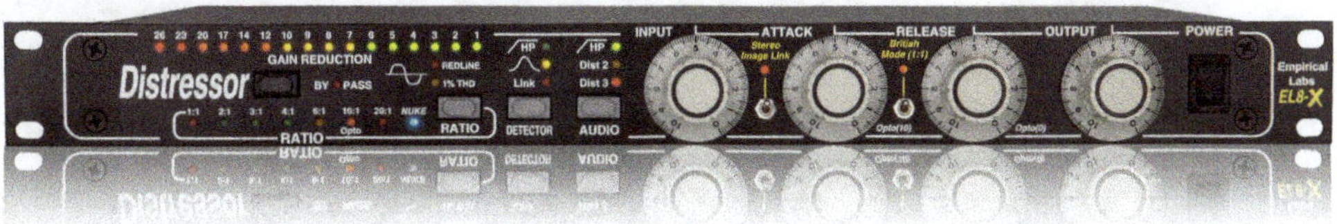

FIGURE 27.1
Modern classic: Empirical Labs Distressor, with selectable distortion colors, various sidechain filters and an optional opto-mode that emulates LA2A-compression.

Source: Photo courtesy of empiricallabs.com.

▶ SNARE

Similar to the bass drum, we'll stick to any modern EQ applied in Chapter 19. Additionally, a beautiful vintage device from API, Neve, SSL or Pultec could add extra character. Again, post-compression of course.

In the snare top 3, most devices are familiar already, however SPL's "Transient Designer" (see Figure 27.3) and Flux's "Bittersweet" require a little explanation. These are not regular compressors but, rather, devices that let you adjust the volume of a note's attack or sustain. With just two knobs, the envelope of a signal can be altered dramatically. Often used alternatives are Waves "Trans-X" and "Smack Attack," Oeksound Spiff, and Eventide "Physion."

Note: When using a transient shaper on snare, always keep an eye on the peak level of the mix, as overloads can occur easily. Also, limiters in the mix bus can go bananas on transient-shaped snares. Although pulling back all the faders by the same amount solves this, it will cost you mix volume.

TOP 3

CHARACTER EQ SNARE

1. API 550, 560
2. Neve 1073, SSL EQ
3. Pultec EQP-1A
4. Free: TDR VOS Slick EQ, Logic Vintage EQ

BASS DRUM FIRST AID

Due to a bad recording, the kick may lack essential low frequencies. In severe cases, even the best vintage EQ can't help you out. Fortunately, bass frequencies can be recreated by using a tone generator. Here's how:

Create a new track that has a tone generator inserted (see Figure 27.2). Set the tone generator to a sine wave with a frequency of 40 to 80 Hz. Then, insert a noise gate (see Section 27.4) post tone generator, and assign it the original bass drum as a sidechain input. Now, the sine wave will only pass when the bass drum triggers the gate. Be sure to dial in the right settings for "Release" (and "Hold"), as it can make or break the bottom end of the mix. A short release will yield insufficient bass, while a long release can cause the enormous energy of the pure sine wave to clutter up the low end of the mix. As a last refinement, the pitch of the sine can be matched to the key of the song. The pitch-to-frequency table in Chapter 9, "Digital Audio Workstation and MIDI," will help you find the right frequency for the sine wave. This trick can also be used on **toms** or **snare**. As an alternative, "Thump" by Metric Halo lets you add sine waves as an insert on the original instrument channel.

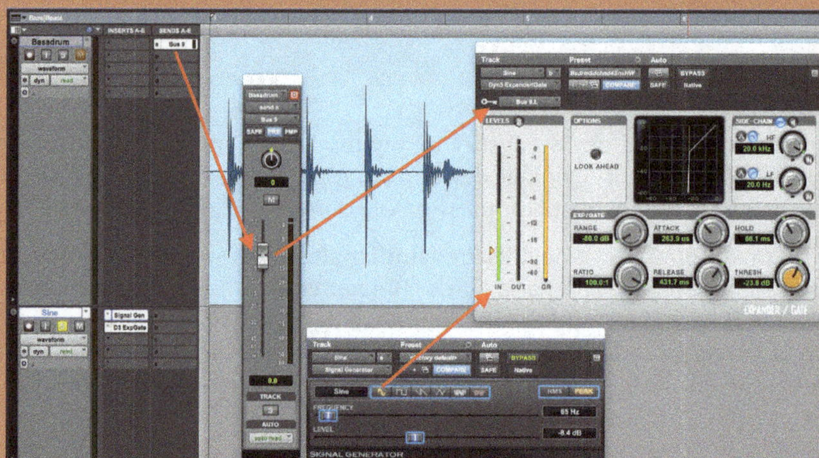

FIGURE 27.2
Sidechain setup in Pro Tools with "Signal Generator" (in Plug In->Other).

TOP 3

CHARACTER COMPRESSION SNARE

1. Empirical Labs Distressor Universal Audio 1176
2. DBX-160
3. SPL Transient Designer, Flux Bittersweet (free)

2. Advanced Effects for Drums

▶ REVERB

Traditionally, reverb for drums and percussion is added with echo chambers, plate reverbs and hardware devices such as the Lexicon PCM60, PCM70 or the AMS RMX16. Nowadays, the Bricasti M7 is popular. The easiest (and cheapest!) way to apply these effects yourself is by using dedicated impulse responses in a convolution reverb. Many impulses can be found on the internet, often for free.

TOP 3

DRUMS AND PERCUSSION REVERB

1. AMS RMX16 (Figure 27.4)
2. Gated reverb (or Logic Enverb)
3. Logic Space Designer, Pro Tools Reverb One
4. Longer reverb: EMT-140, Lexicon 224 and AKG BX20

Gated Reverb

When used in decent amounts, this typical 1980's invention doesn't necessarily make you sound like Phil Collins if you don't want to. In fact, gated reverb continues to be a powerful tool to inflate drums; many reverbs have presets for it. By inserting a gate after the reverb of choice, it's easy to construct a gated effect yourself.

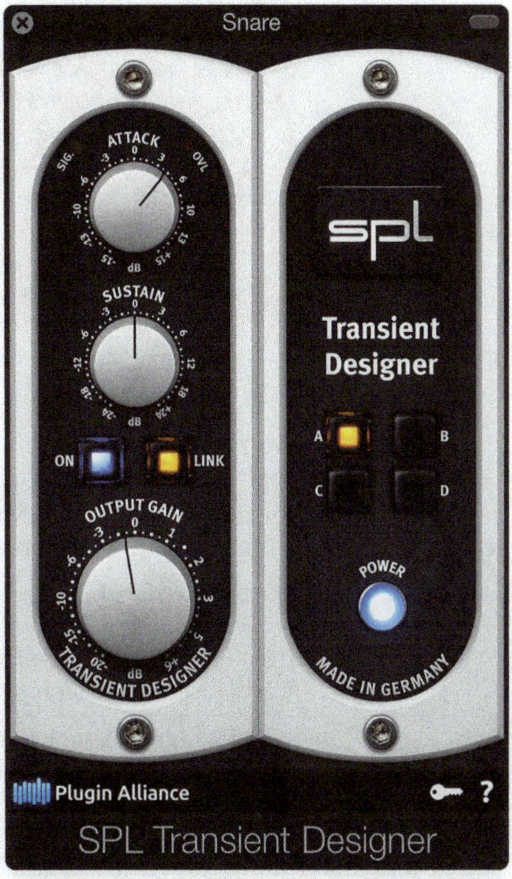

FIGURE 27.3
SPL Transient Designer.

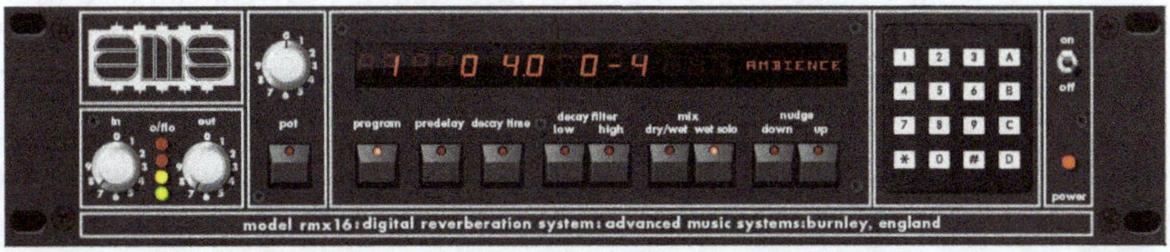

FIGURE 27.4
Although the AMS RMX16 was one of the great reverb units of the 1980s, its reverb doesn't necessarily need to sound retro when used in a contemporary context. Most popular programs must be "Ambience" and "NonLin."

SIZZLE FOR THE SNARE

In case the snare was recorded with a top mic only, you may miss the "sizzle" component. To bring this back in, another signal generator trick can be used. Create a new track with a signal generator and a noise gate inserted. Dial in **"pink noise"** on the signal generator, and select the snare track as a sidechain input for the noise gate. Be sure to **hi-pass** this signal, as pink noise contains large amounts of (unwanted) sub-low frequencies. This technique can be taken one step further by using the noise exclusively as a source for reverb. Altering the EQ curve of the noise channel allows huge changes to the color of the reverb.

Early Reflections

With drums, adjusting the ratio between early reflections and reverb makes all the difference. By using early reflections only, depth and width can be added, without losing definition. It's not unlike gated reverb but less dense.

For **longer reverbs**, classics like the EMT-140 and the Lexicon 224 shouldn't go by unmentioned. Pre-delay (up to 125 ms) will disconnect the direct signal from its reverb. As the effect's audibility will increase, you'll need less of it. This leaves room for other instruments in the mix.

Last, for making a statement with drum reverb, it won't get any easier than using a **spring**. Good examples of its typical lo-fi sound can be heard on albums by Portishead and MGMT.

▶ DISTORTION

Distortion makes an instrument sound more aggressive; as if the musician is playing harder. With drums, the effect can be crucial for the right amount of attitude and character. Although EQ can only boost **existing** frequencies, distortion **adds** overtones. There's many different types of distortion that all have their own color: tube distortion, transistor distortion, tape distortion, stompboxes, waveshapers, clippers and bitcrushers. Last but not least, vintage compressors (or EQs) that are driven, can also produce distortion, especially when attack and release times are set to zero. Basically, every drum signal is suitable for distortion, probably with the exception of cymbals.

Although throwing a few distortion plugins at the drums might get you great attitude in no time, the effect has serious side effects:

- Distortion is like compression on steroids. Soft signals will eventually become as loud as the loud signals themselves. **Crosstalk** might therefore mess up the drum sound.

- Similar to compression, distortion reduces dynamic range. This could result in a **flat** and **lifeless** sound.
- Distortion fills out the spectrum. This might cause the listening experience to be painful when playing loud, especially on a bigger system.
- Added overtones in the mids/highs distract our ear from the low frequencies. As a result, the drums can start to sound "middy" or "undersized." To counteract this effect, there are a few solutions:
    - Cut the offending mids with bell EQ.
    - Boost the lows (and highs) with shelf EQ.
    - Use a "de-harsher." A de-harsher is specifically designed to filter painful components from the signal. Devices with a good reputation are Waves Manny Triple-D, Brainworx BX-Refinement and Oeksound Soothe.

Producers such as John Congleton, Trent Reznor and Dave Friddman are masters at applying distortion in a sophisticated way. Their mixes can generally be played (very) loud, without hurting the ear.

Distortion works like compression on **steroids**: it can make crosstalk as **loud** as the signal itself.

TOP 3

DRUMS DISTORTION

1. Soundtoys Decapitator, Thermionic Culture Vulture (see Figure 27.5)
2. Sansamp PSA1, Izotope Trash
3. Vintage compression (optionally, set attack and release to zero)
4. Analog tape
5. Free plugins: Softube Saturationknob, Audio Damage FuzzPlus 3, Logic "Clipdistortion," Pro Tools "Lo-Fi" or "Air Distortion." Every DAW has multiple distortion boxes on board.

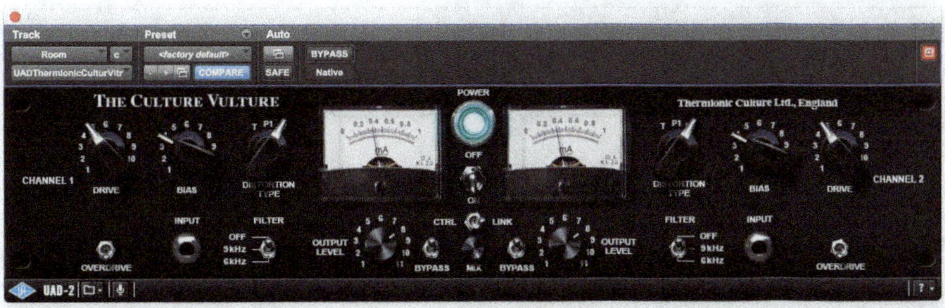

FIGURE 27.5
Thermionic Culture Vulture: a modern classic for adding distortion.

3. Working in Parallel

Parallel techniques allow the original signal to remain unaffected. Because the processed signal is **added**, extreme settings for compression can be used. The effect can be added conveniently and precisely with a fader, without having to open a plugin. Many modern compressors and distortion devices offer a simple dry–wet knob, which saves you from creating an extra channel or aux. Professionals commonly use parallel techniques for their mixes, sometimes extensively. On the original signal they might only use light compression or stay away from the effect altogether. On the parallel channel, they'll bring in the meat with a (driven) compressor.

Both compression and distortion can be added in parallel. You could, for instance, parallel compress the bass drum, parallel compress the snare, and then parallel compress the complete drum kit. Additionally, the snare can be **distorted** in parallel and the whole kit also.

Note

- As always when adding signals, be alert to phase issues, especially when adding distortion thru an aux. Always keep a phase reverse switch at hand, and verify which setting works best for the application.
- As parallel compression on a drum kit boosts softer signals, the hi-hat and overheads might end up too loud in the mix. Again! To solve this, you could send a custom mix to the parallel aux by using effect sends on the drum channels. In this mix, the hi-hat and overheads volume can be decreased or muted.
- As a heavily compressed signal's spectrum tends to focus toward the mids, this can be counteracted by adding both bottom end and top end with a shelf EQ (post-compression, see Figure 27.6). This technique is known as "**New York–style compression.**"

FIGURE 27.6
Typical EQ curve to follow a parallel compressor (New York–style compression).

> **TOP 3**
>
> **OVERHEAD, ROOM AND PARALLEL COMPRESSION**
>
> 1. API 2500, Neve 33609
> 2. Fairchild 670, Universal Audio 1176, Empirical Labs Distressor
> 3. Goodhertz Vulf Compressor, Waves Pie
>
> Free: Roughrider (Audio Damage; Figure 27.7)

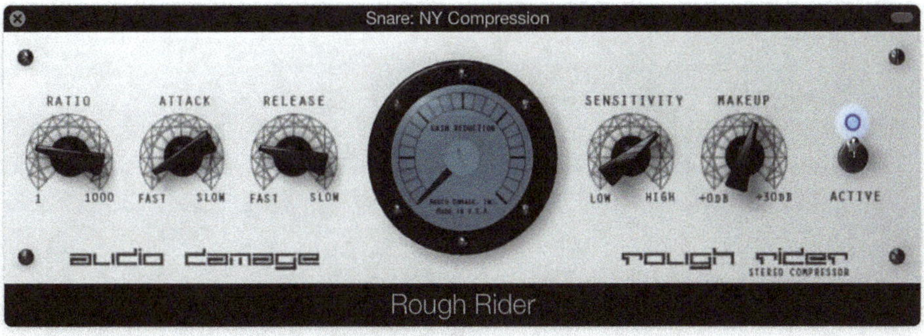

FIGURE 27.7
Drum compression: "Roughrider" by Audio Damage.

RE-AMPING DRUMS

After sending the drum signal to speakers and then record it to a new track, the acoustics of the room can be added to the original signal. Programmed drums might benefit from this technique in particular. By moving the microphone(s) closer or farther away from the speakers, the room signal can be varied between direct and diffuse. Similar to compression, re-amping usually yields best results when **adding** the re-amped signal in parallel to the original signal.

4. Reducing Bleed

Every single mic signal of a drum kit includes a certain amount of bleed. This has consequences when mixing. For instance, when processing the snare, the hi-hat will suffer accordingly. If we could remove spill, we'd be given more control when processing a single instrument. Let's look at a few methods that help attenuate leakage.

a. Noise Gate

A classic tool for removing bleed is the **noise gate**. A noise gate only passes audio when the signal exceeds a certain threshold. Setting this value sufficiently high, causes a loud signal (such as an afterbeat snare hit) to open the gate. In between the afterbeats, the softer hi-hat notes are shut off by the gate. EQing the snare will no longer alter the sound of the hi-hat. **Except** when the gate opens! Then, hi-hat bleed is added to the snare. This could lead to unnatural volume and timbre changes of the hi-hat.

What about ghost notes in the snare track? Well, this is a problem. As these are too soft to open the gate, we're actually messing up the musical performance here. Fortunately, overhead and hi-hat mics have captured the snare's ghost

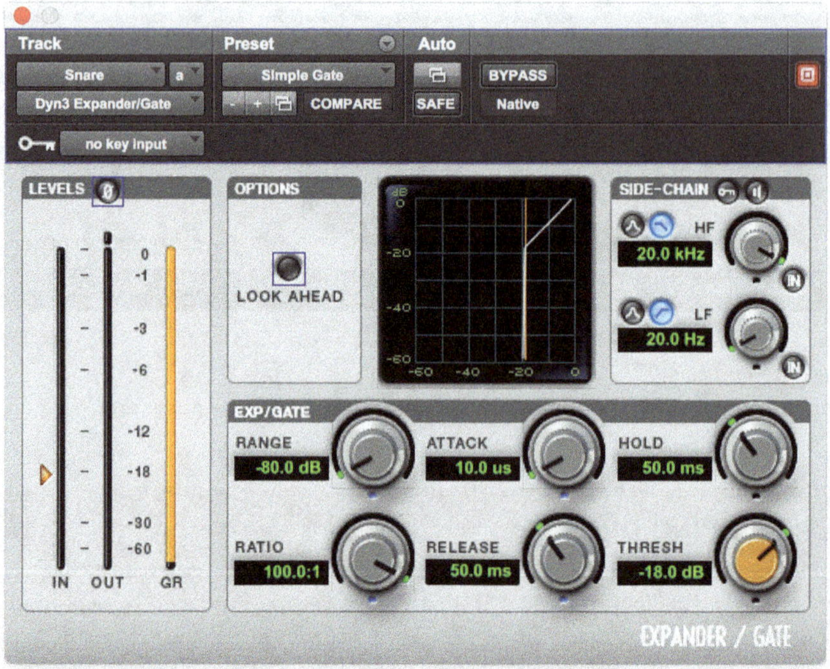

FIGURE 27.8
Dyn3 Expander/Gate in Pro Tools. "Threshold" determines at which level the gate opens, while "attack" and "release" allow for a gradual opening and closing of the gate. To preserve transients, "look ahead" opens the gate just before the actual signal is present—at the expense of some latency, though. Presets allow for quick results. Then, individual controls can be fine-tuned for the application at hand.

notes. Compression will further help the ghost notes to become audible, but their sound will be different from the afterbeats. Although you might get away with it in practice, it's clear to see that noise gates often lead to a compromise.

The usability of noise gates depends on the dynamics of the notes and the amount of spill. Careful tweaking of the controls will certainly help. Then again, the unpredictable on/off behavior of a noise gate causes triggers to be either false or missed. Apart from this, noise gates may also mess with the signals' transients.

4b. Expander

An expander won't suffer from typical gate problems, as it attenuates signals that fall below a certain threshold. How appropriate this is with bleed! In fact, an expander allows for more gradual/musical attenuation of low-level signals than a gate. For snare, kick or toms, start with a ratio of 0.6:1 and zero attack time. "Release" determines how "strict" expansion is applied; a longer release sounds looser and more natural, while shorter release times lead to more controlled results.

4c. Strip Silence

Strip silence (see Figure 16.2) can be used to divide one long region of the kick or the snare into hundreds of individual small clips. Deliberately setting the threshold too low, will result in too many clips, but now you'll be sure that even the softest notes are preserved. Any false clips can be removed by hand. In order to avoid digital clicks, you'll probably need short fade-outs for the clips. Optionally, clips can be set to the same length. All in all, fine-tuning Strip Silence can be labor-intensive but will result in the best controlled "gating." Sonically, it will leave transients intact.

5. Adding Samples

Even after applying advanced techniques, the drum sound might still lack. Your last resort is then **adding samples**. By adding a sample to an acoustic drum hit, you can add punch, or just about any sound characteristic that wasn't there in the first place. In professional productions, drum samples are used more often than you might think. Even to Amy Winehouse's organic and vintage sounding "Rehab," mixer Tom Elmhirst and producer Mark Ronson added a sample to the bass drum! How can you add samples to your own drums?

> Drum triggering is the most effective technique to **facelift** even the worst recording.

▶ REAL-TIME TRIGGERING

Adding drum samples is done by **triggering**. With trigger plugins like Steven Slate Trigger, Drumagog or SPL's DrumXchanger, it's easy to add suitable samples to the kick, snare or toms. Just insert the plugin on the source track, and set the trigger threshold. Then choose a sample, and decide on the balance between the sample and the original drum sound. Similar to gates or Strip Silence, you could end up with either false or missed triggers. If this happens, you can opt for **offline triggering**.

▶ OFFLINE TRIGGERING

Some trigger plugins and DAWs offer functions for converting audio signals into MIDI (see Figures 27.9 and 27.10). The software analyzes the transients and creates MIDI notes in a MIDI track. Deliberately setting the trigger threshold too low prevents soft but relevant notes to be excluded. Afterward, the false triggers can be deleted by hand. For an entire song, this could be a little labor-intensive, but the results may turn out superior to any other solution.

TIP: Using trigger microphones alongside regular mics will provide you with perfect trigger signals that need no editing afterward.

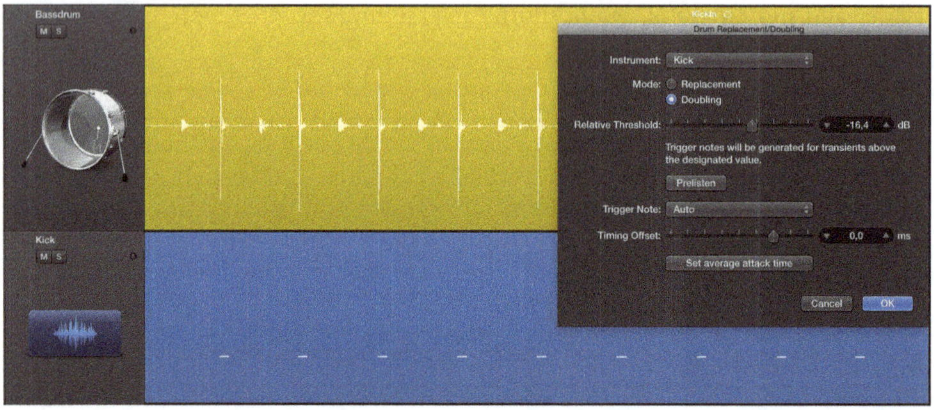

FIGURE 27.9
Logic Pro X offline drum triggering: select the drum track to be triggered. Then, choose Track->"Replace or Double Drum track." Logic will create a snare software instrument plus a MIDI region (blue). After setting the right threshold and instrument (bass drum, snare, tom), you're done.

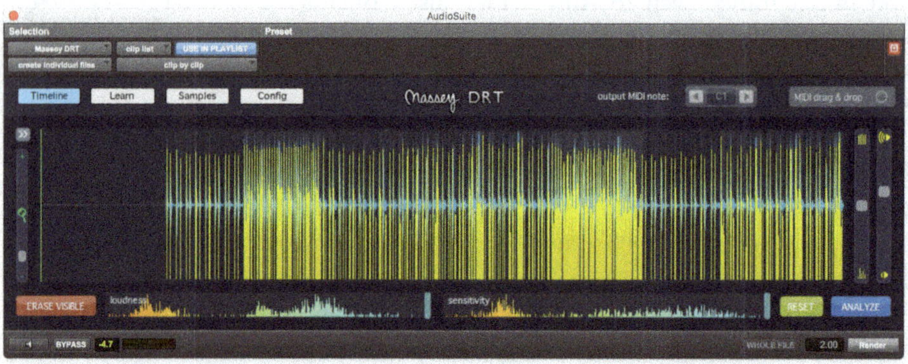

FIGURE 27.10
For offline triggering drums in Pro Tools, the non-expiring free version of Massey's DRT plugin (Audio Suite) can be used.

TOP 3

DRUM LIBRARIES

1. Toontrack Superior Drummer
2. Steven Slate Drums
3. Wavemachine Labs Drumagog
4. Native Instruments drum libraries

▶ WORKING WITH SAMPLES

Which sample will work? This is almost impossible to know beforehand. Browsing through a library of samples while the song plays is probably the best thing to do. In case the drum sample sounds punchier than the acoustic drum, it can be tempting to favor this signal over the original signal. But in practice, the translation of the original dynamics into MIDI notes often lacks precision. This may cause you to lose the human factor in the drum groove. You can verify this by muting the original drum signal. Possibly, editing velocity data might help. The trick here is to add sufficient volume of the sample in order to support the original sound while at the same time preserving the musicality of the performance.

Commercial libraries offer high-quality samples usually. Recording techniques have evolved over the years, and samples are captured dynamically. This means that velocity 98 has a different sample assigned than velocity 99, for example. To further improve realistic sounding drums, manufacturers use the **"round robin"** technique. With round robin, the plugin stores multiple samples for the same velocity. With multiple MIDI notes at the same velocity, the plugin fires **different** samples. This will prevent a 16-s snare fill from sounding like a machine gun.

6. Changing the Timing of the Drums

In certain cases, you may want to change the timing of the drums. Or even quantize them. Which options do we have?

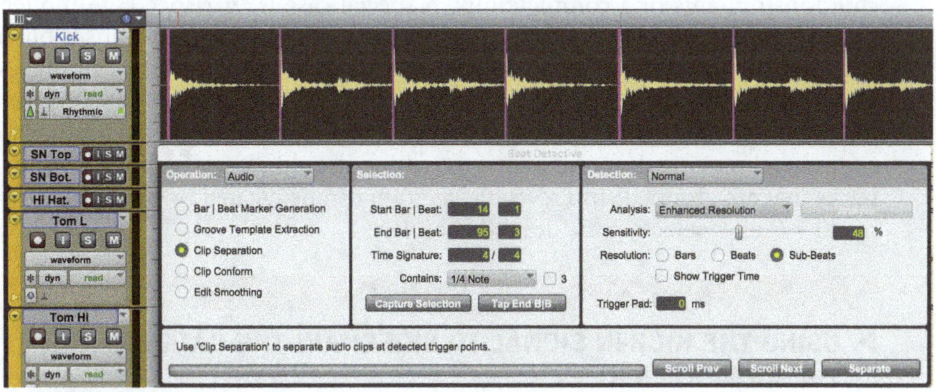

FIGURE 27.11
In Pro Tools, time correction can be done with "Beat Detective." It works by cutting and moving audio:

1. Set Beat Detective (in the "Event" menu) to "Clip Separation," and select the clips that best express the groove, for instance, the kick and snare. Then click "Analyze." After analyzing, increase "Sensitivity," markers will appear. Once you're happy with the results, other clips can be selected, those will be marked too.
2. Clicking "Separate" causes audio to be separated into individual clips.
3. Selected clips can be quantized in "Clip Conform" mode. Select the desired quantize value (for instance 1/8 notes or 1/16 notes), and set "Strength" to 100% (or less). Click "Conform" to quantize.
4. Finally, "Edit Smoothing" removes any gaps, while crossfades can be added by pressing "Smooth."

a. Cut and Move Audio

Preserving audio quality of the drums is vital. Therefore, always try to solve the problem by cutting and moving regions first. However, when moving a snare hit to the left or the right, its crosstalk on other mics won't move. A larger movement will cause a flam, while a small movement will cause phase issues with other mics. That's why you must always commit changes on **all tracks at once.** Both Pro Tools and Logic allow tracks to be **grouped** for that purpose. After activating a group, edits performed on one track will change all other tracks accordingly.

Typically when moving a single hit to the right, a gap at the left will occur. This gap can be filled by resizing the adjacent region. Unnatural transitions and digital clicks can be covered with crossfades.

b. Time Stretching

In case cutting and moving audio cause unnatural results due to gaps between the drum hits, drum timing can be adjusted with time stretching (see Figures 27.12 and 27.14). When stretching, the software speeds up and slows down while reading the audio file. This way, the signal can sound continuous. Although this continuity is a great advantage, you could lose audio quality with it. Therefore, always verify audio quality after committing. Common artifacts are transients that smear, and sustained notes that "wobble." To minimize by-effects, manufacturers include specific settings. For example, "**monophonic**" is optimized for use on a single vocal, bass, sax or trumpet. "**Slicing**" or "**rhythmic**" is optimized for use on drums and percussion, while "**polyphonic**" minimizes artifacts on chord instruments or a complete mix. A special case is "**varispeed.**" Although other algorithms will maintain pitch during stretching, varispeed causes pitch to go up and down, just like a vinyl deck or tape machine. Of course, this makes varispeed useless for pitched sources. But it will, in fact, offer the best quality for any unpitched material, like percussive sources or sound effects. Similar to cutting and moving drum hits, time stretching should always be done to drum tracks that are grouped! In Pro Tools, a group can be created by pressing Command + G after selecting the desired channels.

Tips and Tricks

▶ USING THE KICK-IN SIGNAL TO SIDECHAIN THE GATE ON THE KICK-OUT MIC

The kick-in signal usually contains less spill than the kick-out signal. By using the former signal as a trigger source, the gate on the kick-out mic can open up more reliably.

▶ USE MIDI NOTES TO TRIGGER A GATE

Once you have collected trigger notes of the drums, these can be used as a source for noise gates. Pre-delaying the notes allows the gate to open just before the transient of the signal arrives. Optionally, all notes can be set to the same length.

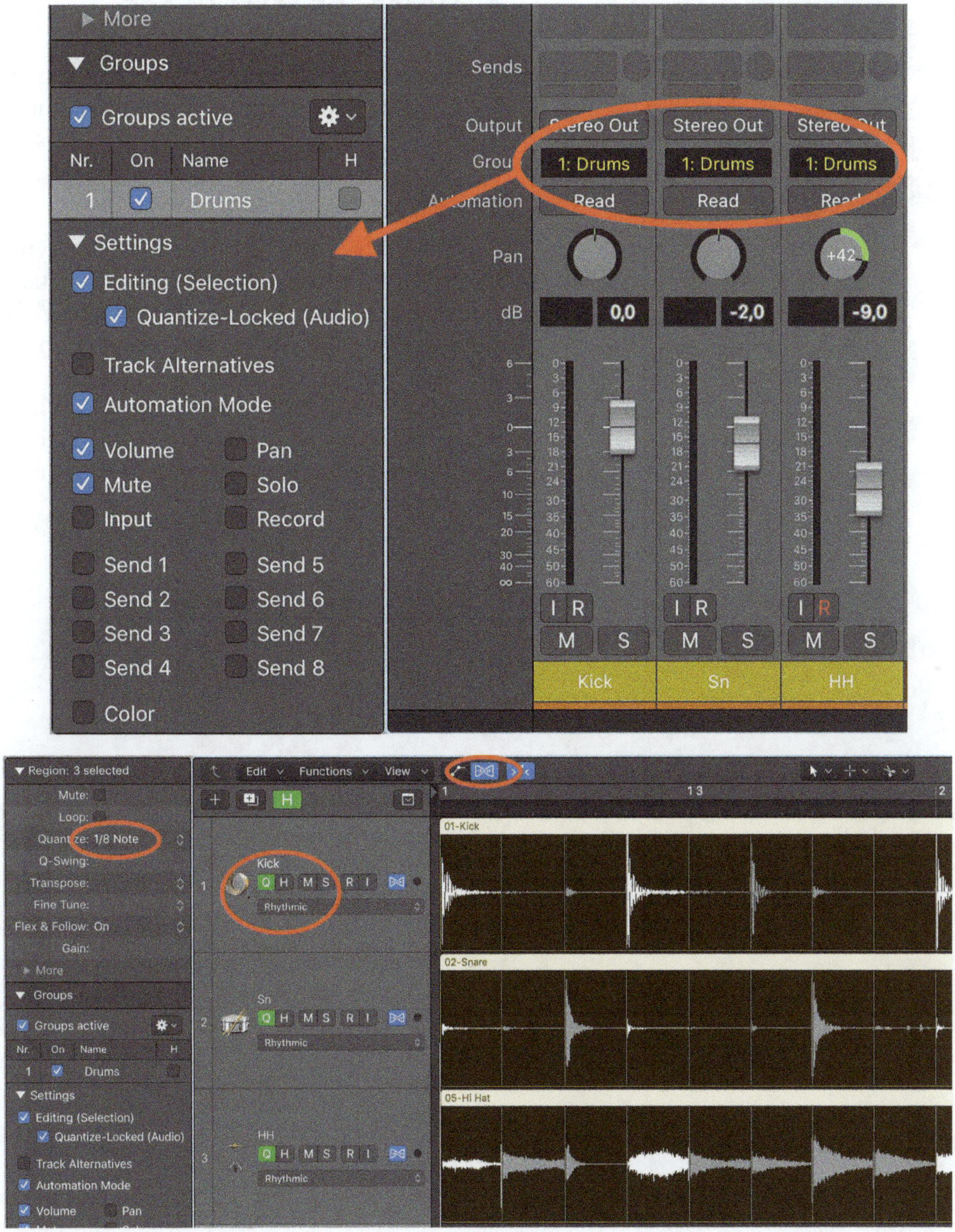

FIGURE 27.12

Quantize drums in Logic. (a) In order to commit changes on all tracks, a group must be created: click+hold the group field in the relevant channels in the mixer and choose a group number. Then, tick "Quantize-Locked Audio" in "Group Settings." (b) After enabling Flex-view (blue button upper right) and choosing the appropriate time-stretch mode in the track headers ("Rhythmic" in this case), regions can be selected and quantized by choosing a quantize value in the "Inspector."

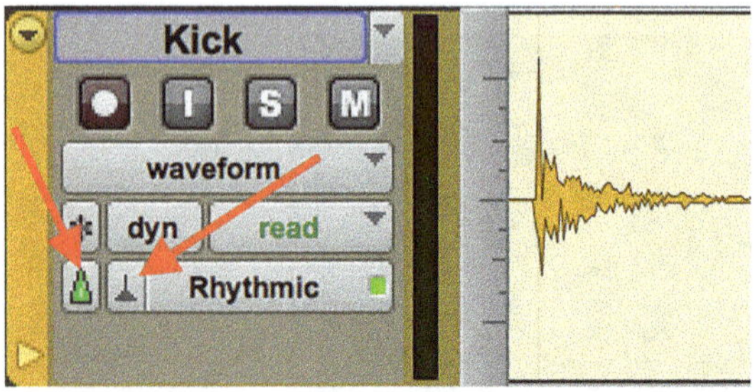

FIGURE 27.13
Elastic Audio (Pro Tools) can be activated by clicking "Ticks" in the track header. The time stretch mode can be selected by click-holding the small "Elastic Audio" symbol. To quantize audio, select "Event Operations" from the "Event" menu and choose "Elastic Audio Events" in the "What to quantize" field.

▶ LOWER THE LEVEL OF THE CYMBALS IN THE ROOM MICS

Try multiband compression as shown in Chapter 30, "Advanced Mixing Techniques | Vocals." Use 8 to 12 kHz for a frequency band.

▶ SMEAR PERCUSSION

Due to large amounts of high-frequency energy in percussion instruments like a tambourine or shaker, they can be "merciless" for the ear and claim their position up front in the mix. What may help is using modulation effects such as chorus, flanger or phaser. Then, transients will smear, which causes these instruments to take a proper place in the mix.

CHAPTER 28

Advanced Mixing Techniques | Bass

Despite our work on the bass in Chapter 20, the instrument might still lack in power, size or audibility. Fortunately, there are some advanced tricks and techniques to further improve sound quality. In this chapter, we'll look at timing, vintage plugins and specific distortion. And what is actually the secret of those thundering basses in reggae and dub or those of James Blake for instance?

▶ ADJUST TIMING: CUT AND PASTE

There are many possibilities for tweaking the sound of a bass. But if the performance lacks, the mix can never sound massive and tight. Temporarily dragging the kick track alongside the bass will easily reveal differences in timing. Any notes that are off can be cut and moved. Digital clicks can be smoothed with (short) crossfades.

▶ ADJUST TIMING: TIME STRETCHING

In case the bass requires extensive editing, time stretching might be your best option. In **Pro Tools,** you'll use "Elastic Audio," in **Logic** "Flextime." Select "**Monophonic**" mode for the best sound quality. Correcting the timing of a single note works by inserting a marker (Pro Tools: Ctrl + click; Logic: left-click) at the start of the note. Now, before dragging the marker to the desired position, surrounding notes must be prevented from moving. This can be done by inserting markers left and right of the section you're about to drag. Then, the middle marker can be safely dragged to the desired position. Time stretching allows correcting parts that either rush or drag in the most musical way. After each edit, always verify if the audio quality is still acceptable.

TOP 3

CHARACTER EQ BASS

1. Pulte EQP-1A
2. Neve 1073
3. Waves R-Bass
   Free: TDR VOS Slick EQ

▶ EQ

Although excellent results can be achieved with stock EQ in **Logic** and **Pro Tools**, third-party plugins might add just that little bit of extra character. This results in a fatter, more saturated bass tone. Similar to kick, Pultec EQP-1A and Neve 1073 are considered exceptional for bass. With Waves "R-Bass," harmonics can be added by means of a fader, resulting in full-bodied bass.

TOP 3

CHARACTER COMPRESSION BASS

1. Universal Audio 1176
2. Teletronix LA2A
3. DBX-160, Fairchild 660
   Free: Klanghelm MJUC jr.

▶ COMPRESSION

After reading Chapter 26, "Vintage EQ & Compression," an 1176 and a LA2A in the compression top 3 can hardly be called a surprise. The DBX-160 can also be very effective on bass, just like it is on kick drum.

▶ DISTORTION

When looking for a distorted bass sound, vintage compression with fast (or zero) attack and release settings might work well. For more dirt, an 1176 with "*all buttons in*" can be used. There are also loads of dedicated dirt boxes, all offering their own specific distortion colors. Last, re-amping the bass can be very effective.

TOP 3

BASS SATURATION/DISTORTION

1. Thermionic Culture Vulture
2. Soundtoys Decapitator
3. Tape-plugins
   Free: Softube Saturator

▶ THUNDERING BASS

Many people love the thundering basses of, for example, Massive Attack, James Blake, FKA Twigs or those in hip-hop, dub and reggae. Contrary to what you might expect, the total amount of bass in these productions is roughly the same as other music! How can that be achieved?

Often, the secret is both an empty arrangement **and** a dull bass. In an open arrangement, the bass is no longer masked by other instruments. Even dull, sine-like basses can become audible and play a mystical role. Sine-like basses can be constructed by using a lo-pass filter (24 dB/oct at 80–125 Hz). On a synthesizer, a sine-like sound can be dialed in by simply choosing a sine wave or triangle as a waveform. Cleaning up other instruments in the low mids (100–400 Hz) might help considerably for audibility and size of the bass. As soon as guitars, organs or pianos enter the arrangement, it will be harder for the bass to sound big, as these instruments all have their root notes in the lower mids.

▶ IMPROVING AUDIBILITY OF THE BASS

The weak harmonic structure of a bass guitar often prevents audibility in the mids. As we've seen, saturation and distortion can greatly help with this, but there's more:

1. Similar to the bass drum trick in the previous chapter, **split the bass signal over two channels**, one with a **lo-pass filter** (100–200 Hz, 12–24 dB/oct) and one with a **hi-pass filter** (200–800 Hz, 12–24 dB/oct). Insert an amp, tape machine or saturation/distortion unit in the high channel. Now that the lows and the highs each have their own fader, the instrument can be easily reconstructed with just two faders. Start off with the bass component only. This determines the amount of lows in the mix, which is more or less a given. Then, bring in the high signal to taste, this determines the bass's articulation. Cutoff frequencies of the hi-pass and lo-pass filters account for the total amount of low-mid energy.
2. **Bass in octaves.** An old trick for improving audibility of the bass is to double it one octave higher with a guitar, synth or other instrument. Although audibility improves, the bass itself might now appear thinner or "smaller." This is because our ear's attention is directed toward the **high** octave, as it is less sensitive to low frequencies. To solve this, the octave part can be **hi-passed**, for instance with a 24 dB/oct hi-pass filter set at 1 kHz or higher. Now, the separated spectra will cause both registers to be perceived as two different instruments. This way, the low bass can retain its primal function. This idea can be taken one step further by doubling the bass **two** octaves higher, the high part will only need just a little bit of level in the mix before being audible.

Audible bass may prevent the instrument from sounding **beefy**.

Be careful, **audible** bass may prevent the instrument from sounding **beefy**. As soon as overtones of the bass become more prominent, the dark instrument may cease to play its **mystical** role.

Tips and Tricks

▶ PARALLEL COMPRESSION ON KICK AND BASS

For more synergy between kick and bass, send both channels to an aux. Compression on this aux generates a signal with **consistent** bottom end and added overtones. The compressed sound can then be mixed with the original signals.

▶ AUTOTUNE ON BASS

Although you might associate the name AutoTune with vocals, it can be used on bass too. In case the bass is off-pitch, you can tune it with "Pitch Correction" in **Logic** and the free "MAutoPitch" by Meldaproductions in **Pro Tools**.

CHAPTER 29

Advanced Mixing Techniques | Guitar

For the average guitar sound, third-party compression and EQ plugins might get you better results than stock plugins, like more character and more tone. Then, saturation can be used to increase the size of the instrument and add extra overtones. In case all the fiddling doesn't seem to get you anywhere, re-amping is not only the most radical but also the most effective solution. The good news: virtual amps are getting better and better!

▶ ADJUST TIMING

We might throw 20 plugins on a guitar track, but if the performance isn't right, we'll play a match that is lost beforehand. If necessary, timing corrections can be done by either cutting and moving audio regions or by using time stretching. "*Polyphonic*" mode usually works best for guitar.

▶ VINTAGE EQ

Similar to other instruments, *tone shaping* with vintage EQ could turn out slightly more appealing than modern/clean EQ. Apart from the familiar Pultec and Neve EQs, many mixers use API and SSL EQ for guitar.

FIGURE 29.1

TOP 3

CHARACTER EQ ELECTRIC/ACOUSTIC GUITAR

1. Neve 1073
2. API 550, 560
3. SSL Channel, Pultec EQP-1A, Pultec MEQ-5
   Free: TDR VOS Slick EQ

▶ VINTAGE COMPRESSION

Even if you don't want to mess with dynamics, vintage compression may help the guitar tone to come alive, due to added overtones and coloring. Popular choices are both the Teletronix LA2A and LA3A. As the LA3A is transistor-driven, it sounds slightly cleaner than its older brother while reacting a little faster. Compressors that might react just as well to guitar are the ubiquitous 1176 and the Empirical Labs Distressor.

TOP 3

CHARACTER COMPRESSION ELECTRIC/ACOUSTIC GUITAR

1. Teletronix LA3A, LA2A
2. Universal Audio 1176
3. Empirical Labs Distressor
   Free: Klanghelm MJUC jr.

FIGURE 29.2

▶ ADVANCED CLEANUP OF NASTY PEAKS IN THE SPECTRUM

Sometimes, nasty peaks in the signal are just too complex to be taken down with modern EQ. Fortunately, there are tools that could do a better job:

1. **De-esser.** A de-esser can actually suppress painful mid-high guitar frequencies in a pleasant way. Engaging "sidechain monitor mode" will help you find the offending frequency. "Threshold" determines how much energy is cut.

2. **De-harsher**. Plugins such as "Manny Triple-D" (Waves), "BX Refinement" (Plugin Alliance) and Oeksound "Soothe" are dedicated tools to take down certain spectral components.
3. **Pultec MEQ-5.** This EQ can clear up harshness in the mids unlike any other EQ.

▶ RE-AMPING

Re-amping really is both the most drastic and powerful technique to make a guitar timbre fit a track. To keep options open, it's therefore a good idea to always record a DI-track of the clean guitar. Then, the guitar can be re-amped, even during the mix. Once other instruments have taken on their final shape, it will be much easier to find the exact mic and amp settings for the guitar. Apart from this, the clean DI-track can sometimes be useful for adding brightness to an amped signal. Be careful with the volume of this signal though, it contains large amounts of high frequencies, and could easily overpower the amped signal.

▶ VIRTUAL RE-AMPING

Guitars, either clean or amped, can also be re-amped with **virtual** amps, speaker cabinets, stompboxes and other gear. Digital models are getting better and better over the years. Apart from **Logic's** "Amp Designer" and **Pro Tools'** "Eleven," there are third-party hardware amps from Kemper ("Profiler"), Positive Grid ("Bias") and Universal Audio ("Ox"). and software amps from Universal Audio, Brainworx and iKMultimedia (Amplitube). Last, convolution reverbs are suited for re-amping too; impulse responses of famous cabinets can be found on the internet, often as a free download.

Tips and Tricks

- **Compressing or (subtly) distorting a guitar's reverb** can warm up the signal and make it sit more stable in the mix. Moreover, other reverbs in the mix may benefit from compression, saturation or distortion as well.
- **Transient Designer for sustain.** In Chapter 27, we used SPL's "Transient Designer" for adding attack to the drums. Why not use this effect on guitar? Not only can you add attack but also more tone and "meat" by increasing the sustain knob.

CHAPTER 30

Advanced Mixing Techniques | Vocals

Hopefully, Chapter 24 gave you sufficient clues for getting a good basic vocal sound. At the same time, it may not yet be up to par with your favorite records. How can you bridge that gap, and arrive at the refined sound of the pros? And how do you approach deviations in timing and tuning?

To further build the vocal sound, we'll dive into the following topics:

1. **Vintage EQ for vocals**
2. **Vintage compression for vocals**
3. **Advanced effects for vocals**
4. **The order of effects**
5. **Adjusting pitch and time**

1. VINTAGE EQ FOR VOCALS

Chapter 24 has shown you how to remove any unnecessary energy and unwanted resonances from the vocal. For an advanced approach, we'll now use vintage EQ to further shape the vocal's timbre. Even if the frequency spectrum seems "finished," it's worth trying to squeeze out that last bit of quality. With vintage EQ, there's more happening than just boosting a certain frequency. Many vintage plugins add sound, even with all controls at zero!

Similar to the drums, you can use vintage EQ either as the finishing touch to already-applied EQ, or you can null the boosted bands in the modern EQ and replace them with similar boosts from vintage EQ. Applying character EQ **post**-compression lets you add energy without affecting compression. This also allows compensating for any losses caused by compression.

The Pultec EQP-1A and Neve 1073 are probably the most famous EQs for adding either a big bottom end or an "airy" top end. A mid boost on the MEQ-5 or the 1073 can give extra bite and make the vocals sound forward. With slightly different colors, the API or the EMI/Abbey Road EQs could prove just as valuable, experimenting is key here.

TOP 3

CHARACTER EQ VOCALS

1. Pultec EQP-1A, Pultec MEQ-5
2. Neve 1073
3. API, EMI/Abbey Road
   Free: Logic Vintage EQ

▶ EQing With Multiband Compression

Let's say, a female vocal is harsh sounding in the mid-high register. To solve this, you would normally carve out the 1- to 2-kHz area with bell EQ. But in quieter passages, that same mid dip could cause the vocal to lack intelligibility and appeal. Multiband compression to the rescue!

With multiband compression, the signal is split into several frequency bands that can be compressed individually. First, bypass all but one of the frequency bands. Then, "Solo" that band, and carefully choose the lowest and highest frequency you want to work on (1259–3414 Hz in Figure 30.1). Next, set the "range" parameter to the maximum amount of compression (–7 dB). By lowering the "Threshold," the compressor will only attenuate notes with too much mids. As you want the process to work almost like a limiter, fast attack and release times and a hard-knee setting are often needed.

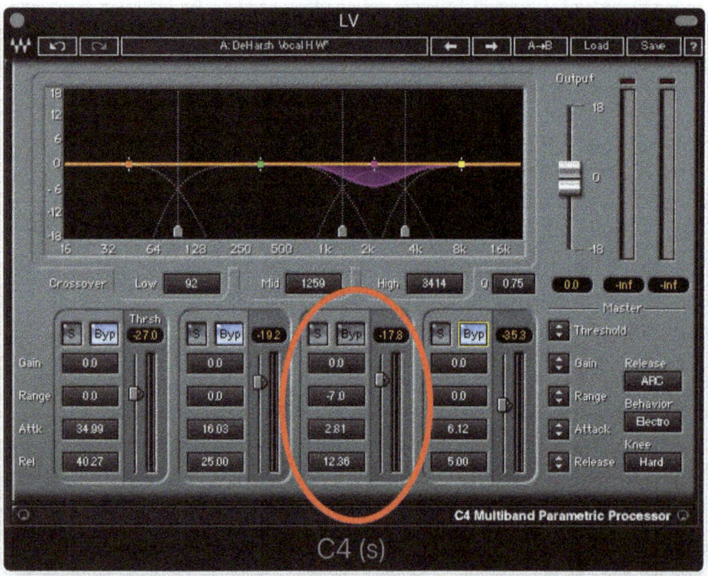

FIGURE 30.1
Waves C4 with settings to remove the harsh mids from a vocal.

As you see, a multiband compressor is actually a **dynamic EQ**; it can be used to process any problem area in a signal. A multiband compressor could also work well for vocals that have been recorded with too much of the proximity effect. Then, the low-shelf band of a multiband compressor allows reducing the bottom end of low notes without affecting higher notes.

TOP 3

MULTIBAND COMPRESSION

1. FabFilter ProMB, Waves C4/C6, McDSP ML4000
2. Budget tip: TDR Nova
3. Free: Logic Multipressor, Pro Tools Pro Multiband Dynamics

WRONG PAIRING

Randomly throwing a top 3 plugin at the vocal doesn't automatically lead to a good sound. In practice, certain devices can work adversely on a specific signal, specifically compressors. When mixing, you should always be alert for such a "wrong" pairing. In case you don't get a satisfying sound from a certain device within a reasonable amount of time, quickly replace it with alternatives until you get better results.

2. VINTAGE COMPRESSION FOR VOCALS

We'll replace the stock compressors from Chapter 23 with third-party vintage compressors. These allow for more compression, exhibit fewer side effects and offer better tone shaping. A vintage compressor can add a "sticky" quality to the vocal. A popular setup for vocal compression is the Teletronix LA2A for global leveling, followed by a Universal Audio 1176 (Bluestripe) or DBX160 for adding punch and aggression.

TOP 3

CHARACTER COMPRESSION VOCALS

1. Teletronix LA2A
2. Universal Audio 1176
3. Fairchild 660
4. Empirical Labs Distressor

Parallel Compression

Parallel compression is suited for vocals too. Similar to drums, it can boost the soft elements in a vocal very efficiently, so that the vocal's impact never ceases. For more punch and aggression, you can use higher ratios and larger amounts of gain reduction. An 1176 in "Bluestripe" mode could work well, and don't forget to try New York–style compression (as noted in Chapter 27, "Advanced Mixing Techniques: Drums").

PAINTING WITH COMPRESSORS

What's better than one compressor on the vocal? Right, six! On the vocal of Coldplay's "Viva La Vida," Michael Brauer inserted an Empirical Labs Distressor for light compression. That signal was used to feed five other compressors: an 1176, a second Distressor (in "Nuke" mode), a Gates STA-Level, a Federal and a Fairchild 666. These compressors were sent to separate faders, with Brauer adjusting the balance per song section. Even in the professional world, this is exceptional. But the example clearly shows the sheer amount of creative techniques to arrive at a good sound.

TIP

FREE THOSE FADERS

Sometimes, when you want to grab a channel fader, it appears that you can't use it, as it is locked by automation. Changing its level will cause the fader to jump back to the recorded value as soon as you hit "play." This renders the fader useless for controlling volume. A good workaround is to automate the gain of any plugin that's already inserted in the channel, such as an EQ or compressor (see Figure 30.2). In case there are no plugins in the channel, any simple plugin will do, as long as it doesn't interfere with sound, such as a gain plugin or simple EQ. Now you'll have the faders available for what they are intended for, namely, intuitively adjusting volume. Beware to not automate gain *before* a process that reacts dynamically, such as compression or distortion, as this will alter the sound. A rule of thumb is automating gain of the last plugin in the channel.

By automating the output volume of the channel's last plugin (BF76 in this case), the channel fader remains available for adjusting the final volume.

Limiting

In rare cases, the vocal exhibits fast peaks, even after using one or more compressors. This can cost precious level in the mix bus. To counteract peaks, a brickwall limiter or tape plugin can be used. As a nice by-effect, limiting can push the vocal forward in the mix. Be careful, though; too much of this good will cause the vocal to sound lifeless, a few dBs are generally the maximum.

3. ADVANCED EFFECTS FOR VOCALS

When it comes to vocal effects, conventional tools like slapback delay, ping-pong delay and (longer) reverb remain the most popular tools. Often, better results can be had by trading in stock plugins for third-party offerings, as the artifacts of certain classic devices may have been modeled more precisely.

FIGURE 30.3
Ping-pong settings with Waves H-Delay.

> **TOP 3**
>
> **VOCAL DELAY**
>
> 1. Soundtoys Echoboy
> 2. Soundtoys Primal Tap
> 3. Waves H-Delay (see Figure 30.3)

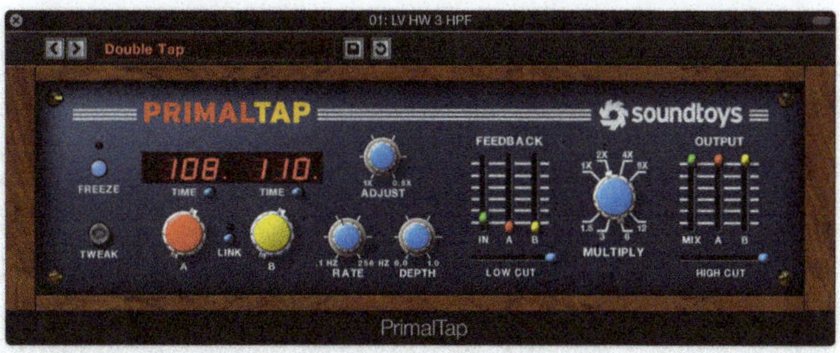

FIGURE 30.4
Slapback echo with Soundtoys' "Primal Tap." This plugin is inspired by the Lexicon Prime Time, an early digital delay from 1978. By doubling delay time, the unit's frequency response is divided by two, resulting in filtered and more interesting repeats.

TIP

OFF-BEAT DELAYS

Most delay plugins have a switch for tracking the tempo of the song (host tempo). By default, this is usually on, resulting in beautiful rhythmic echoes. In case you're not necessarily looking for rhythmic delays, off-beat echoes could work too. That's because rhythmic delays often coincide with loud hits of the drums or other instruments, causing them to be masked. To compensate for this, you can, of course, increase the delay's volume, but there's only so much space in the mix. As an alternative, try switching off "host tempo" and scroll through the milliseconds scale to find a delay time that **feels** good.

▶ Reverb

Arguably, most used reverbs in pop music ever are the EMT140 (plate) and Lexicon 224 (X) ("Rich Plate" and "Rich Chamber"). Emulations of these vintage devices can be found with many different manufacturers. In case you lack the budget for additional plugins, it will not be hard to find (free) impulse responses of these effects on the internet. The IRs can be loaded into your convolution reverb of choice.

TOP 3

VOCAL REVERB

1. EMT140
2. Lexicon 224
3. Bricasti M7, Valhalla, Audio Ease Altiverb
   Free: Pro Tools Reverb One or Air Reverb, Logic Space Designer or Chroma Verb

How to Decide on the Best Reverb

When browsing through the various reverb presets, it may be hard to hear differences or even tell them apart. What exactly do you listen for? This trick could help: temporarily exaggerate the reverb amount. Or even listen to the reverb-only signal, both in solo **and** in context of the mix. This can be done by changing the vocal channel's post-fader send effect into pre-fader and pulling back the fader. Then, similar to regular instruments in the mix, reverb parameters can be adjusted until the color of the reverb sounds good in the track. If needed, use EQ to cut any frequencies that aren't helpful, such as sub lows, top highs or ugly-sounding mid peaks.

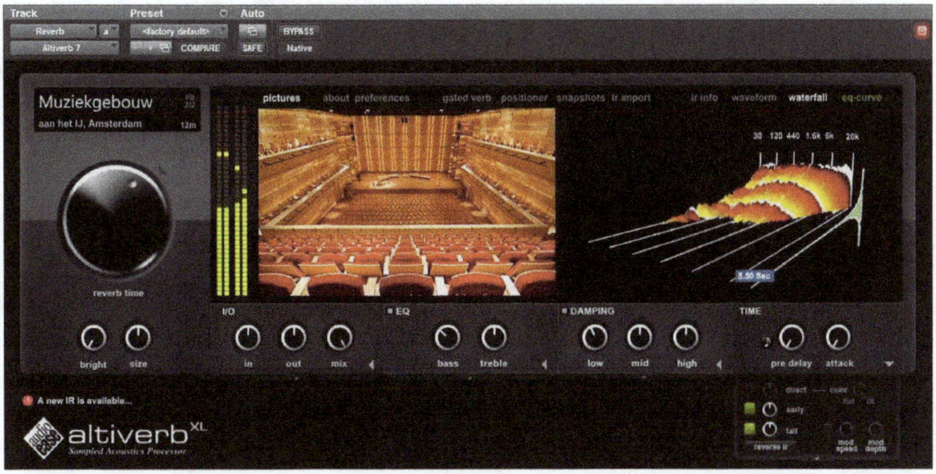

FIGURE 30.5
Audio Ease "Altiverb" convolution reverb.

TIP

SUDDEN MUTES OF REVERB AND DELAY

In pop songs, a sparsely arranged section may sometimes follow a loud and dense section. In such an intimate section, most instruments may have ceased playing, and it's only the lead vocal and accompanying chords that are left. If you mute the vocal's reverb and delay at such a spot, you'll notice that the vocal suddenly comes forward. The absence of effects adds to the vocal's expression, even if reverb and delay weren't loud in the first place. This trick could help the vocal's delivery in a full stop right before the chorus too.

▶ Harmonizer: The Secret Weapon of Many Mixers

Harmonizers produce multiple versions of the input signal that are detuned, for example, octaves or fifths. For mixing applications, harmonizers are often used for detuning in cents (one cent being 1/100th of a semitone) for thickening and widening. Now there are two brands that introduced legendary harmonizers in the 1970s: Eventide and AMS. The Eventide H949 was responsible for the piano sound in David Bowie's "Ashes to Ashes," while the AMS DMX-1580 (see Figure 30.6) was used for Phil Collins' vocal on "In the Air Tonight" and "Mama."

A popular setting for vocals is –6 cents, 5 ms delay for the left channel and +6 cents, 7 ms delay for the right channel (see Figure 30.7). This can make the vocal

FIGURE 30.6
AMS DMX-1580S harmonizer.

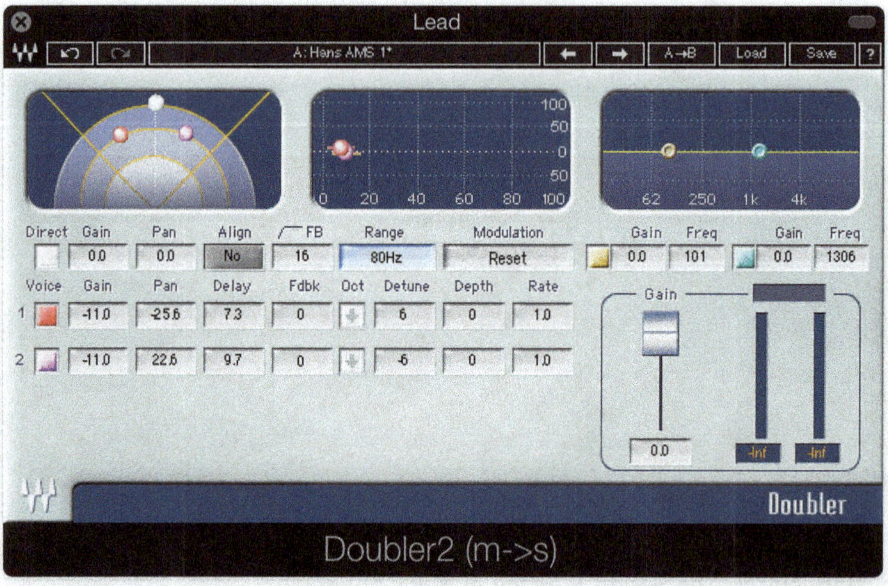

FIGURE 30.7
Waves' "Doubler 2" as an insert effect: the "Direct" signal passes unchanged (white box top left), while Voice 1 and Voice 2 are added at –11 dB. Both are detuned by 6 cents, delayed by a few milliseconds and panned at 9 o'clock and 3 o'clock.

sound as if it were double tracked, albeit with a mechanical twist. The sound of a harmonizer could remind you of a chorus, although it lacks the typical cyclic variation of a modulation effect. A harmonizer sounds solid, and that's why it has become a secret weapon for many mixers. Just a little bit of the effect on the vocals can subtly add magic, without being noticed consciously.

TOP 3

HARMONIZERS

1. Eventide H3000
2. Soundtoys Microshift
3. Waves Doubler, DMG Pitchfunk, Sonnox VoxDoubler, iZotope Vocal Doubler (free)

4. THE ORDER OF EFFECTS

The order of effects greatly influences the sound, especially with devices that react to volume, like compressors and distortion. What is a good sequence for effects on the vocal?

1. **Tuning** (AutoTune, Melodyne, Waves Tune): these processors do their work best on clean and untreated signals, so they should be positioned on top of the chain.
2. **Surgical EQ**: in "cut before, boost after" fashion, the signal is cleaned from any unnecessary energy before hitting following processors.
3. **Compression**: one compressor can be used for global leveling, optionally followed by an additional compressor for adding aggression.
4. **Multiband compression** (optional): this could be positioned either before or after compression. An advantage of an early position in the chain is that regular compression can now work on an almost "finished" signal. At the same time, regular compression could counteract the effect of multiband compression. That's why you could also position this device post regular compression.
5. **Character EQ**: after compression, the vocal can be given its final tone.
6. **Limiting** (optional): short peaks can be removed with a limiter (but be careful though!).
7. **De-esser:** sibilance is at its loudest due to compression. Post-compression, it is easier for a de-esser to detect *s*'s.
8. **Delay, reverb and other effects**: these appear late in the chain, as they should be fed with a "finished" signal. They can be applied as an insert effect or in an aux.

FIGURE 30.8

Every recording calls for its own specific approach, this is just an example of an order that could work. More important than just carbon copying this chain for every mix is the thought process that preceded it. Instead of asking, "**How** can I do this?" you should ask yourself, "**Why** am I doing this?"

5. ADJUSTING PITCH AND TIME

Talent has been the key element for the exemplary vocal performances of artists like Brian Wilson, Freddie Mercury or Whitney Houston. But talent alone wasn't enough. In the analog days, vocal sessions comprised endless take recording and meticulous comping. Only when the producer, artist and engineer were satisfied was the vocal called final. Although this labor-intensive method is still a good recipe for getting the ultimate vocal, there are also other ways to arrive at a good vocal quality. In this section, we'll go through the options for correcting pitch and time.

AutoTune and Melodyne

In 1997, US company Antares released an early version of "AutoTune" (see Figure 30.9). For the first time, a vocal could be tuned automatically and in high quality. A few years later, German company Celemony released "Melodyne." This software allowed altering pitch, timing and vibrato on a per-note basis. Both these inventions have caused a revolution, not only for the workflow in the studio but also for the sound of pop music.

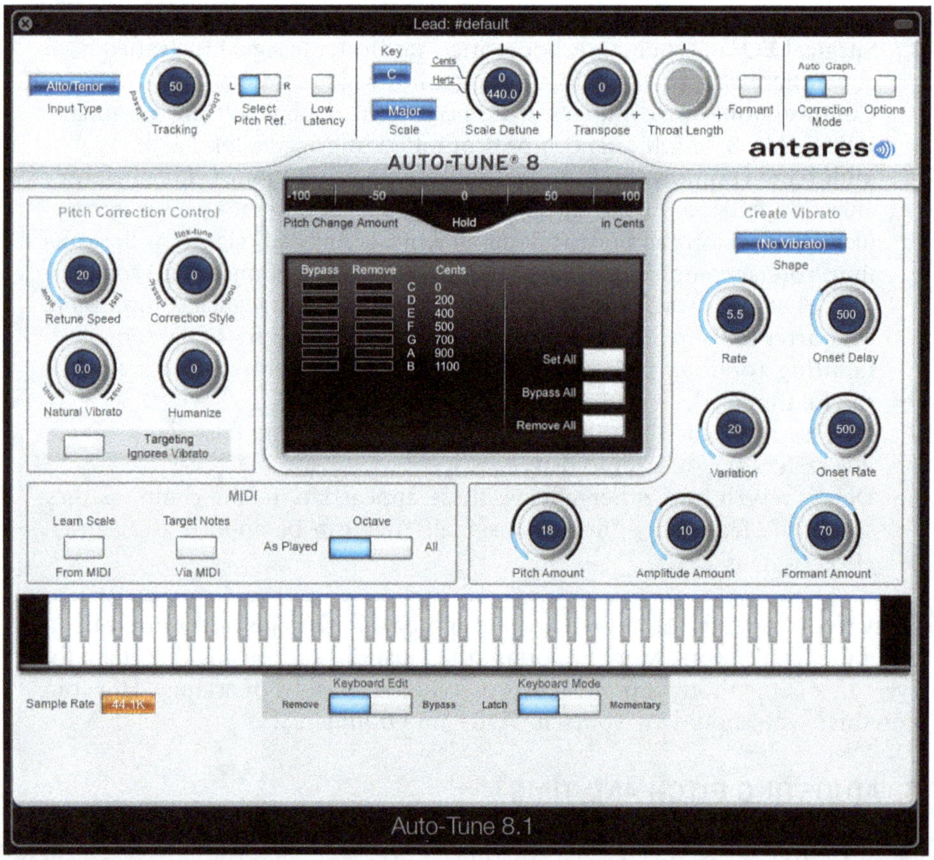

FIGURE 30.9
Antares AutoTune: "Retune Speed" determines audibility of the effect.

Tuning for Sound

Most people recognize the AutoTune effect from Cher's "Believe," rapper T-Pain's music or EDM (Electronic Dance Music) vocals. Less apparent but ever so effective is its use by artists like Adele, Ke$ha, Michael Bublé, Katy Perry, Adam Levine, Bon Iver, Sufjan Stevens and Robert Plant. They may not use Auto-Tune for its gimmicky sound but, rather, for pitching *almost* perfect notes to *just* perfect. As a by-product of tuning, overtones get arranged more evenly, which provides for a tight and cutting vocal **sound**. So the sound argument can be just as important as the tuning argument.

Pitch plugins can be divided into two categories: **real time** and **offline**. Auto-Tune is in the real-time category, just as Izotope Nectar, Waves Tune and Logic's "Pitch Correction." With offline pitch plugins, such as Melodyne, Logic's "Flex" mode or AutoTune's graphical mode, the vocal notes are shown in a piano-roll type of display and can be edited individually.

The original incarnation of pitch and time software allowed tuning **monophonic** sources only, such as a single vocal, saxophone, trumpet and so on. But since Celemony invented the "DNA technique," it has now become possible to tune notes **within** chords. In case one of the vocalists has sung a wrong note in a backing vocal track, simply dragging it to the desired pitch will fix the issue.

TOP 3

VOCAL TUNING (REAL TIME)

1. Antares AutoTune
2. Izotope Nectar, Waves Tune
3. "Pitch Correction" in Logic (free)

▶ *Set and Forget: Real-time Pitch Correction*

Working with real-time pitch plugins couldn't be easier. Just insert the plugin on the vocal track and chose the right key. Every note will now automatically be corrected to the nearest tone in the scale. "Retune speed" determines how fast the correction process takes place. The fastest setting leads to the characteristic AutoTune sound, while slower settings will cause the process to take place relatively inaudible. This delayed tuning may cause any off-pitch notes to slip through, however.

BLUE NOTES

Blue notes float between adjacent tones in the scale. In the key of C, for example, they glide between D# and E, and F, F# and G. Blue notes are not only used in blues but also in pop. Since a pitch plugin will correct blue notes toward the "right" tone in the scale, this causes the melody to change. If you don't want that to happen, then the blue notes can be bypassed in the keyboard image in the plugin.

TIP

THAT SOUND

There are many good real-time pitch plugins on the market today. For subtle corrections, many of them will do the job nicely. However, if you're looking for the original, extreme effect, the original version 5 software of Antares is the one with **that** sound, not unlike other early digital gear like the Fairlight, samplers from Casio and Ensoniq and the reverbs from Lexicon and AMS. By popular demand, Antares have built in a version 5 "classic mode" in newer versions of the plugin. Never forget to dial in sufficient top end on an AutoTuned vocal. This will expose the artificially arranged overtones very effectively.

How much of this tuning technology you use is up to personal taste. But producing a record that is literally "out of tune" with its genre **can** be avoided.

TOP 3

VOCAL TUNING (OFFLINE)

1. Celemony Melodyne
2. Antares AutoTune (Graphical Mode)
3. Flex Pitch in Logic (free)

▶ *Total Control: Offline Pitch Correction*

If you want precise control on a per-note basis, you'll need **offline** pitch software. Not only allows this altering the basic pitch of individual notes, but also the onset pitch, pitch drift, vibrato, volume, formant, timing and length. By changing the timing and the length of a note, adjacent notes automatically stretch or shrink while the rest of the notes will stay in place.

Classic Pitfalls

For vocalists, singing a diatonic sequence of notes (the adjacent note in the scale) is relatively easy. Larger intervals can be harder, however. As the vocalist (subconsciously) knows that the large interval is more difficult to sing, she may strain and therefore overcompensate. This causes the target note to turn out **sharp** (overshoot), and the upbeat note to turn out **flat**. If unprepared for the big jump, the opposite can happen: a sharp upbeat note and a flat target note (undershoot).

Notorious intervals are triads (C–E–G or A–C–E), fifths (C–G) or octaves. Generally, we "understand" the intention of the vocalist, and we're inclined to accept imperfections. But on second sight, off-pitch notes can be disturbing. Corrections can help the performance, cause the vocal to sound tight and solid and make it cut through the mix more easily.

Note: pitch perception is relative. Therefore, a note can be perceived as out of tune when, in fact, the preceding note was off.

PERFECT PITCH, LESS-THAN-PERFECT SOUND QUALITY

Correcting pitch and time is a relatively new and technically complicated process. Therefore, it can (and will) mess with audio quality. How bad that is depends on the situation. Sometimes, the software analyzes certain notes incorrectly, or notes may exhibit artifacts, even before changing. Other times, a relatively large adjustment sounds okay, while a small adjustment elsewhere causes digital artifacts. These artifacts can quickly become apparent in a sparsely arranged section, while in a dense section you might get away with notes that sound slightly unnatural. As the vocal is crucial for the production, always carefully verify audio quality after each edit.

Checking and correcting a complete vocal track might seem like quite a job, but you'll quickly become adept at recognizing and correcting notes that are off. Correcting is always worth a try, even if there wasn't a problem at first sight. Never forget, it's the lead vocal, so no stone should be left unturned.

That being said, not every musician will be enthusiastic about "correcting" his performance (. . .). Therefore, always consult the vocalist about making adjustments. Apart from this, there is also the risk of "correction frenzy." You may get caught in the correction process, wanting to correct more and more. But that could leave you with a result that's lifeless. To prevent this, try **leaving** notes that are less questionable. Or correct an individual note that's 40 cents flat by 20 cents. In case such a note sticks out like a sore thumb later on in the mix, the fix is only a few clicks away.

By using offline pitch software, the musical quality of a performance is no longer a question of intuition only. Along the way, your ears get better, and you'll develop both skills and taste. Not unimportantly, a trained ear will also lead to better decisions during recording.

ARTIFICIAL HARMONIES

Offline pitch software is also handy for constructing harmony vocals. Copying the lead to a new track and dragging the notes up or down a few semitones will get you a harmony in no time. Of course, the farther you move away from the original pitch, the more artificial the sound. But this is not necessarily a problem, as harmony voices are usually softer than the lead. In case the exact same timing of the harmony vocal sounds too perfect, you could use a spare take from the recording session. Or you could steal the lead's audio of a different song section.

▶ Changing Timing

Artists like Brittany Howard (Alabama Shakes), Katy Perry, Beyoncé and D'Angelo share an exceptional timing accuracy. Their records clearly demonstrate how the superior positioning of notes helps the performance. Every single syllable is exactly there, where the vocalist (and producer) wanted it to be. But there's more: an often-overlooked aspect of timing is **note length**. Some vocalists may tend to stretch notes nice and beautiful. But "nice and beautiful" could at the same time sound sluggish or pleasing in the wrong way. It could also lack a statement. A good example of an artist with an exceptional feel for note length is Michael Jackson. He has proved that the right duration of words can add to the groove tremendously.

While recording, you'll try to capture the best timing and groove. But even then, there's a good chance that timing and note length can be improved afterward with technical means. Cutting and moving audio is the preferred technique for correcting timing, as it will preserve audio quality. Only if this technique falls short can offline pitch software be called to the rescue.

Backing Vocals

Similar to the lead, the timing of backing vocals can be examined and changed if necessary. Dragging the lead vocal track alongside the backing vocals makes it easier to assess variations in timing. Depending on how far you want to go, every single syllable of every track can be aligned. Because firing up 24 different pitch editors isn't exactly practical, you'll most likely want to perform the edits with scissors and mouse in the main screen of your DAW.

Note:

- Backing vocal takes that end differently can easily become audible when they are (hard) panned. In case a word is too long, do not cut off its ending, but rather, remove its middle section (see Figure 30.10). Shuffle (drag) the resulting

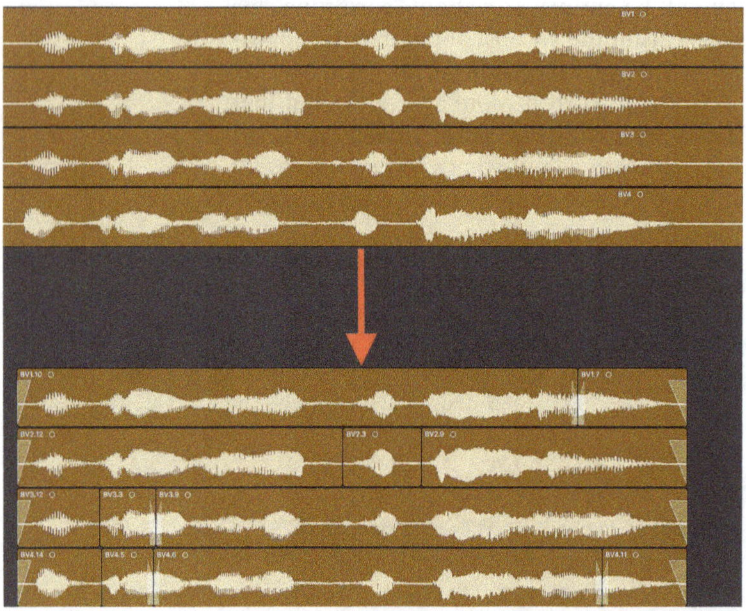

FIGURE 30.10
Backing vocals, aligned without the use of time stretching. Top: original; bottom: edited and faded. The last word of the first and fourth takes was too long. Instead of fading the words, the end of the word was separated and dragged to the left. The third and fourth take exhibit a typical phenomenon with backing vocalists; the later the take, the more they tend to rush. That's why these takes were nudged right ever so slightly.

right clip to the left clip, and apply a short crossfade. Now the word can end properly, while the note's slightly unnatural sounding middle part is (hopefully) covered by other choir voices.

- Any untimed *s*'s and *t*'s in the backing vocals can easily lead to a machine gun salvo, especially at the ending of a line. In most cases, just a few *s*'s or *t*'s are sufficient. Or maybe just one, of the lead vocal.
- For an organic sounding choir, leave in (some) timing variations. More variations make up for a choir that's wider and thicker.

Finally

How much of this tuning technology you use is up to personal taste, it largely depends on taste and genre. As a mixer-producer you are the last authority to watch over the musical material, and protect the artist from sounding off. As the notes' timing and tuning directly affect the perception of the music, these aspects are more important than any mix process.

Tips and Tricks

Move Ss to a Separate Track

Even when *s*'s are attenuated with Region Gain, they'll travel through compressors, EQs and other devices. All this processing will prevent them from sounding natural. By copying the *s*'s to a separate track (without compression and EQ), they'll sound natural. Their level can then be controlled conveniently with a fader. As a last refinement, the amount of reverb and delay on that track can be decreased or muted altogether. Although this method may seem labor-intensive, it could save you time in the end. That's because the alternative method, namely, every time revising gain on all the vocal's *s*'s, could take longer.

De-essing Reverb and Delay

Esses bouncing back and forth in reverb and delay might be disturbing. By simply inserting a de-esser before the reverb, this problem can be solved.

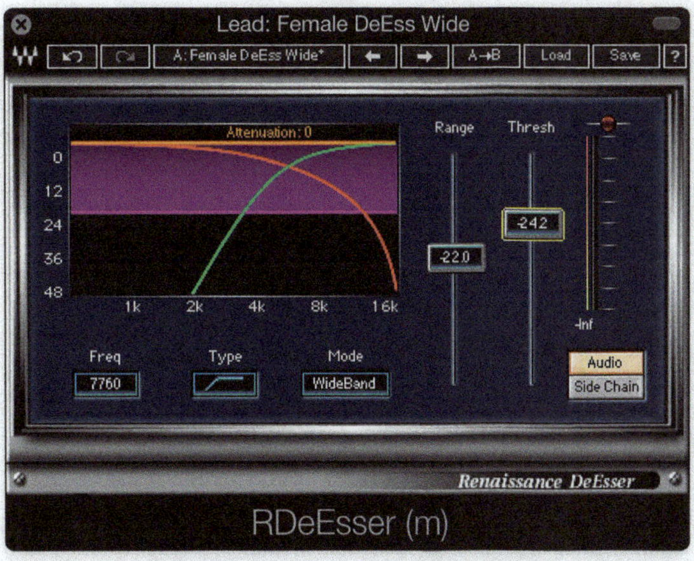

FIGURE 30.11
Popular de-esser: Waves R-DeEsser.

Reduce Stereo Width of a Ping-Pong Delay

Although the left–right echoes of a ping-pong delay provide for a beautiful, wide stereo image, they may feel disconnected from the vocal. To counteract this, pan the output channels toward the center.

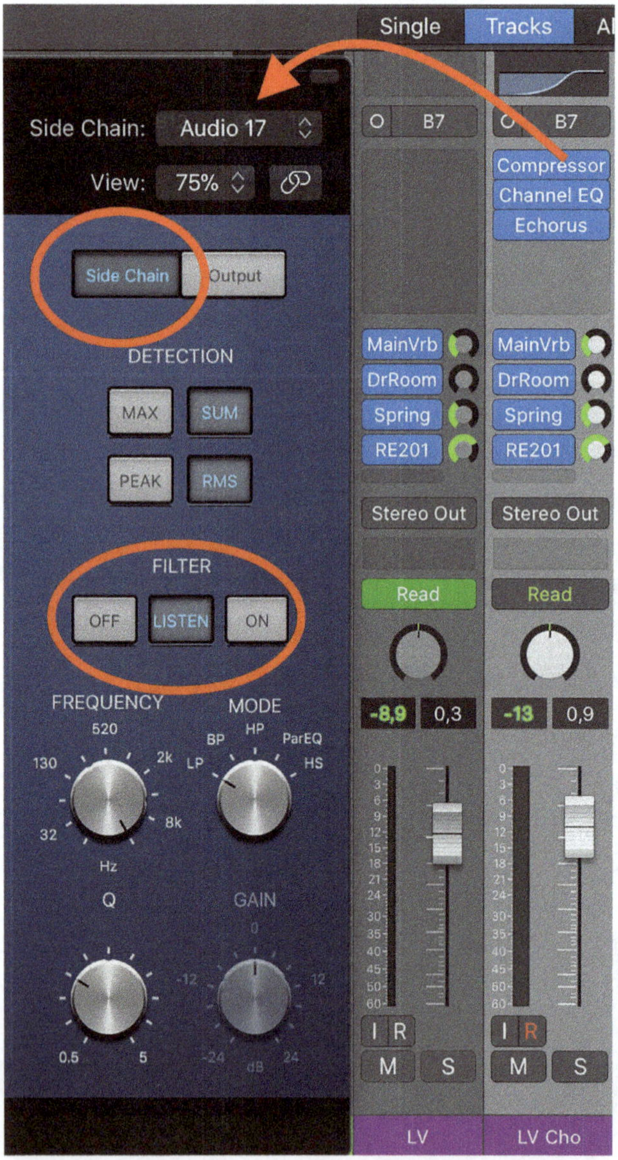

FIGURE 30.12
Hi-passed chorus on the lead vocal through a parallel channel.

Sidechaining Delay

Sidechaining a quarter note delay on the lead vocal will add to the effect's rhythmic quality.

Gating Vocal Reverb

Insert a noise gate post-reverb. By assigning the vocal itself as a sidechain signal for the noise gate, reverb will only sound when the vocalist sings. This thickens the vocal, while any "retro-sounding" reverb tail is avoided.

Hi-passing Chorus on the Lead

Although adding chorus to the lead vocal can give a nice widening, the effect can quickly become apparent and cause the vocal to become unstable. For a more subtle effect, try adding chorus through an aux, after hi-passing the signal with EQ (500–4 kHz). Optionally, this signal can be used as a send for the reverb (Figure 30.12).

Easy Parallel

In case you want to parallel-process an individual audio track without copying the audio regions, create a new audio track and insert a plugin that has a sidechain option (like a compressor or a gate) in the upper slot. Now enable "Side chain monitor" (or "Listen"), and choose the relevant input as a sidechain. This will cause the audio to be piped to the parallel channel so that it's ready for processing. Logic users: instead of using buses as a sidechain input, you can directly choose audio tracks (or instruments) from the drop-down menu.

Swap Outputs of a Stereo Harmonizer

Doing so makes the backing vocals sound bigger, as a detuned version of the left channel is positioned opposite of the original.

Vocal Doubling With AutoTune

Duplicate an AutoTune'd lead track twice. Tune one track a few cents down and the other track a few cents up. To widen the lead, pan both tracks to taste.

Vocal Doubling With Melodyne

In case you want to double the lead, but there are no additional vocal takes available, a mechanical double can be constructed by capturing the original vocal in Melodyne. After choosing "Add Random Deviation" (see Figure 30.13), every note can be given its own unique pitch and timing.

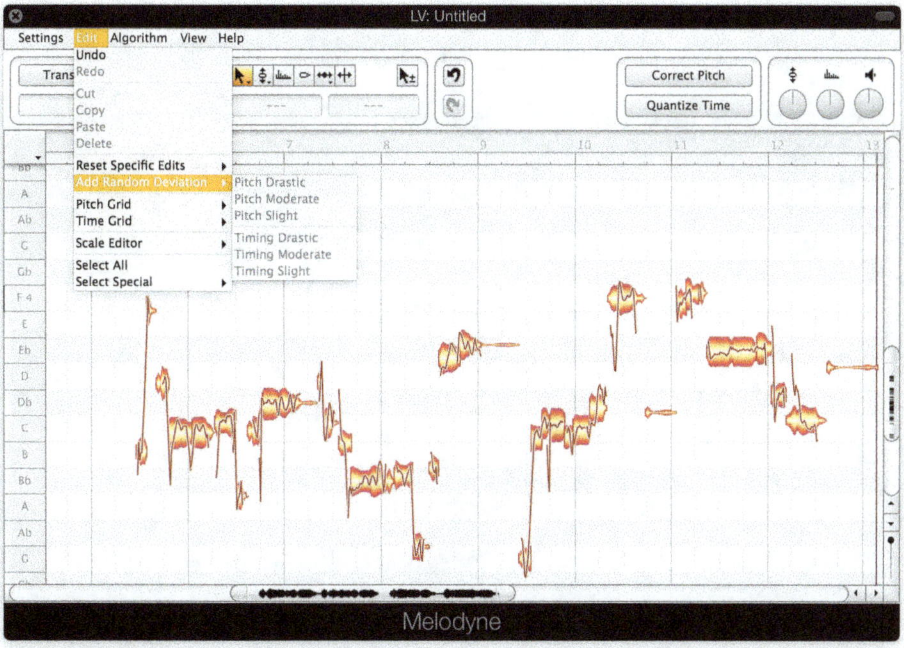

FIGURE 30.13
Vocal doubling with Melodyne.

PART IV
Mastering

CHAPTER 31
Mastering

In everyday life, people use many different formats to listen to music, like vinyl, CD, film, games and digital streaming (such as iTunes and Spotify). During the specialized process of mastering, music can be given such a specific format. The resulting *production master* can then be sent to a CD factory, a vinyl pressing plant or online distributors. The mysterious part of mastering is the sound aspect. Just before hitting the ear of the consumer, a finished mix can be made to sound better and get that typical "sounds-like-a-record" polish. What exactly is involved with mastering? How can you anticipate for professional mastering with your own mixes?

FIGURE 31.1
Mastering room at Wisseloord Studios, Hilversum, The Netherlands.

Source: Photo courtesy of wisseloord.nl.

THE MASTERING ENGINEER

A mastering engineer can be seen as an executive engineer with super trained ears. He uses specialized EQ, compression and other equipment to fine-tune the tonal balance of the mix. Although drastic interventions can sometimes be necessary, the art of mastering is often about subtle but strategically aimed corrections. For an album, the mastering engineer will sequence the songs in the right order, match them in volume and balance their timbres. This can be challenging in case the mixes were produced by different engineers in different studios.

As the mastering engineer works with a finished stereo mix, there are limitations to what can be achieved. Although a specific EQ setting can be beneficial for one instrument, other instruments might suffer. At the same time, a small tweak that's well executed can make all the difference for the mix to sound good. This is a paradox of mastering.

Not **everything** in music is **audible**.
—*Charles Rosen*

Famous mastering engineers are Bob Ludwig and Howie Weinberg (Masterdisk), Ted Jensen, Tom Coyne, Greg Calbi and George Marino (Sterling Sound), Bernie Grundman, Brian Gardner, Stephen Marcussen, Mandy Parnell and Dave Kutch. Some of them have mastered more than 200 albums a year for several decades!

Mix engineers often work with the same mastering engineer. Due to a mutual understanding, the resulting master will come out better. The mastering engineer knows about the mixer's specific preferences, while the mixer can anticipate for mastering by leaving out certain processes, or stick to certain levels. Although mix engineers (and artists or producers) sometimes attend the mastering session, it is more common for the mastering engineer to work alone. When mastering is finished, the resulting files are sent to the artist, producer or record company. If there are any comments, the mastering engineer will produce new revisions.

FIGURE 31.2
Standard procedure when mastering: is the mix mono-compatible?

Since a few years, there are also automated online mastering services. Using a form of artificial intelligence, a computer program makes decisions about the sound of a given mix. Although, at present, results are still inferior to human mastering, sound quality will definitely improve as technology advances. A great advantage of automated mastering services is their instant delivery and reduced pricing.

THE MASTERING STUDIO

Acoustically, a mastering studio meets the highest standards. It is often built as a "box-in-a-box," preventing outside noise to penetrate. Mastering speakers are often large and expensive and designed to reveal the smallest details. Linearity is a requirement not only for the frequency response but also for the room's **reverb**

time per frequency band. In case certain frequencies take longer to decay than others, this could result in resonances that blur the image.

▶ MASTERING EQUIPMENT

In the old days, boutique devices, or *in-house*-designed analog equipment (for instance, at EMI) was used. This vintage equipment continues to be popular because of its unique sound. But it's not only about tubes, transformers and vintage anymore; **digital** devices offer processing that wouldn't be possible in the analog world, like linear-phase equalizers (see Figure 31.3), brickwall limiters or devices that add specific harmonics to the signal or software that shows the spectrum of a signal over time, which allows removing certain spectral components by the eye. Almost like "Photoshop" for sound. Last, a nice feature of digital equipment is its precision and "recall-ability." This is why most mastering studios have a hybrid workflow nowadays.

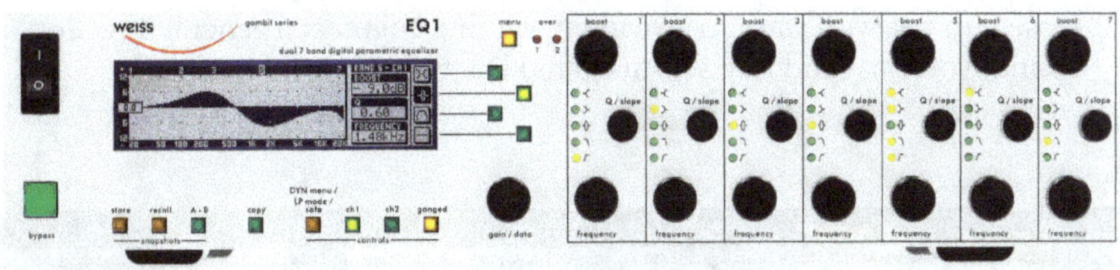

FIGURE 31.3
Weiss 7-band digital mastering EQ.
Source: Photo courtesy of weiss.ch.

▶ STEM MASTERING

In consultation with the mastering engineer, a so-called *stem mastering* can be a good alternative to regular mastering. The mastering engineer will import the separate files of, for instance, drums, bass, guitars, keys and vocals into his workstation and make a mix. This arrangement allows for more drastic changes than ever achievable with a stereo file.

LISTENING COPY AND MASTER

After finishing a mix, it can be confusing for an artist if the volume of his music appears to be lower than other music. That's because the mix hasn't been mastered yet. Therefore, mix engineers often bounce two versions of the mix: one listening copy with limiting and the original, untreated version that's intended for mastering.

▶ HOW SHOULD A MIX BE DELIVERED TO A MASTERING STUDIO?

There are slight differences between different mastering studios. Before sending files, always verify their exact delivery requirements. The following are common requirements:

- **High-quality audio files.** This means bit depth should be as high as possible (usually 24), while the sampling rate should be equal to the original project. Common rates are 44.1 kHz, 48 kHz, 88.2 kHz and 96 kHz. Some mastering houses accept analog tape as a delivery format.
- **No overloads.** Most mastering studios prefer a headroom of –6 to –3 dB for the mix. This allows the mastering engineer to start processing right away without the risk of overloads.
- **Unprocessed mixes.** Never use EQ, compression or limiting on the mix bus when bouncing the mix. That's because mastering equipment has a better quality and more precision than standard mixing gear. Apart from this, mastering processes are more effective when the mix signal is clean and unprocessed. In case you want to give the mastering engineer a general direction for sound, you can send in a self-mastered listening copy as a reference.

FIGURE 31.4
SPL mastering console.

CHAPTER 32

DIY Mastering

"Never use plugins on the mix bus. Only a mastering engineer can make good decisions about the total sound of the mix. Anything you do yourself could potentially do damage." **Yes indeed, mastering is a specialized craft. Of course, a professional is likely to get the best results when treating that precious mix signal. But in practice, you may lack the budget for professional mastering. And, more importantly, when releasing online, your music must compete with music that has been mastered! Therefore, why not try to process the mix signal as good as you can yourself? This chapter shows you appropriate tools and guides you through the thought process. Along the way, mastering will reveal the flaws in your mixes too!**

Sometimes, mastering calls for specific operations, like removing noise, clicks, or pops, or adding reverb. Using special tools, a mastering engineer can work wonders in such a situation. With the average project though, it's "just" about improving sound by using EQ, compression and limiting. That's why this chapter focuses on these processes. But before we start, we must define the goals for our master-to-be.

Online, **your tracks** must compete with music that has been mastered professionally

▶ SETTING GOALS FOR MASTERING

What exactly are we heading for, and how will this influence the listener's perception? As you generally wouldn't want your music to stand out like a sore thumb in an online playlist or on the radio, the ultimate goal of mastering is to arrive at a sound that's **in tune** with other music in the same genre—or at least not "out of tune." This is where reference tracks can help. As every recording is unique, carbon copying the characteristics of one specific record is pointless. Reference tracks should rather be used to arrive at the tonal balance and volume of the genre in general.

Tonal Balance

Getting the right balance between lows, mids and highs is where we strike on the heart (and art) of mastering. By emphasizing relevant frequencies and overtones, individual instruments can flourish, which causes the mix to **appeal**. A **denser** and **wider** spectrum has a better chance of triggering all available hair cells of the listener's auditory organ.

What happens if we add energy in the various frequency bands? Well, if we max out the low frequencies, the earth will shake when the mix is played on a large system. By maximizing the mids, we'll achieve maximum aggression, excitement and loudness. Maximizing the highs provides clarity and emphasizes transients. Undeniably, these are all great and important qualities for our master to be. But adding energy cannot be endless, of course. Maxing out the lows without sufficient support of mids and highs causes the music to lack aggression and excitement. Too much mids and highs may make the mix sound loud on the radio, but without sufficient support of the lows, the mix will lack warmth or can even be painful when played on a proper system. So for our ear, frequency bands are **related**. After changing one band, other bands may need to be adjusted.

Tonal Balance and Volume

There are two ways to look at volume in mastering: **absolute volume,** being its volume in comparison to other music, and volume **within** a track (aka dynamics). In order to increase energy and impact, dynamic range is often reduced in mastering. Although this improves the audibility of softer elements in the mix, the spectrum will become denser and more linear. This causes focus to shift from the lows toward the mids/highs. A mix that sounded warm beforehand may sound aggressive now. So here we see, dynamics and tonal balance are related.

Balanced and Predictable Sound

Ugly peaks or resonances in the mix's frequency spectrum will cause it to sound unbalanced and unpredictable. In case the mix is played on speakers that happen to emphasize such a frequency, the listening experience could even be disturbing. In order to correct peaks and resonances, search and destroy can be used. After corrections, the listening experience will be **less straining**. Unconsciously, this is what you may have experienced by listening to music that has been mastered professionally.

▶ THE MASTERING CHAIN

Now that we know a little bit about the thought process and general approach, it's time to get our hands dirty. Which devices are suitable, in which order should you use them, and which settings are applicable? Unfortunately, there is no such thing as *the* solution to all mastering problems. With each mastering session, you'll need to find the best order of the best devices with the best settings. The thought process is just as important as listening. You first select the devices and then position them in an order that prevents them from counteracting. Based on what you hear, devices, settings or order can be altered. Time to look at a possible chain!

1. **Surgical EQ (for corrections)**
2. **Compression (neutral)**
3. **Compression (character)**
4. **EQ (character)**
5. **Limiting**
6. **Spectrum analyzer**

- **EQ:** before the signal is sent to a compressor, we'll remove any unnecessary energy. After compression, character EQ is used for plus curves.
- **Compression:** one compressor is used for **compacting** and **gluing** the mix, a second device can be used for **tonal shaping**.
- **Limiting:** this is used to arrive at a certain loudness.

In case **large** amounts of EQ are needed, there's a problem with the mix. Always aim to solve problems in the mixing stage, if possible.

1. Surgical EQ

The mix signal has a full frequency response and contains the combined energy of many instruments. As such, it is sensitive and precious. Often, a small boost or cut at the right frequency is just what's needed to improve tonal balance.

Hi-pass filters: Frequencies below 20 to 30 Hz don't contain any musical information but may freak out gear that's farther down the line, like compressors, amps and speakers. That's why a hi-pass filter (12–24 dB/oct) can be used to filter sub lows. As steeper filters cause larger phase shifts, it's better to choose a curve that's *just* steep enough. Sub lows don't need to be erased completely but, rather, be attenuated until in reasonable proportion to the rest of the bottom end. Note that bass frequencies are already down by 3 dB at the frequency shown!

Lo-pass filters: similar to sub lows, unnatural amounts of top highs can be tamed with a lo-pass filter that's not too steep (6–12 dB/oct).

Minus curves: broader frequency dips may clean up the mix in a pleasant way. Sweet spots can be found by scanning through the spectrum with a bell EQ. As noted before, the signal might also contain narrow frequency bands that resonate or make up for a straining/painful listening experience. Similar to what we did with single instruments, search and destroy will help you cure this.

Plus curves: although it is perfectly fine to boost frequencies with modern, surgical EQ, you don't get the character that vintage EQ or passive EQ could add. That's why we'll boost frequencies in the fourth step ("Character EQ").

Reference tracks are of great help when EQing. Switching back and forth between your own project and other music lets you meticulously investigate the differences per frequency band. Be sure to compare at the same volume!

Which EQs are suitable? With surgical EQ plugins, differences in operation are generally bigger than differences in sound quality. Popular types are BX Digital V3, FabFilter Pro-Q or stock EQ from **Pro Tools** and **Logic**.

SOLO THAT BAND

With EQs from BX Digital, FabFilter (see Figure 32.1) and others, a single frequency band can be set to **solo**. This will temporarily mute the rest of the spectrum. In **Pro Tools**, solo mode in **"EQ3"** and **"EQ7"** is activated by Shift + Ctrl + Clicking on the relevant frequency band. Be careful, as listening to a single frequency band for a longer period of time will cause you to lose your reference point.

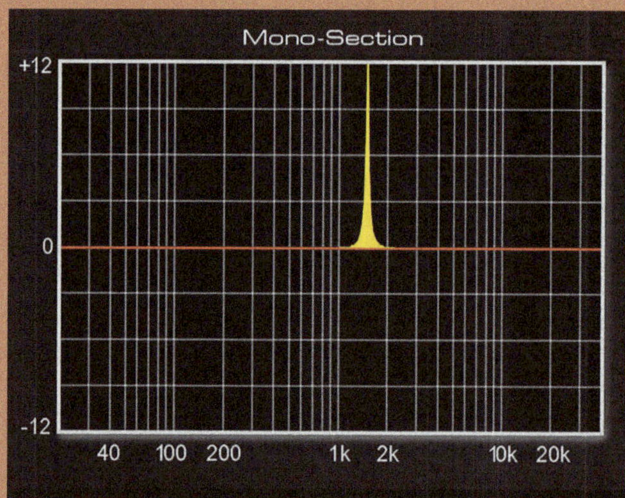

FIGURE 32.1
BX Digital V2: EQ solo mode.

2. Clean Compression

Compression may provide for either mix glue, tightness, punch, urgency or excitement. Or maybe all at the same time. As artifacts will quickly become noticeable, don't expect to compress more than 2 to 3 dB at this stage. Check Chapter 15, "Effects | Compression and Limiting," once again on how to find the best settings.

Note that the compressor will eagerly react to kick and bass, as those instruments contain most energy. Pumping might be the result, while the vocal hasn't been able to touch the compressor yet. If the compressor has a sidechain option, pumping effects can be reduced by enabling the (hi-pass) sidechain filter (see Figure 32.2).

FIGURE 32.2
Waves API 2500: With "Tone" at "Med" or "Loud," the compressor's bottom end sensitivity is decreased.

> **TOP 3**
>
> **CLEAN COMPRESSION MASTERING**
>
> 1. API 2500
> 2. Softube Weiss DS1-MK3, SSL Bus Compressor
> 3. Focusrite Red-3 (d3 in Pro Tools)
> 4. Budget tip: TDR Kotelnikov

3. Character Compression

After applying clean compression, it's worth a try to see if a vintage compressor can bring something to the table. Not only for compression, but also for shaping the tone. Even without gain reduction, the signal passes nonlinear circuitry, thereby adding color. By increasing both threshold and input gain, compression decreases while distortion increases. By adjusting the dry/wet balance, this coloration can be added in parallel. In case the compressor doesn't have a dry/wet knob, compression can be added through an aux.

Different compressors distort the signal in different ways. How such a color fits your specific mix signal depends on many factors. In order to investigate this, temporarily exaggerate both the compressor's gain reduction and ratio. This will point your ears toward any negative by-effects. After it's become clear what the compressor is doing, dial in attack- and release settings that minimize by-effects. Last, bring back both the amount of gain reduction and ratio to the desired setting.

Be extra alert to distortion on kick and bass, as it can spoil the sensation of big bottom end in the mix. Kicks or basses that sound purer, that is, like sine waves, can be very sensitive to distortion.

TOP 3

CHARACTER COMPRESSION MASTERING

1. Manley VariMu, Fairchild 670
2. Shadow Hills Mastering Compressor (see Figure 32.3)
3. LA2A, Focusrite Red 3
4. Free: TDR Kotelnikov (see Figure 32.7)

FIGURE 32.3
The Shadow Hills Industries Mastering Compressor is one of the priciest compressors around; it has both an optical and a VCA compressor, which can be combined. Universal Audio has a software version.

Source: Photo courtesy of shadowhillsindustries.com.

MULTIBAND COMPRESSION IN MASTERING

In mastering, multiband compression can be used to correct mix problems. Let's suppose we have a mix with a bass that is too soft. By simply boosting the lows with EQ, this would cause the kick drum to become louder too. Multiband compressor to the rescue! Enable one band of a multiband compressor, let's say between 30 and 125 Hz. Lower the threshold of the band until the compressor responds sufficiently to the kick. As the compressor has now lowered the kick's volume, this needs to be compensated for with makeup gain. As a result, the bass will be louder. Aggressive settings can be used, like a high ratio (up to 100:1), hard knee and (very) fast attack and release times. Other problem areas in the mix can be processed likewise.

The question arises if multiband compression should be applied pre-compression or post-compression. Well, that depends. As regular compression could counteract on areas that you've attenuated with multiband compression, inserting the multiband device post-compression seems logical. But in our example, quite some mix energy gets rearranged. Therefore, inserting the multiband device pre-compression has the advantage of regular compression working on a mix that has a proper balance of kick and bass. Chapter 27 has a list of popular multiband compressors.

4. Character EQ

This step allows for compensating any losses caused by compression. And, our master-to-be can be given its final timbre with plus curves. By boosting certain frequencies, we can bring forward the good elements in the mix. For example, 30 to 100 Hz provides power, 100 to 300 Hz improves audibility of the bass, 800 Hz to 2 kHz adds aggression, 2 to 6 kHz adds excitement and 6 kHz and beyond provides for "air." But hey, this is mastering! Other aspects in the mix may suffer. For instance, boosting 150 Hz may bring out the snare nicely but at the same time cause the bass to sound honky. Boosting 1200 Hz might add aggression to the guitars but will color the vocal along the way and so on. At every step, you'll have to ask yourself if the advantages outweigh the disadvantages.

Due to its transparent sound and exemplary phase behavior, **passive** EQ is popular for this application. Another option would be to use **linear-phase EQ.** Linear-phase EQ is very precise, transparent and doesn't introduce distortion. Only small quantities are needed: a shelf of 0.8 dB at 10 kHz may provide for better transients and just the right amount of air.

TOP 3

CHARACTER EQ MASTERING

1. Manley Massive Passive (see Figure 32.4)
2. Elysia Museq, SPL Passeq, Native Instruments Passive EQ
3. Pultec EQP-1A
4. Free: TDR VOS Slick EQ, d2 in Pro Tools, Logic Vintage EQ

FIGURE 32.4
The Manley Massive Passive 4-band EQ must be the most popular mastering EQ ever. It is an "improved" version of the Pultec EQP-1A. The "Pultec trick" can not only be executed on the lows but also on the highs, allowing high-frequency boosts with fewer sibilance problems.

Source: Photo courtesy of manley.com.

SONIC CANDY

Other than EQ, there are devices that can shape the spectrum in a special way. For example exciters from Aphex, SPL and BBE, both in hardware and software. Kush Audio has "Clariphonic" (see Figure 32.5), which, in fact, is a parallel equalizer. Other examples are "Identity" from Zynaptiq, "Cobalt Saphira" from Waves and "Soundbrigade" from SKNote. Although none of these devices can be called a standard in mastering, they could bring useful colors to the table. No stone should be left unturned when searching for that last bit of extra harmonic content. Insertion post-compression seems obvious.

FIGURE 32.5
Kush Audio Clariphonic.

Source: Photo courtesy of thehouseofkush.com.

▶ TAPE

In case you're curious about what tape can do to the mix: now is the moment! The saturation of tape can bring warmth, take off a bit of that digital edge and make the total sound more pleasing. Increasing the input level to exaggerate the saturation effect allows finding out what tape does to the mix. At the moment you hear distortion, you've probably gone too far.

A great advantage of tape is that it will flatten peaks. Therefore, the brickwall limiter that we'll use later in the chain can be set to work gentler. A tape machine can either be inserted **pre or post** our character EQ. Pre-character EQ has the advantage of being able to compensate for possible losses, post-character EQ has the advantage of the tape machine limiting the final signal.

When tape is used in addition to other vintage gear, the total amount of distortion can easily exceed certain limits. It may be necessary to back off compression in such a case.

There are great choices of tape plugins with good reputations: Universal Audio (Studer A800, Ampex ATR-102, Oxide, Fatso), Waves (Kramer Tape, Abbey Road J37), Slate Digital (Virtual Tape Machines), U-He (Satin), Softube (Tape), SKNote (Roundtone) and Massey "Tapehead" (only for Pro Tools).

▶ DE-ESSING

The more compression and distortion on the mix, the louder the sibilance of the vocals. Again! The only advantage of loud *s*'s is that they can be easier recognized at this point in the chain (. . .). Be careful with de-essing in mastering: the process can quickly become audible when the device grabs other high-frequency sources, such as cymbals or snare.

TOP 3

DE-ESSING MASTERING

1. Softube Weiss DS1-MK3
2. Eiosis E2 De-Esser
3. Sonnox Oxford SuprEsser, FabFilter Pro-S
4. A multiband compressor (Waves C6, FabFilter Pro-MB)

5. Limiting

At this point in mastering, the total volume is probably still lower than other music. This can be compensated for with a brickwall limiter. How can the optimum setting for a brickwall limiter be determined? Similar to what we did with compression, temporarily exaggerate the amount of limiting will reveal its negative side effects. Then, try to find attack and release settings that generate the least amount of distortion. Generally, 1 to 5 dB of gain reduction is the maximum before side effects become apparent. A brickwall limiter automatically maximizes its output level, leaving you with a signal that's touching the 0-dB mark, without overloads. Remember, to prevent digital distortion, a brickwall limiter is **always** the last device in the chain. Most popular brickwall limiter ever must be Waves' L2. Since its release in 2001, a great deal of pop music has been mastered through this device, be it in software or in hardware. Its musical character has contributed to a classic status.

As stated before, compression and limiting **will** alter tonal balance. This is not necessarily a bad thing, as long as you're conscious about the effect. But it may require changing EQ and/or compression.

A brickwall limiter is **always** the last device in the chain.

And there's a last thing to consider: compression and limiting **will** alter the macro dynamics of the mix! In case the soft sections have now become as loud as the loud sections, this can be counteracted by automating the **threshold** of the brickwall limiter. **Increasing** the threshold in the soft sections reduces limiting and, as an aftermath, loudness. In fact, this technique restores some of the original dynamics, which is always an advantage for delicate song sections.

TOP 3

BRICKWALL LIMITERS

1. Softube Weiss MM-1, FabFilter Pro-L (see Figure 32.6)
2. Izotope Ozone, Sonnox Oxford Limiter
3. Waves L2
4. Budget options: brickwall limiters in Pro Tools ("Maxim") and Logic ("Adaptive Limiter")

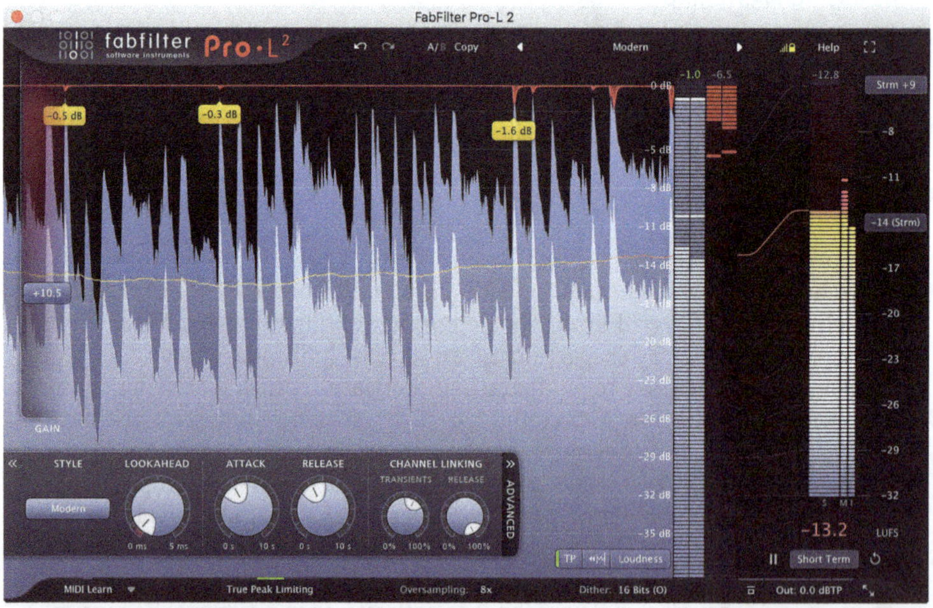

FIGURE 32.6
FabFilter Pro-L2 brickwall limiter.

6. Spectrum Analyzer

In order to properly compare the mix with other music, a spectrum analyzer should be inserted last in the chain. How can a spectrum analyzer help with mastering?

Lows: This is the area where a spectrum analyzer can help best. Within a certain genre, the lows may exhibit less variation than other frequency bands. Moreover, different tracks might display (almost) identical bass spectra!

Highs: Although we can hear frequencies up to 20 kHz at birth, the upper limit decreases with age (and hearing damage). So before adjusting the 10- to 20-kHz area, check your own hearing by using a tone generator in your DAW. In case you notice a reduced sensitivity for the highs, this could be due to hearing damage, but the ear canal could be blocked by ear wax as well. This can easily be removed by a doctor. Anyone with a reduced upper limit can prevent making mistakes in the highs by looking at the spectrum analyzer.

Mids: from this area, it may be harder to draw conclusions. There's simply too much variation within different records. Fortunately, our ears are doing a good job in the mids!

Overall spectrum: Within a current genre, you may find the course of the spectrum to be similar between different records. For example, after peaking in the 40 to 125 Hz area, the spectrum often decays toward 4 to 10 kHz, with an even quicker decay when approaching 20 kHz. The decay rate between the lows and highs is of particular interest here, as it gives you an idea of the tonal balance we discussed earlier. Slower decays indicate mixes that sound (too) dense, painful or aggressive. Faster decays indicate mixes that are bass-heavy or unexciting. It would be impossible to describe the "ideal" picture, as this differs per genre and evolves over time. Rock might gravitate toward a horizontal line, while R&B might expose a mid dip within the decaying line. Different analyzers will show different responses, so each device requires learning its specific picture.

▶ ALMOST DONE!

Now before bouncing the master, there are a few things to check:

- **Distortion:** "vintage" may seem like the magical solution for all problems, but distortion increases with every vintage device added. Too much distortion could make the mix sound as if it is played on multimedia speakers.
- **Dynamics:** as a sum of all processes, dynamics might have suffered too much. Therefore, bypass every dynamics device and verify its contribution to the total gain reduction. Also verify if transients of kick and snare are sufficiently in shape.

Now bypass the complete chain, and compare the master to the original mix. At the same volume, of course. What are the gains, what are the losses? In case a certain improvement has caused too many side effects, you'll have to go back to the drawing board and try to come up with solutions.

▶ FAQ

What Happens When You Master on Speakers That Aren't Flat?

In that case, taking down an ugly frequency could mean that you're actually trying to compensate for the nonlinear response of your speakers. This will mutilate the master. That's why it is important to use speakers with a flat frequency response when mastering. A flat speaker response allows the spectrum of a mix to be corrected until it is balanced. This is the best chance for the mix to sound predictable on **all** speaker systems.

Should You Use Special Mastering Software?

Not necessarily. Although professional mastering software offers a streamlined workflow for common mastering tasks, there is no difference in sound quality with popular DAW's like Cubase, Logic or Pro Tools.

Can You Master While Mixing?

It depends. Professionals separate mastering from mixing. Either side of the production process takes the most of an individual, and wearing two hats at the same time can be confusing. With mastering, the engineer should take on the **consumer's perspective**. In case you're also the mix engineer, a pause might be needed in order to detach yourself from the intricacies of the mix.

In the **ideal** world, negative side effects of mastering can be counteracted in the mix.

On the positive side, running a mastering chain on the mix bus is just fine, technically. Mastering in the mix allows you to undo certain by-effects of mastering. For example, when applying compression and limiting, low-level signals in the mix will be boosted, like reverb, s's or cymbals. As every negative aspect can be counteracted in the mix immediately, mastering will be most effective. Professional mastering engineers would die for the possibilities!

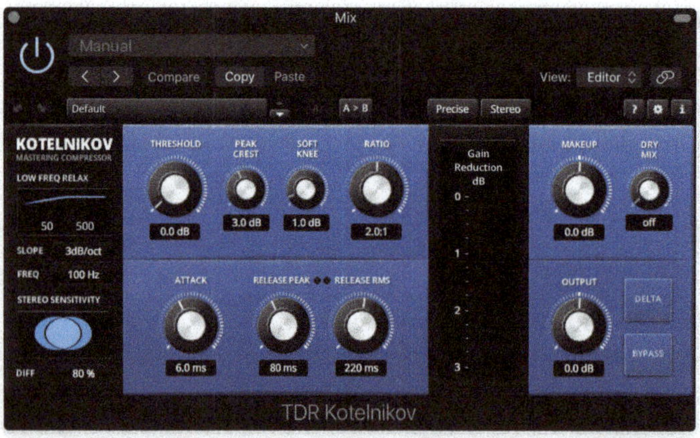

FIGURE 32.7
TDR Kotelnikov (free). Refreshingly, this mastering compressor does not imitate the behavior of vintage compressors. Release can be set for high and low frequencies independently.

Can You Use Headphones for Mastering?

Yes! Good-quality studio headphones have a (very) flat frequency response, even in the sub-low frequencies. On top of that, the fine detailing of headphones prevents noises, ticks or pops to go by unnoticed.

Other Aspects of Mastering

▶ INTER SAMPLE PEAKS

A brickwall limiter protects the mix bus from overloading, no matter what signal you throw at it. At least, that seems so. Actually, overshoots **can** occur, and we call them *Inter Sample Peaks*. "ISPs" cause digital distortion when D/A converters in cheap CD players and laptops produce rounding errors when converting the signal. Distortion caused by ISPs can also occur when converting a high-quality file to a lossy format, like MP3. A normal peak meter won't show ISPs; they must be monitored with a *True Peak Meter*. By taking down the *ceiling* (or output level) of the limiter, the ISPs will disappear. Common figures are –0.1 dB, –0.5 dB or –1 dB. Switching on *oversampling* in the limiter might also help reduce these peaks.

"MasterMeter" in **Pro Tools** (see Figure 32.8) lets you monitor True Peaks, while **Logic Pro** has a True Peak setting in "MultiMeter."

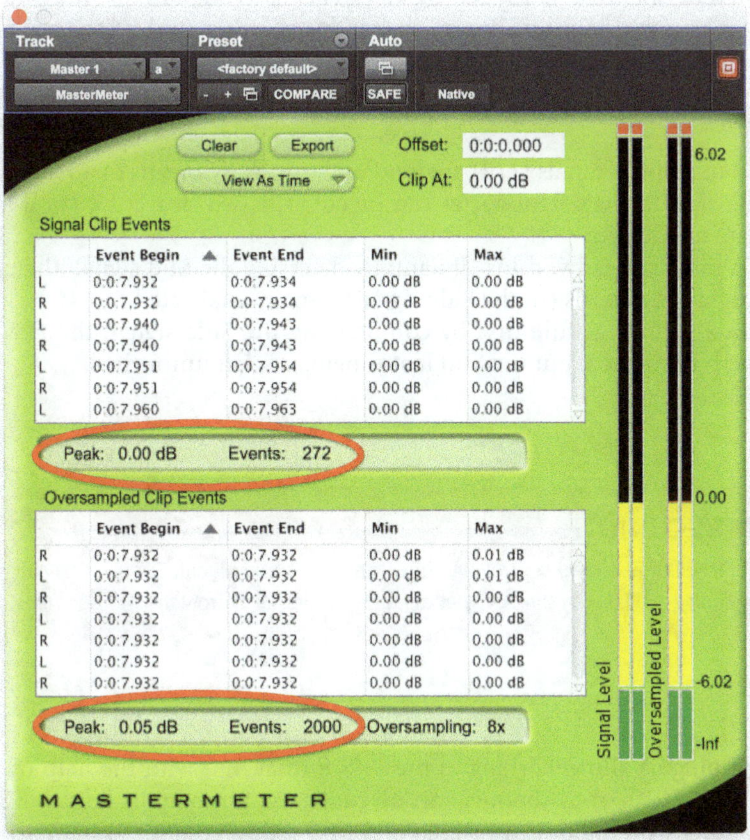

FIGURE 32.8
Pro Tools: Mastermeter.

▶ DITHER NOISE

As noted in Chapter 10, "Recording on the Computer," D/A converters have trouble converting low-level signals. As there are fewer bits available for describing the signal, the converter can become indecisive about the signal having either one value or the next value. This results in rapid switching, which sounds like distortion. To help the converter, the signal level can be increased by adding noise. As the converter has now more substance to work with, switching is reduced. At the expense of noise entering the signal, of course. With normal monitor levels though, dither noise will be inaudible.

Dither noise should **only** be applied to the final production master or when converting files from a higher bit depth to a lower bit depth (24 bit ->16 bit). Most brickwall limiters have a dither option.

▶ M/S MASTERING

As noted in Chapter 24, a stereo signal can also be coded in the Mid-Side (M/S) format. In mastering, this comes in handy, as either mid instruments (often kick, snare, bass, lead vocal) or side instruments (like drum overheads, backing vocals, synths) can now be processed independently. Let's look at a few examples:

- In a mix, the cymbals (L/R) are loud and brittle. By means of an M/S EQ, the Side signal's high frequencies can be cut. This will leave the vocal and the snare unaffected.
- In a mix, the lead vocal is sibilant: by de-essing the Mid signal, the cymbals and backing vocals (in the Side signal) remain unaffected.
- A mix sounds a little dry: by compressing the Side signal, the (low-level) reverb is boosted, but the Mid instruments are left untouched.

STEREO WIDTH

The M/S format allows controlling the **width** of a stereo signal. Simply increasing or decreasing the level of the S signal will either widen or narrow the stereo image.

Not all plugins allow working in the M/S format. But with the help of Voxengo's "MSED" (free), a channel can be run in M/S mode. In **Logic,** "Channel EQ" allows working both in stereo and M/S (see Figure 32.9). Other popular plugins that offer M/S compatibility are H-EQ (Waves), Pro-Q (Fabfilter) and most plugins of Brainworx. The Fairchild 670 compressor works in M/S mode by enabling "Lateral/Vertical" mode.

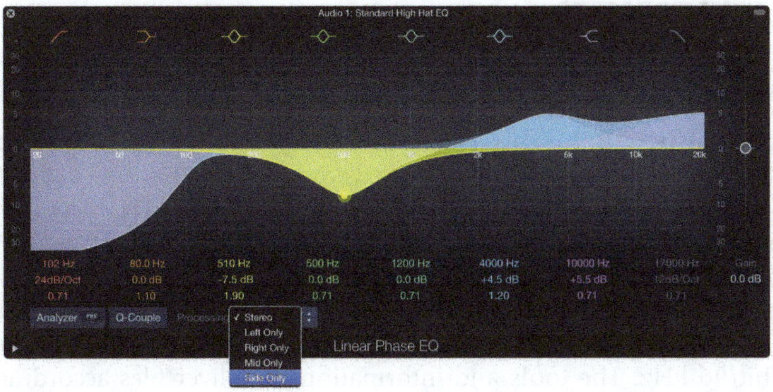

FIGURE 32.9
Logic's EQ can be set to work at any part of the stereo signal (left, right, Mid, Side).

MONO BASS

With too much stereo information in the lows, the bass may sound less tight or even unstable. As our hearing is not equipped to discern directional information in the bass area, left and right channels can be summed to mono below a certain frequency, for example, between zero and 80 to 120 Hz.

FIGURE 32.10
Brainworx Digital V2 mastering plugin. At the left, the EQ section for the Mid signal can be found, at the right, there's an EQ section for the Sides. In the middle section there's the control for summing L/R bass frequencies ("Mono Maker").

▶ TARGET FORMATS

Now that sound is finalized, the final step in mastering is printing production masters for each delivery format:

- **Digital distribution.** This is the most popular and convenient distribution format: a master can be delivered worldwide in no time at full quality. Most portals work with WAV files (16–24 bit/44.1–48 kHz). The portal may convert this file into the final distribution format, for example, MP3, AAC (Apple) or Ogg Vorbis (Spotify).
- Apple has the "**Mastered for iTunes**" label. This has nothing to do with altering sound but rather refers to the high quality of the original master, namely, 24 bit/96 kHz. The tools and information to deliver files according to the "Mastered for iTunes"-standard can be found at www.apple.com/itunes/mastered-for-itunes/.
- **Mastering for CD.** To master a CD according to the "Red Book" standard, dedicated software is needed for adding "PQ codes." PQ codes store information about track numbers and pause times between tracks. You can also add the artist name, song title and album title. Similar to ISBN codes for books, ISRC codes can be added, that help performing rights societies collect royalties. When the mastering is finished, a so-called **DDP** file must be rendered. This file is used by the factory to press CDs. Once a CD is stored in the DDP format, sound can no longer be altered. Popular software for CD mastering is Sonoris DDP-Creator, Steinberg Wavelab, Magix Sequoia and Hofa CD Burn. DDP.Master.
- **Mastering for vinyl.** Vinyl is a very "physical" format, meaning that the characteristics of both sound waves and movement of the stylus present us with certain consequences and limitations. For instance,
  - the lower and louder the bass frequencies, the larger the movement of the stylus. Eventually, the stylus could lose its track and jump out of the groove. That's why you have to be careful with the amount of sub-low energy when mastering for vinyl. Too much stereo information in the low frequencies may also lead to tracking problems.
  - as louder bass causes grooves to be wider, the playing time on one side is reduced.
  - because rotation speed is continuous, there's less room for accurately describing the signal when the stylus approaches the end of one side. For songs that appear later in the sequence, quality of the bass will suffer accordingly.
  - vinyl might have trouble reproducing certain (in-harmonic) high-frequency sounds.

Once the listener is playing his record, sound waves have been **transduced** twice: once in the factory when cutting and once on playback. This is the reason why vinyl has a fundamental amount of distortion and a limited dynamic range. To anticipate for this, the mastering studio will often produce a dedicated vinyl master with less, or even no, limiting. There's also another reason for less limiting with vinyl: vinyl lovers will generally favor good quality sound over loudness.

CHAPTER 33
Just One Louder

Producers want to sound as "hot" as possible, bands want attention and record companies want their artists to sound as loud as the competition—or louder. There are massive advantages to compression and limiting: it makes music compact and injects it with excitement and urgency. Loud mixes are appealing and translate better when played softly, especially in a bar, a shopping mall or when hovering. This chapter looks at techniques for making the master as loud as possible. But we'll also raise the big question: Isn't 10 better than 11?

▶ THE LOUDNESS WAR

Since the 1980s, pop music has become louder and less dynamic. A prelude to this trend was orchestrated by SSL when it released mixers that featured a compressor on **every channel**, plus one on the mix bus. A further boost to the volume of pop music was given by digital compressors and limiters in the 1990s. It was the times of "hey, we have the tools, let's make our tracks as loud as possible, that's fun." This resulted in notorious loud albums such as Oasis's *What's the Story Morning Glory?*, The Stooges' *Raw Power*, Avril Lavigne's *Avril Lavigne*, Kid Rock's *Rock and Roll Jesus*, U2's *How to Dismantle an Atomic Bomb*, Arctic Monkeys' *Whatever People Say I Am, That's What I'm Not* and Franz Ferdinand's *You Could Do so Much Better*. Blinded by the sheer amount of compression available, everybody jumped on the loudness train without realizing that piling up compressors could cause the music to sound harsh, cold and tiring. This is what has come to be known as the **Loudness War** (see Figure 33.1). Similar to what digital reverb did in the 1980s, or 'stereo' in the 1960s, this is another example of technology dominating the sound of pop music temporarily.

Without **soft** notes, **loud** notes will cease to have impact.

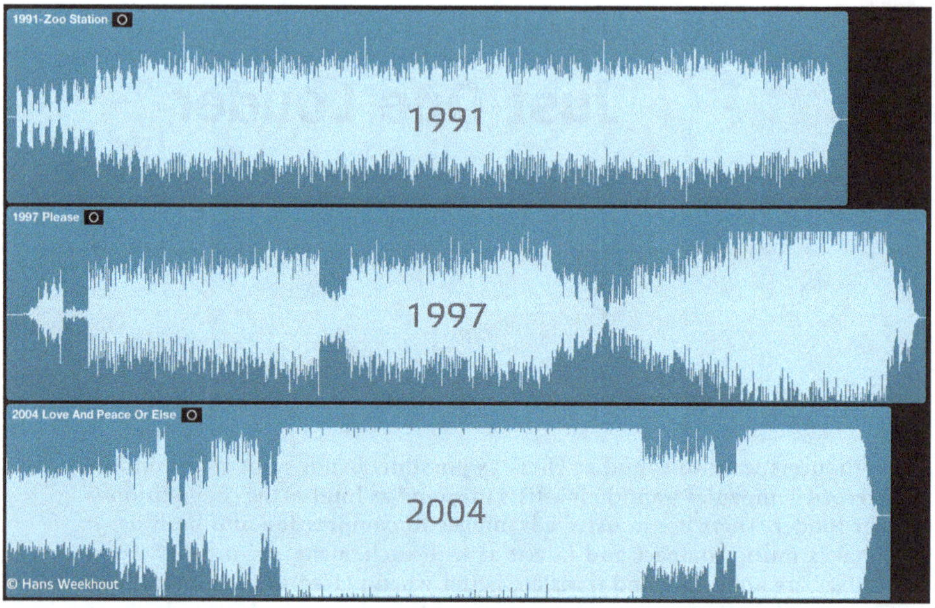

FIGURE 33.1
Despite their peaceful intentions, U2 have contributed to the Loudness War too.
Top: waveform of the song "Zooropa" (1991)
Middle: "Please" (1997)
Below: "Love and Peace or Else" (2004)
All songs were mixed by Flood (The Killers, Foals, Goldfrapp, Editors). Arnie Acosta mastered in 1991 and 2004 while Howie Weinberg mastered in 1997.

IS 10 BETTER THAN 11?

To answer that question, we'll compare Johnny Guitar Watson's "A Real Mother for Ya" (pre–Loudness War) with Snow Patrol's "Run" (2003). With Snow Patrol on a low volume, every single detail can be heard: the arrangement sounds complete. Plus, the mix sounds energetic; it is tempting to turn up the volume. However, when you turn up the volume on a good pair of speakers, the listening experience is less rewarding. As the frequency spectrum is filled out, the mix is slightly painful and tiring to listen to for a longer period. As more and more instruments are introduced during the song, the actual volume hardly increases. Note that this is an effect unknown to live situations.

Now, let's play Johnny Guitar Watson softly. Although the vocal is audible, the band fails to make an impression. As the mix seems to contain less energy, it's not inviting to turn up the volume. However, if you do so, it suddenly seems like you're standing right in between the musicians. Every hit on the snare sounds like a sledgehammer, while the vocals are truly dynamic. As a listener, this is a rewarding experience, especially if you have invested in good speakers.

Loudness War 2.0

Around 2005, certain producers and artists started to oppose against the volume competition. Contrary to the trend and favoring sonic integrity over loudness, their music contained no excessive limiting. They reasoned, if listeners can't hear the music properly, they will reach over for the volume knob, which is very valid, of course. Some notable examples include Norah Jones *Little Broken Hearts*, Belle & Sebastian *Girls in Peacetime for Dance*, Arcade Fire *Reflektor*, Daft Punk *Random Access Memories*, Bon Iver *Bon Iver*, Alison Krauss & Union Station *Paper Airplane*, Jack White *Lazaretto*, Mark Ronson *Uptown Special* and Tool *10,000 Days*.

▶ LUFS

Support for this movement came in 2010 with the introduction of a new technical standard, called *LuFS* ("Loudness Units relative to Full Scale"). LuFS is an intelligent method to express loudness as perceived by humans. The LuFS value of a track can be stored as a *tag* in the audio file.

Since the introduction of LuFS, most online streaming platforms have adopted the standard to play music at a lower volume, the **target loudness**. This obviates the need for excessive compression and leaves room for short/loud peaks. When played at a lower volume, overcompressed material will sound inferior in comparison to music that has left some dynamics in. The leading platforms use a slightly different standard: YouTube works with –13 dB LuFS, iTunes Radio uses –16 dB LuFS and Spotify uses –14d B LuFS.

Waves (see Figure 33.2), TC-Electronic, NuGen and others offer software that allows measuring LuFS. **Logic** has a LuFS reading in "MultiMeter," while Pro Tools users can install the free "dpMeter II" by TB Pro Audio.

FIGURE 33.2
Waves: WLM Loudness meter.

> **INNOVATING APPLE**
>
> Already with the introduction of iTunes 3 in 2003, Apple came up with a method similar to LuFS, called "Sound Check." When importing a track in iTunes, the perceived loudness is measured, and stored in the file as metadata. Switching on Soundcheck in the settings of iTunes or an i-device causes all tracks to play at the same volume. iTunes Radio and Soundcheck share the same LuFS value (–16 dB).

Loudness War 3.0

Now that the leading platforms have standardized volume, there aren't many reasons for a master to be louder than –13 dB LuFS. Loudness is only relevant for music on CD or files that are bought online. As a logical consequence, one would expect the loudness of music to drop. But actually, the opposite is true. By the end of the 2010s, leading pop producers and mastering houses continue to produce loud music. As if there hasn't been discussion and as if LuFS wasn't invented. It's apparent in all genres: the average volume for either Top 40, rock, R&B, EDM, hip-hop or alternative is easily louder than –10 dB LuFS, with the loudest records approaching the –5 dB LuFS mark!

Producer Butch Vig (Nirvana, Foo Fighters): "Compression is a necessary evil. The artists I know want to sound competitive. You don't want your track to sound quieter or wimpier by comparison. We've raised the bar and you can't really step back" (Rollingstone.com, December 26, 2007, "The Death of High Fidelity").

Apparently, loudness continues to be important. But simply dragging down the threshold slider of a brickwall limiter would cause too much distortion and impose the device's character on the signal.

▶ HOW CAN YOU PRODUCE A LOUD MASTER WITHOUT TOO MANY SIDE EFFECTS?

By dividing the compression load over several stages, the artifacts of a single stage or processor are minimized. Every half dB must be cherished here, as it will all add up to a master that's considerably louder. These are the processes to be considered:

- **Multiband compression.** Compressing each individual frequency band of the mix will even out dynamics in the most efficient way, resulting in a denser spectrum with more energy. Multiband compressors may be somewhat intimidating, as there are so many parameters to set. Plus, it's hard to oversee the effect of three (let alone six) different compressors, all working on a specific part of the spectrum. A good strategy is to start with one band first. After problems are solved, a second band can be added for treating other parts of the spectrum. A few notes on the always difficult attack and release times:
  - Tempo and density of the notes dictate release time.
  - Compressing a single band works similar to compressing a full range signal (Chapter 15). Short attack times will attenuate peaks, causing the note's sustain to get louder. Longer attack times allow transients to pass, and the note's sustain to attenuate.
  - Low notes tend to decay slower than high notes; release time can be set accordingly.
  - Temporarily exaggerating compression allows finding the setting that has the least negative side effects. Then the amount of compression can be taken back to the desired amount.

General tips for working with a multiband compressor:

- To verify that you're working on the right part of the spectrum, solo a single band.
- Makeup gain per band allows for EQing the mix efficiently.
- Do not set a band's crossover frequency to the sweet spot of an important instrument (for instance, the root note of the snare).

Multiband Limiting

Similar to multiband compression, a multiband limiter divides the spectrum in separate frequency bands. Limiting can then work efficiently on an isolated part of the spectrum. Some multiband limiters offer "profiles": these are optimized for certain styles of music.

Brickwall Limiting

Fine-tune settings, experiment and try other brands/types. Eventually, use two limiters successively. **Wave clipping** smooths the round natural shape of sound waves. As opposed to compressors and limiters, there's no attack and release involved in this process, so there are no pumping artifacts. Apart from this, waveform decapitation generates harmonics. This causes the mix to sound brighter, differently to what an EQ could do. Professionals deliberately clip the inputs of their A/D converters when recording the signal from their analog mastering equipment ("ingress"). Good processors for this purpose are LVC ClipShifter (see Figure 33.4), Sonnox Oxford Inflator, DMG Limitless, TDR Limiter 6 GE and Logic's Bitcrusher.

FIGURE 33.4
LVC ClipShifter.

Fix it in the Mix

Last but not least, reducing dynamics in the mixing stage will probably yield the greatest gain in volume. Defeating the peaks of individual instruments eases the load on mastering compressors and limiters farther down the line. Watch the meters for single instruments and instrument groups, especially those for drums and vocals. Excessive peaks can be counteracted with any or all the methods mentioned earlier.

▶ SIDE EFFECTS TO BE AWARE OF

More compression and limiting causes more artifacts:

- Reverb/ambience gets louder, which causes the mix to appear wetter.
- Attacks will be attenuated.
- *S*'s and *t*'s will get louder.
- In the frequency spectrum, focus will shift from the low frequencies toward the mids/highs.
- Distortion will increase.

Mixing and mastering in the same project allow most by-effects to be compensated immediately.

▶ THE FINAL QUESTION

Still, one question remains to be answered. If the major platforms stream music at a reduced loudness (and they do), why are we still producing loud masters? Well, it's hard to come up with an answer that's all encompassing. There are a few factors that contribute, though:

- CD is not dead. There are more unsigned artists than ever; they continue to release music on CD.
- There's many online platforms (such as Soundcloud and Bandcamp) that don't use any standard for loudness.
- Music is still downloaded as audio files, either illegally, free of charge or in online stores;
- For equal loudness to work, iTunes requires Sound Check to be activated by the user. In Spotify, the option "Set the same volume for all tracks" in "Preferences" must be enabled. The question is, How many consumers actually know of any "preferences," let alone the one that enables equal loudness?

On the production side, things have changed too. Old loudness is not new loudness. **Awareness** and dislike of the typical Loudness War 1.0 sound among producers, engineers and mixers have urged them to find ways to increase loudness without its associated problems. That's why we've started reducing dynamics with sophisticated techniques like multiband compression, spectral enhancement, and more extensive parallel compression. Apart from this, there's also a tendency to avoid excessive energy in the aggressive mids (1–3 kHz), which allows playing a loud record on big speakers without being painful.

As we've seen many times in this book now, if you want to change things, it is most effective to do so at the source. In the quest for loudness, a good solution is clearing out the **arrangement**. With fewer instruments playing at the same time, there's space for the music (and our ears) to breathe. The art of deduction. In an empty mix, every single note can be loud and make maximum impact. And indeed, it seems that in some genres (notably Top 40), arrangements have become emptier than they were before.

All in all, it seems that the loudness issue will continue to keep us busy for some time. Although it is hard to predict when the competition will end, it has at least taught us to be wary of distortion, the importance of overtones and harmonic content, and how to get that last bit of quality from our equipment. How great it is that this is just one of the aspects that makes music production so fascinating!

PART V

Appendices

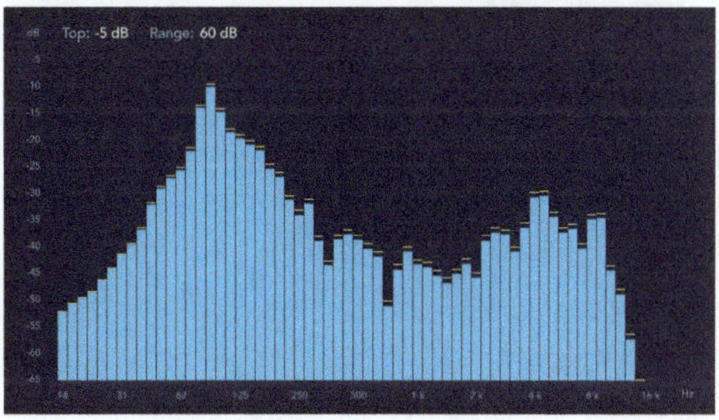

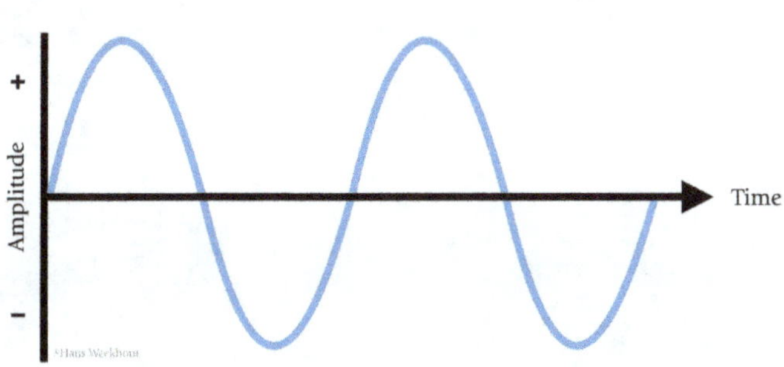

APPENDIX 1
Characteristics of Sound

In the middle of a production, it is not obvious to think about the physical movement of air molecules. However, a better understanding of the behavior of sound waves will lead to better sounding productions. What about decibels, frequency and phase? What can we read from the frequency spectrum? How come songs in a certain key sound better than others? And bass frequencies may be left out without a musical penalty!

When an instrument or a speaker produces sound, air molecules are pushed forward and backward. Although they basically stay in place, it's the molecules' vibrations that cause energy to be transferred. This happens at the speed of sound (1,125 ft/s). Once the eardrum is set into motion, the auditory nerve is stimulated and an electrical signal is sent to our brain. This is how a change of air pressure can lead to a sensation of sound.

▶ THE FOUNDATION OF SOUND

The most basic form of sound is a sine wave (see Figure BM1.1). Everybody knows the sound of a pure sine wave, it is the beep that's used for censoring TV programs. Although pure sines are rare in nature, they are the physical foundation of sound. Every sound that we know of, is the sum of multiple sine waves, each at a different frequency (pitch), amplitude (volume) and phase (timing).

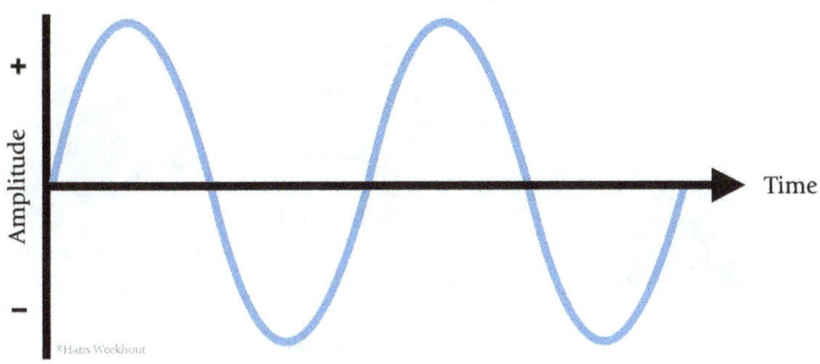

FIGURE BM1.1
Pure sine wave.

▶ PHASE

Phase indicates the starting point of a wave and is expressed in degrees. The starting point is determined at 0°, and a full period (one node and one antinode) is called 360°. Once a second wave starts half a period later than the first, their phase difference is 180°. We call this **out of phase**. In case these waves meet each other, they cancel, which results in perfect silence.

▶ AMPLITUDE, SOUND PRESSURE AND DECIBEL

Human hearing has a huge range in terms of volume; from a leave falling from a tree to the immense volume of a live concert (see Figure BM1.2). Volume (or sound pressure) is measured in µPa (micro Pascal). For a young human, the

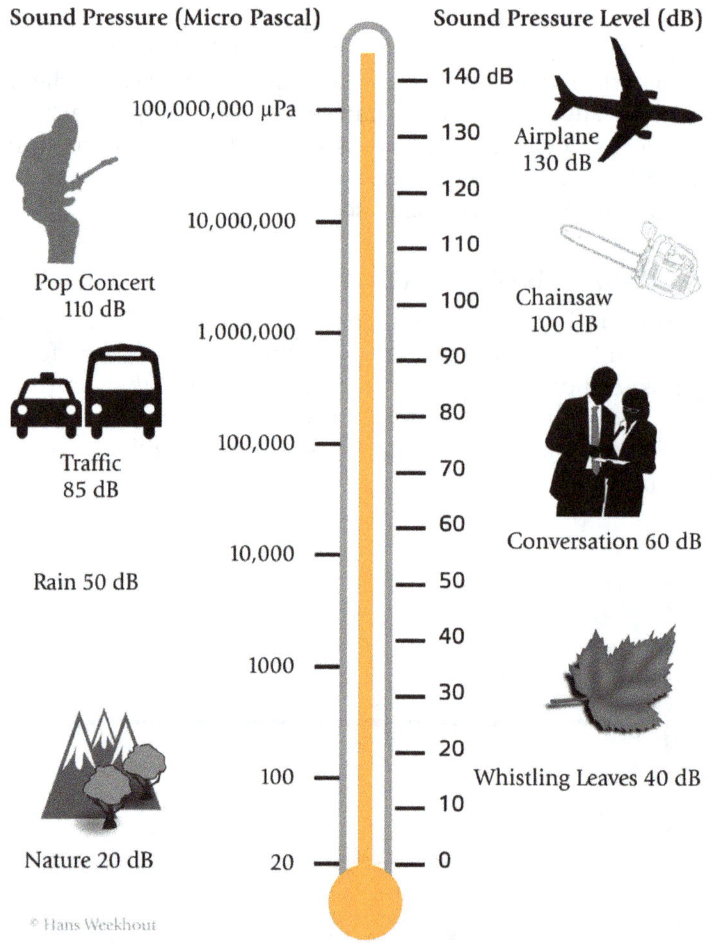

FIGURE BM1.2
Dynamic range of the human ear.

hearing threshold is about 20 μPa; this equals an eardrum's movement as small as a hydrogen atom's diameter! The threshold of pain is at 100,000,000 μPa.

As the Pascal scale, with its large numbers, is impractical, we usually work with the (logarithmic) **decibel** scale (see Table BM1.1). With dBs, there are two figures important to remember: **1 dB** is thought of as the **smallest audible** change, while **3 dB** represents a **doubling** of energy.

Table BM1.1 Air pressure versus decibel

Relative Increase of Energy (micro Pascal)	Number of Decibels (dB)
1	0
2	3
10	10
100	20
100	50
10,000,000	100

▶ FREQUENCY

Frequency equals the number of complete sound waves that fit one second. It is measured in cycles per second, or **Hertz** (Hz). At birth, our hearing has a frequency range of about 20 to 20,000 Hz. Lower or higher frequencies exist in nature but are inaudible to humans. Different frequencies are perceived by our ear as a change of **pitch**.

As of our musical interest, it is important to learn about the musical relationship between frequencies. If a tone has a frequency of 440 Hz (middle A on the piano), the A one octave higher has a frequency of 880 Hz. The next octave upward equals 1760 Hz. With this in mind, it follows that our hearing has a range of about 10 octaves. The exact frequencies of notes can be found in Figure 9.15. Of all acoustic instruments, a pipe organ produces the lowest frequency (C0 at 16 Hz), a piccolo the highest (C8 at 4186 Hz). The spectrum beyond 4186 Hz consists of **overtones**.

▶ FREQUENCY RESPONSE

In the studio, the maxim is *"what goes in, must come out."* Every frequency going into a device must sound equally loud on playback. This applies to the microphone, the microphone pre-amplifier, the recording medium (tape or computer) and the speakers. If one of these devices falls short, playback will be **colored**. That's why manufacturers aim their devices to have a frequency response as linear as possible (from 20–20,000 Hz). Although this is not too hard to achieve with audio interfaces nowadays, speakers can be (very) nonlinear, especially in combination with bad acoustics. Depending on the type, microphones may also color the sound, especially sound that arrives at the back or the sides.

Manufacturers indicate the linearity of a device as "20–20,000 Hz ± 3dB" (see Figure BM1.3). This means that deviations from the linear goal are no greater than 3 dB. As a reference, 1000 Hz is used because even the worst device is generally able to produce this frequency properly.

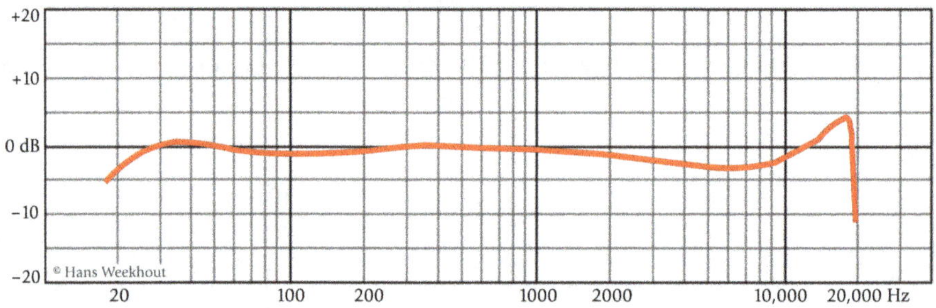

FIGURE BM1.3
Example of the frequency response of a tape machine, 20–20,000 Hz ± 2dB.

▶ DIFFERENCES BETWEEN LOW AND HIGH FREQUENCIES

Once an air molecule sets its neighbor into motion, energy is lost. This is especially true for high frequencies; these waves are small and contain only little energy. That's why they decay quickly and are obstructed by objects easily. Everybody has experienced the dull sound when being far away from an open-air concert; after approaching the speakers, the sound has suddenly become brighter.

Low frequencies consist of large, long waves; they contain large amounts of energy, and it's not easy to tame them. They'll even travel through walls or windows. While visiting the bathroom in a club, everybody has experienced the ever-present boom of the bass, while high instruments have almost disappeared (Table BM1.2).

Table BM1.2 Frequency versus wavelength

Frequency	Wavelength
20 kHz	0.68 inch
10 kHz	1.36 inch
100 Hz	11.3 ft
20 Hz	56.5 ft (!)

Now that you know about the characteristics of sound waves, it's possible to draw some practical conclusions. As low frequencies contain more energy than high frequencies,

- a *tweeter* (high-frequency unit in a speaker) can be small in size, while a *woofer* (low-frequency unit) must be large;
- tweeters need a less powerful amplifier, whereas woofers need a large amp;
- only small or thin materials are needed for absorbing high frequencies. Low frequencies need (much) more material and space for absorbing;
- a compressor is more "grabby" on low frequencies than it is on high frequencies.

Last, **directionality**, or the way sound waves radiate from a source changes with frequency: the higher the frequency, the more directional. Low frequencies radiate in all directions. There are some practical consequences of this:

- Tweeter units should always be pointed precisely at your ear.
- Just one subwoofer is needed for reproducing low frequencies. Its position is less relevant for our ear (though it is for the acoustics).

▶ THE SPECTRUM OF SOUNDS

With a spectrum analyzer, it's possible to view the **spectrum** of a signal. For example, the spectrum of a piano's middle A shows us that the **fundamental** frequency (440 Hz; see Figure BM1.4), is the loudest partial. Apart from the

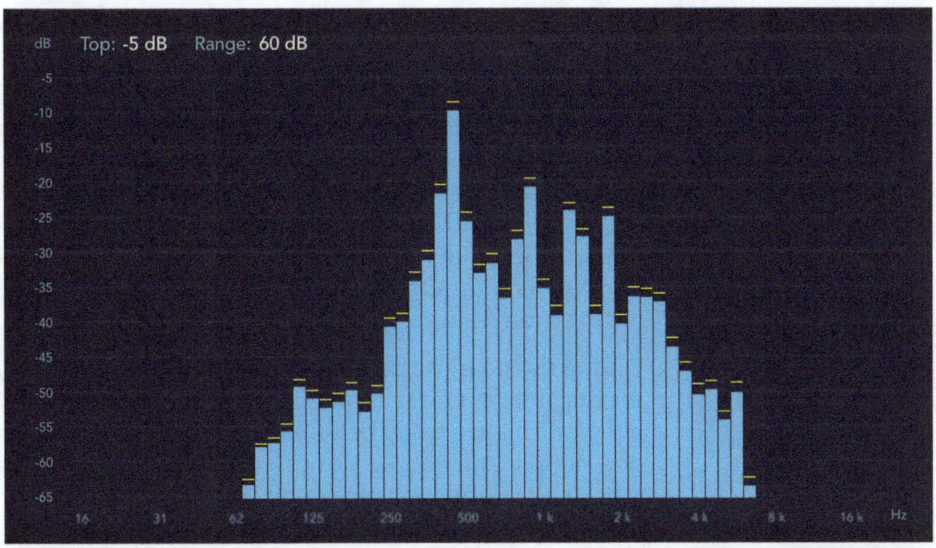

FIGURE BM1.4
Spectrum of an A4 (440 Hz) on piano. Harmonic overtones, such as the octaves at 880 Hz and 1760 Hz, are clearly visible.

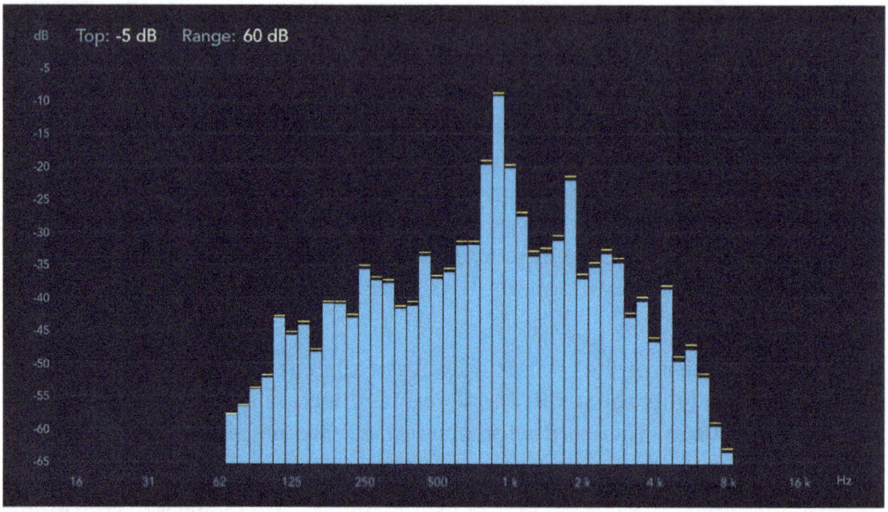

FIGURE BM1.5
Spectrum of an A5 (880 Hz) on piano. Notice the low frequencies below the fundamental tone; these are caused by resonances of the wooden casing.

fundamental, **harmonic** frequencies (octaves and fifths) show up, but also **enharmonic** frequencies. These are caused by the hammer hitting the string, and vibrations of the frame, the case and other strings. Playing the same note on another instrument will generate its own, unique spectrum. Our brain recognizes an instrument by the specific shape of the spectrum.

▶ ATTACK AND SUSTAIN

The volume of a musical note evolves over time. By hitting a drum, or picking a guitar string, a short, loud **attack** is followed by a soft, longer **sustain**. During the attack phase (**transient**), an instrument produces (many) more harmonics than it does during its static sustain phase. That's why it's easier for our brain to recognize an instrument by its attack than it is during sustain. With more overtones, we know that an instrument is played more aggressively. Only with sufficient high frequencies, an instrument (or mix) can sound defined, while transients can flourish.

It's in the **attack** that an instrument is recognized.

During the note's **sustain** phase, the spectrum is simpler, making it harder for our brain to tell one instrument from another. This can easily be verified by removing the attack of a sampled piano in a software instrument. In the low register, the instrument can sound pretty similar to a cello.

▶ WHEN LOW FREQUENCIES CAN BE LEFT OUT

Our brain is smart at recognizing instruments by their spectrum—actually so smart that if you take out the bottom frequencies, it will **automatically make up for any**

missing information. Even without any low end being present, a musical person will be able to point out the corresponding root note precisely. In music production, this information is vital. By cutting off the bottom end of instruments that don't necessarily need bass (maybe the piano or guitars), space can be made for **true** bass instruments, like kick drum and bass.

After cutting an instrument's **bottom end**, our brain will make up for any missing information **automatically**.

Now it becomes clear why multimedia or other small speakers can still provide you with a reasonable impression of a mix's bottom end—despite the fact that these speakers may not reproduce sound below 100 to 150 Hz!

Unfortunately, our brain cannot make up for missing information in the upper part of the spectrum. In fact, we knew this already since the exact structure and loudness of overtones are essential for our brain to recognize instruments.

▶ PRACTICAL CONSEQUENCES OF FUNDAMENTALS AND OVERTONES

- With instruments that have a narrow spectrum, such as piccolo, glockenspiel, triangle, falsetto voices, certain kick drums, basses and organs, there are limitations to what you can achieve with an equalizer. The frequency range of an equalizer is so wide that it affects **volume** rather than the spectrum.
- Percussion instruments such as bells, cymbals, gongs and drums contain large amounts of **enharmonic** frequencies (see Figures BM1.6 and 1.7). This

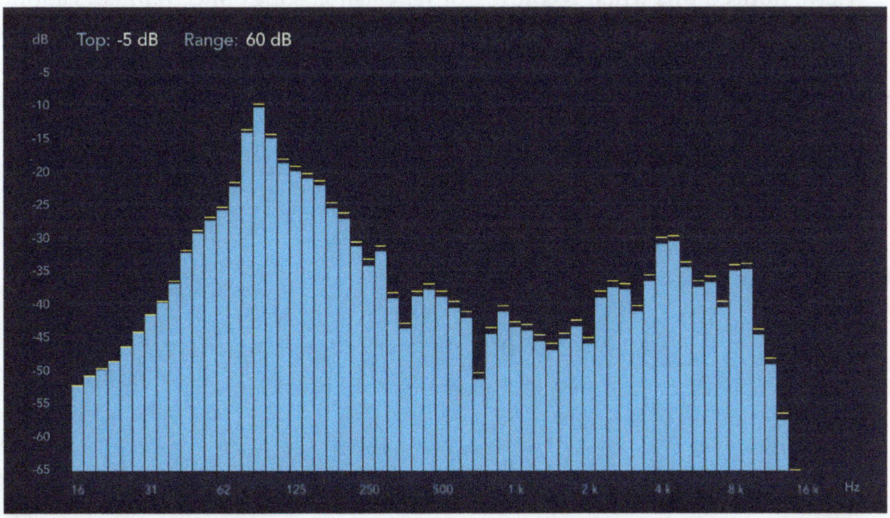

FIGURE BM1.6
Spectrum of a bass drum. This low instrument contains even more energy in the higher frequencies than the piano notes in the previous examples.

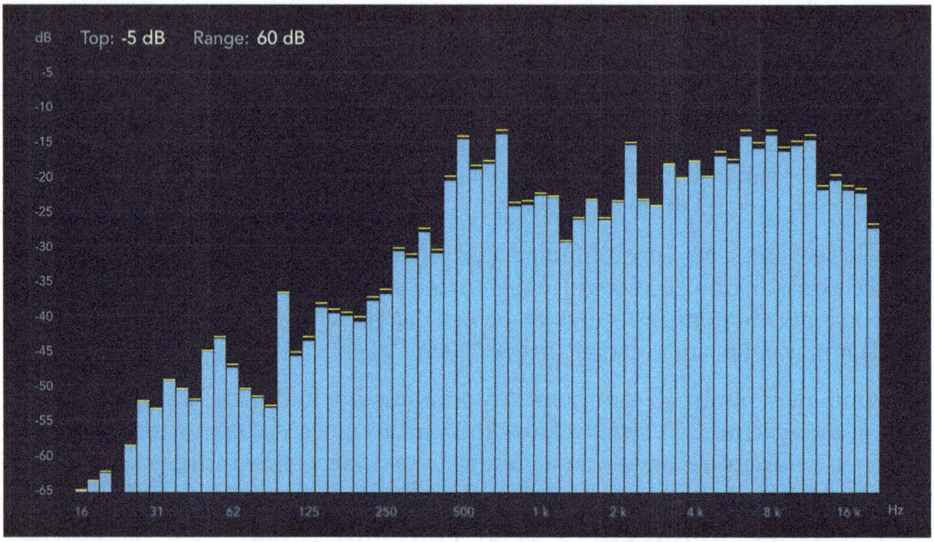

FIGURE BM1.7
Spectrum of a crash cymbal. Note the energy in the lower frequencies.

makes detecting their pitch harder. Then again, one specific frequency may be the loudest, for instance, with a kick drum. In case this doesn't match the key of the music, it will cause a harmonic clash with other instruments. Especially in the bottom end, this can be problematic. Unconsciously, you may have noticed this when recording or rehearsing: songs in certain keys sound better than others. Tuning the drums, or changing the song's key, will result in a sound that's tighter.

APPENDIX 2
Our Hearing

In the studio, our ears are the weakest link. Their frequency response is far from linear, and the perception of specific frequencies changes with volume. How can we compensate for this effect in our quest for the best possible sound? Why do we like music to be loud? And how can you stay away from hearing damage?

Loudness

▶ HEARING LOSS

Sound vibrations enter our ear through the eardrum. Behind the eardrum, the hammer, anvil and stirrup bones transmit vibrations to two sets of cilia (hair bundles). Each hair resonates at a certain frequency and transmits a signal to the auditory nerve once it is set into motion. Before loud sound can hurt the ear, certain muscles in the middle ear offer short-term protection by contracting. However, this protection cannot prevent breaking the connection between cilia and nerve when loud sound continues for a longer time. As a result, people may experience a reduced sensitivity for certain frequencies, mostly in the 4- to 6-kHz area. This is the frequency band where speech intelligibility resides. As other frequencies may still be clearly audible, this disorder often fails to be recognized.

Hearing damage can also result in **tinnitus**. Similar to the phantom pain of people with an amputated arm or leg, our overactive brain starts generating signals that correspond to damaged cilia. People may hear a continuous ringing tone or noise for the rest of their life. In extreme cases, normal hearing is impossible.

Playing music for 8 hr at 80 dB SPL (*Sound Pressure Level*) is considered the maximum before hearing loss occurs (see Table BM2.1). By listening twice as loud (an increase of 3 dB), only 4 hr are considered to be safe.

As music production often means long hours of work in the studio, it is imperative to get used to **low monitoring levels** as quickly as possible. This also applies to listening on headphones and in-ear monitors (on stage).

Table BM2.1 Maximum duration before hearing damage may occur

Sound Pressure Level	Duration
85 dB	8 hr
88 dB	4 hr
91 dB	2 hr
94 dB	1 hr
97 dB	30 min
100 dB	15 min

At 106 dB and above, acute hearing damage could occur (Source: www.osha.gov)

VINTAGE EARS

Although humans can hear frequencies from 20 to 20,000 Hz at birth, this range narrows as we age. From 30 years onward, our sensitivity to low frequencies decreases with approximately 3 dB per 10 years. Perception of high frequencies decreases by 8 to 13 dB per decade.

▶ LIVE

Live concerts are invariably loud and easily exceed the 100-dB mark. At the same time, almost everyone, from the audience to musicians and technicians, wears earplugs in order to prevent hearing damage. One would argue that a simpler solution can be found (. . .). Why do we like sound to be loud? For one, the **physical** sensation of big bass waves vibrating our skeleton is nothing short of spectacular. The physical pressure on our eardrums feels good in a way. Last but not least, loud music stimulates **more nerves**, which simply causes **more excitement**.

Loud music causes more excitement, as **more** nerves are stimulated.

On the other hand, **not** protecting your ears is irresponsible. Many people use drugstore earplugs. Although protection is usually good, they'll close off high frequencies almost completely, thereby leaving you with an unexciting experience. A better solution is provided by tailor-made earplugs from specialist shops that attenuate frequencies in a linear fashion. Fancy models let you even set the amount of attenuation. Considering that your ears are the most important assets for music production, the expense for a pair of decent earplugs is negligible.

▶ MONITORING LEVELS IN THE STUDIO

Similar to live concerts, it is tempting to turn up the volume in the studio. But loud monitoring has disadvantages:

- The acoustics of the room will play a bigger role; sound will change accordingly.
- As distortion of amplifiers and speakers increases, so will the impression of the mix.
- Loud music causes listening fatigue.
- Hearing damage may occur; listening for many hours at a moderate volume could already harm our ears.

Most professionals work at low monitoring levels. Although that may seem boring, it makes them work harder in order to bring forward the energy and emotion that's contained within the music. If that works at a low volume, it will definitely work at a high volume. The ideal monitor volume in the studio

APP CHECK

There are smartphone apps that measure sound pressure levels (see Figure BM2.1). As an aspiring pro, it is a good idea to start measuring audio levels during concerts and in the studio. This could prevent hearing damage and provide you with an objective reference.

FIGURE BM2.1
Studio Six Digital SPL meter.

is similar to the consumers' volume at home or in a car. SPLs between 65 and 75 dB are considered good and safe.

On the other hand, with a creative product, not everything can be judged by the head. Turning up the volume now and then is fun. It will keep everybody

> Once all elements in a mix can be heard at a **low volume**, they can certainly be heard at a **high volume**.

inspired and inform you about the mix's **physical** characteristics. **Don't do this too often though.**

Why not? Well, everybody knows the effect of walking into a dark room after being in a light room (or vice versa). Before being able to see properly, our eyes need time to adapt to the new situation. The same is true for our ears. Directly after changing monitoring volume, it will be hard to interpret a mix, as our ears need time to recalibrate. Therefore, **changing monitoring levels often and abruptly will make you lose your reference point.**

Frequencies

▶ HEARING FREQUENCIES

About 100 years ago, American researchers Fletcher and Munson (1993) examined our ear's sensitivity for different frequencies.

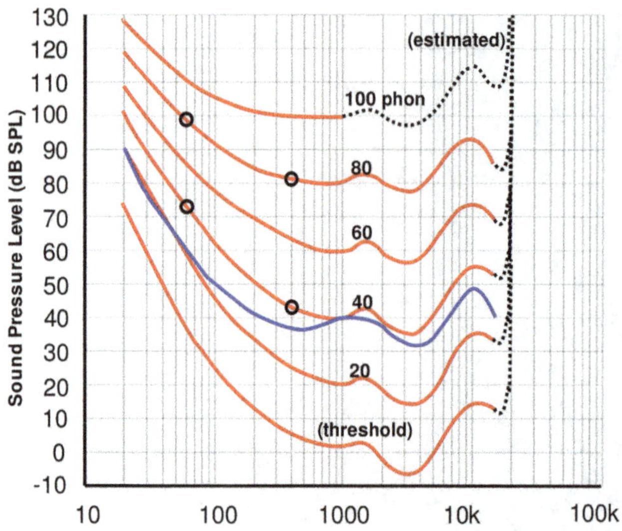

FIGURE BM2.2
Fletcher and Munson's "Equal Loudness Contours" indicate the frequency response of our ears. The curves can be read as: "how loud should a certain tone be presented to the ear to make it sound as loud as the reference tone of 1000 Hz?".

The lowest curve indicates the softest possible audible tones or hearing threshold. The highest curve is the threshold of pain. In order to illustrate the ear's dependency on volume we compare two tones, 60 Hz and 400 Hz, at a volume of 80 phon. As can be read from the graph, 400 Hz is perceived roughly 15 dB louder than 60 Hz. After turning down the listening volume to 40 dB, 400 Hz is now a whopping 30 dB louder than 60 Hz!

They asked test persons to listen to different sine tones. Starting at 1 kHz, subjects were asked to adjust their headphone's volume in order to hear every new frequency equally loud as the reference tone. Since perception of volume is subjective, Fletcher and Munson averaged a large number of results, and put this in a graph called the "Equal Loudness Contours" (Figure BM2.2). What can be concluded from this?

1. **Our ears' frequency response is far from linear.**
2. **Our ears are most sensitive around 3.5 kHz** (as illustrated by the dip in the curves in that area). This is thought of as a result of evolution: in prehistoric times, people's safety largely depended on the cry of a baby or a scream in the distance. In music production, this sensitivity becomes apparent when EQing a fuzz guitar for example. For our ears, a change of only 1 dB in the mids seems to have a relatively large effect on sound.

3. **The lower the frequency, the louder it must be presented to our ear.** Albeit subtler, at the high end of the spectrum (around 10 kHz), the same is true. Simply put, our ears are least sensitive to bass. When listening to music, the volume must be turned up in order to hear the bass properly. Likewise, when EQing the bass, small changes are hard to perceive.

4a. **At high volume, frequencies are perceived more evenly.** Fletcher and Munson performed their research at different volumes. Each graph in the picture represents a certain subjective loudness, or *phon*. Clearly can be seen, that higher volumes lead to a (much) more linear perception. In other words, at a high volume, the lows and highs seem louder to our ears. So here we have another explanation for us liking loud music!

b. The fact that our ears' frequency response is dependent on loudness is useful knowledge when working in the studio: when changing monitoring level, instruments will seem to take on a different place in the mix, especially kick drum and bass.

REFERENCE

Fletcher, H. & Munson, W. A. "Loudness, its definition, measurement and calculation", *Journal of the Acoustical Society of America* 5, 82–108 (1933).

Index

For Product Safety Concerns and Information please contact our EU representative GPSR@taylorandfrancis.com
Taylor & Francis Verlag GmbH, Kaufingerstraße 24, 80331 München, Germany

www.ingramcontent.com/pod-product-compliance
Lightning Source LLC
LaVergne TN
LVHW080805011026
849160LV00015B/1537

* 9 7 8 1 1 3 8 6 2 6 1 0 2 *